Foundations of Algebra

Active Learning Textbook

First Edition

published by
Pacific Crest
Lisle, IL

Foundations of Algebra: Active Learning Textbook

First Edition

by Wade Ellis Jr., Janet E. Teeguarden, Dan Apple, and Denna Hintze

Product & Cover Design, Editing: Denna Hintze
Editing: Anne Scanlan-Rohrer, Breanna Apple

Pacific Crest
P.O. Box 370
Hampton, NH 03843-0370
Phone: 603-601-2246
www.pcrest.com

September 2013

Acknowledgements

As authors, we are indebted to and acknowledge our dads' contribution to the passion and love we have for learning mathematics. We hope to pass on this passion and love to the students who use this book. It is also our hope that other parents instill in their children this same joy for learning mathematics so that others discover the excitement and wonder that we each hold so dear.

We would like to thank Deirdre Brennan and Breanna Apple for their careful construction of answers to problems, questions, and challenges. We would also like to thank Anne Scanlan-Rohrer for her editorial work. The errors that remain are clearly our own.

Wade Ellis, Jr.

Janet E. Teeguarden

Dan Apple

Denna Hintze

ISBN # 978–1–60263–650–7

Table of Contents

Preface

Algebra as it is taught in many colleges is the ability to use the language of mathematics to model the real-world. In order to do this, students should learn to simplify, expand transform or translate expressions and equations using a set of mathematical properties, rules, techniques, methodologies, key concepts and a set of conventions to purposefully analyze, evaluate, solve and graph various types of equations and functions. An algebra course should give students the opportunity to improve their habits of mind: their ability to reason mathematically, to use mathematical tools, and to learn how to learn mathematics.

Students who successfully complete an algebra course should have increased their algebraic skills, both in quality and quantity. These skills include their ability to:

1) re-write an expression by expanding, factoring, or simplifying for analysis or evaluation,

2) re-write an equation(s) in equivalent forms to

   a. solve an equation or system of equations

   b. graph an equation in any two variables

3) analyze or graph any function to describe its behavior over its domain and range.

And students should be able to use these enhanced habits and set of skills to more fully investigate and understand their chosen discipline and the real world around them.

This activity book improves student self-reliance in developing their understanding of mathematical concepts and in mastering transference skills. The Critical Thinking Questions component of each activity strengthens students' ability to apply their reasoning and sense-making skills. Demonstrate Your Understanding and Identify and Correct Errors components allow them to develop their problem solving ability.

The book was developed using the principles of Process Education that are based on and informed by Bloom's Taxonomy as updated by Anderson and Krathwohl. Each activity allows students to reach the problem solving level of learning for the topic covered.

A web-based Instructor's Guide Edition of the book supports instruction in active learning modalities. The Guide also provides answers to Reading Questions, Critical Thinking Questions, Demonstrate Your Understanding and Identity and Correct the Errors problems.

The book is intended to function as a standard college developmental algebra textbook.

The basic structure of the activities in this book are shown here.

Why? EXAMPLE

This portion of the activity explains why it is worth your time and effort to learn the content of this activity. When there are immediate real-world applications, they are generally mentioned.

What Do You Already Know? EXAMPLE

Here you will find several questions, each of which you should be able to answer before you begin the activity.

Goals EXAMPLE

This part of the activity lists the proficiency goals you should be able to achieve once you have completed the activity.

Building Mathematical Language EXAMPLE

New Terms: *This is where new vocabulary terms and concepts are listed. Each of them are defined and discussed in the activity itself.*

Key Concepts EXAMPLE

As the name suggest, this portion of the activity is devoted to the introduction of concepts and ideas that are of central importance to the activity.

Methodology EXAMPLE

What the Methodology Teaches

The methodology (steps for performing a task) for solving the problems in Examples 1 and 2 is shown in the following table. You should complete the Your Turn problem in the same table, following each step in the methodology.

	Example 1	**Example 2**	**Your Turn**
Multiply:	$(2 + i)(3 + 5i)$	$(3 + 2i)(4 - 3i)$	$(2 - i)(5 + 3i)$

Steps	**Discussion**
1 Multiply the real number components	**Pull out the real number components and multiply them.**

Ex 1		**Ex 2**		**Your Turn**	
	$2 \bullet 3 = 6$		$3 \bullet 4 = 12$		

Model 1 — EXAMPLE

Isolate y^2: $3y^2 - 4ax = 7ax - 5 + ay^2$

Step 1 **Rewrite the equation** $3y^2 - ay^2 = 7ax - 5 + 4ax$

Step 2 **Simplify** $3y^2 - ay^2 = 11ax - 5$

While not every activity includes a model, many do. These demonstrate the application of the methodology in different contexts, helping you deepen your understanding.

Addressing Common Errors — EXAMPLE

We all tend to make the same kinds of errors when we first learn something. In this section of every activity, we share some of the most common errors that learners tend to make with the kinds of problems covered by the activity.

ERROR: **Distributing in a multiplication expression**

Incorrect Process	Resolution
$2 \cdot (3 \cdot 7) = 2 \cdot 3 \cdot 2 \cdot 7$ $42 \neq 84$	*Use the Associative Property of Multiplication correctly.*
Correct Process	**Validation**
$2 \cdot (3 \cdot 7) = 2 \cdot 21$	$42 = 42$ ✓

Incorrect Process *demonstrates the error.* ***Resolution*** *is how to correct the error or keep from making it. The* ***Correct Process*** *shows how the problem should be solved without making any error. And* ***Validation*** *demonstrates how to ensure that the answer found through the* ***Correct Process*** *really is the right answer.*

Preparation Inventory — 1.1

I can...	OR *Here's my question...*
☐ recognize the types of numbers.	☐
☐ give examples of the properties of numbers.	☐

This is a checklist of what you should now be able to do, based on what the activity has covered to this point. If you feel you are not yet able to perform an item, you should draft a question about that item so that you can ask your instructor for precisely the help you need.

Critical Thinking Questions EXAMPLE

1. What are five examples of different types of numbers?

These questions require you to step back from doing to think about the hows and whys of what you're learning, improving your metacognitive abilities. The more you are able to do that, the deeper will be your understanding and the greater your proficiency with the content.

Demonstrate Your Understanding EXAMPLE

Simplify each of the following expressions.

a) $4\sqrt{7} \cdot 2\sqrt{7}$

*In this portion of the activity, you are given the opportunity to demonstrate that you have learned the information from **Key Concepts** as well as the problem solving skills taught through the **Methodology** and **Model(s)**.*

Identify, Describe, & Correct the Errors EXAMPLE

In this part of the activity, you are presented with a problem as well as a solution that is incorrect (a mistake has been made). Your challenge is to solve the problem correctly and to describe the error, each in the appropriate place. Some students prefer to describe the error after solving the problem; you are welcome to do it in whatever order you like.

Problem 1	Correct Process	Describe the Error
Is the following number: $\sqrt{36}$ Rational or irrational? Why?		
Worked Solution *(Find the error!)*		
$\sqrt{36}$ is irrational because it contains a square root symbol.		

What Did You Learn About Learning Math? EXAMPLE

1. What role do Critical Thinking Questions play in helping you to learn the material more effectively and efficiently?

Each activity ends with a few through-provoking questions about your learning. Remember that the more aware you are of how you learn and how you can improve your ability to learn, the greater will be your overall performance as a student.

Concept Map — EXAMPLE

As you go through this course, you will be asked to create concept maps at the end of each chapter, linking the concepts that are introduced in that chapter. At the beginning you might start with a very basic concept map, like the one below.

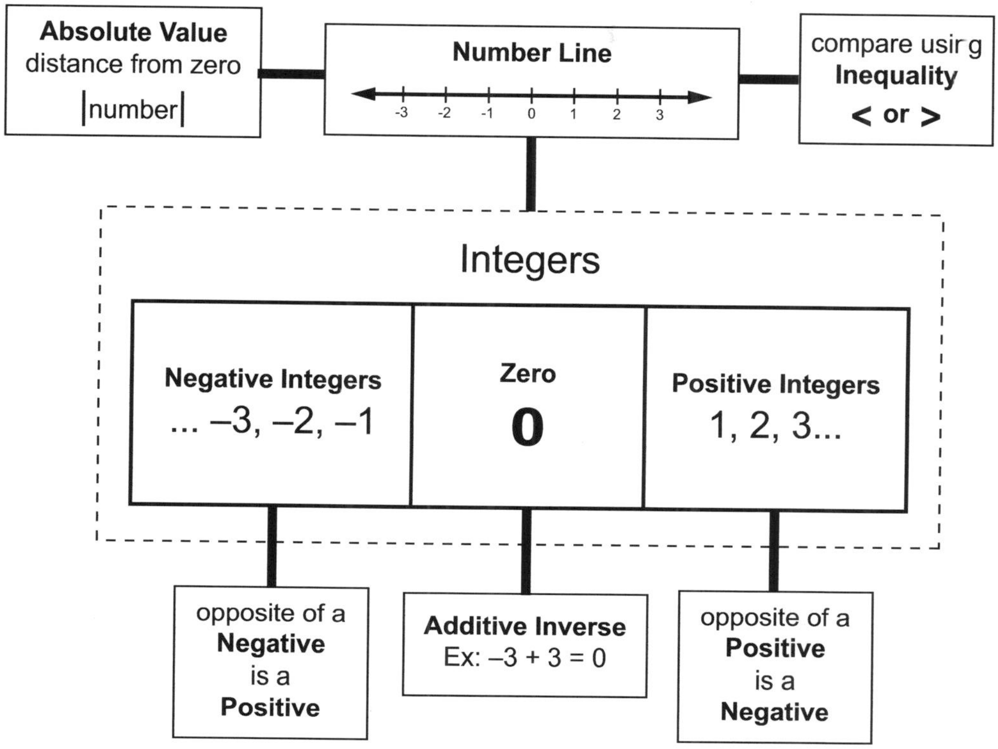

As you move through the course it will be important to make sure you document the links between words and be able to link more of the terms. An example of a more detailed concept map is shown below and should represent a goal for you as you progress through the course.

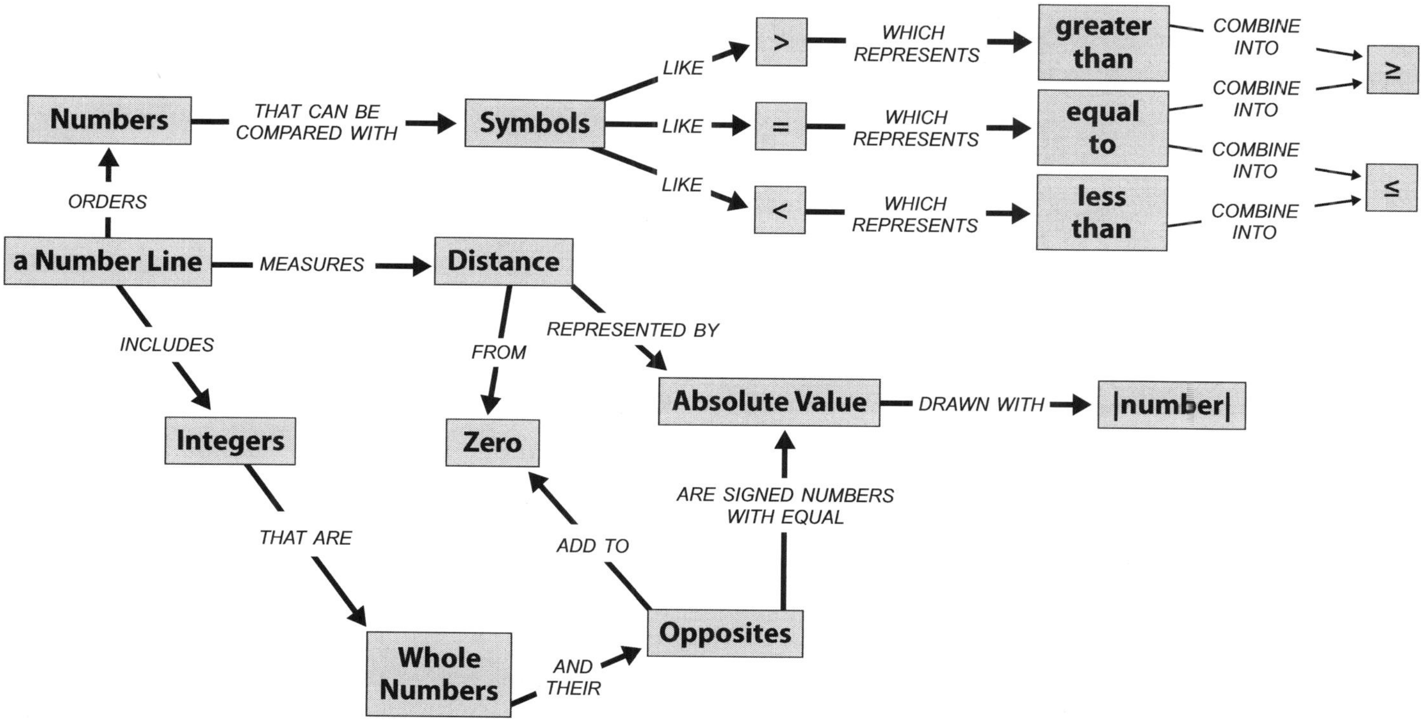

Student Notes

CHAPTER 1

An expression can be thought of as a set of instructions for computing a number from other numbers. This chapter deals with the kinds of numbers that can appear in an expression and then to determine if there are equivalent expressions — sets of instructions that give the same result. We can use various ways of rewriting expressions based on the rules and conventions of arithmetic to determine if two expressions are equivalent. Finally, we will work on ways of interpreting verbal statements symbolically.

Section 1.1 Expanding the Number System

Why? 1.1

You know about integers and fractions from your work in arithmetic. In this section, you will learn about other types of numbers that have the same properties and for which the order of operations still holds. These numbers will be used in all topics throughout the course. For example, the cost per linear yard of a diagonal asphalt path through a town square which is 100 yards on each side is \$20. Using the Pythagorean Theorem, the town computes that the path will be $100 \cdot \sqrt{2}$ yards long and the cost will be $\$20 \cdot 100 \cdot \sqrt{2} \approx \$2828.43$. You will learn to identify numbers like these.

What Do You Already Know? 1.1

What types of numbers are you aware of?

What properties are you aware of?

What is the Distributive Property?

What is the Order of Operations?

Goals 1.1

When you complete this activity, you should be able to:

1. List the several types of numbers and number systems
2. Identify the properties of all number systems
3. Recognize a number's type
4. Describe the difference between a rational and irrational number

Building Mathematical Language 1.1

New Terms: complex numbers, irrational numbers, pure imaginary numbers, identities, number systems, rational numbers, inverses

You have worked with **whole numbers**, which include:

- counting numbers along with zero (0, 1, 2, 3 . . .)
- positive and negative whole numbers or integers (. . . –3, –2, –1, 0, 1, 2, 3 . . .)
- rational numbers like $\frac{2}{3}, -\frac{5}{1}, \frac{73}{41}, -\frac{21}{11}$ (that is, numbers that can be expressed as the quotient of two integers, with the denominator not zero)

In evaluating numerical expressions, you have used the Associative and Commutative Properties of Addition and Multiplication and the Distributive Property, which applies to both addition and multiplication.

Associative *Means you can add numbers or multiply numbers in any order*

Addition:	$a + (b + c) = (a + b) + c$ $2 + (3 + 7) = (2 + 3) + 7$	Multiplication:	$a \bullet (b \bullet c) = (a \bullet b) \bullet c$ $3 \bullet (5 \bullet 2) = (3 \bullet 5) \bullet 2$

Commutative *Means you can change the order of a single addition or a single multiplication*

Addition:	$a + b = b + a$ $5 + 7 = 7 + 5$	Multiplication:	$a \bullet b = b \bullet a$ $5 \bullet 7 = 7 \bullet 5$

Distributive *Means you can change multiplication into addition*

$$a \bullet (b + c) = a \bullet b + a \bullet c$$
$$2 \bullet (5 + 3) = 2 \bullet 5 + 2 \bullet 3$$

Identity Elements

An Identity Element for addition or for multiplication is a number that does not change the value of a number that it is added to or multiplied by. Zero (0) is the Identity Element for addition and one (1) is the Identity Element for multiplication.

Addition:	$a + 0 = a$ $0 + a = a$	$5 + 0 = 5$ $0 + 5 = 5$	Multiplication:	$a \bullet 1 = a$ $1 \bullet a = a$	$2 \bullet 1 = 2$ $1 \bullet 2 = 2$

Inverse Elements

An Inverse Element for addition or for multiplication is a number that, when added to a number or multiplied by a number, yields the Identity Element for Addition or for Multiplication. The number $-a$ is the Additive Inverse of a.

The number $\frac{1}{a}$ is the Multiplicative Inverse of a. The number 0 does not have a Multiplicative Inverse.

Addition:	$a + (-a) = 0$ $3 + (-3) = 0$	Multiplication:	$a \bullet \frac{1}{a} = 1$	$3 \bullet \frac{1}{3} = 1$

Key Concepts 1.1

The rational numbers form what is called a ***number system***, since the four basic operations, when performed on a rational number, always yield a rational number, especially division (except for zero). The ***real numbers*** are also a system, but the integers are not because, for example, 3 divided by 4 is not an integer.

The ***irrational numbers***, like $\sqrt{2}$ and $\sqrt{\frac{17}{31}}$, are numbers that cannot be expressed as the quotient of integers (with the denominator not zero).

Sometimes the product of two irrational numbers is not an irrational number.

For example, $\sqrt{3} \bullet \sqrt{12} = \sqrt{36} = 6 = \frac{6}{1}$ is a rational number, not an irrational number.

The ***algebraic numbers*** contain numbers like $\sqrt{2}$ and all numbers that are the solution to algebraic equations. An algebraic equation involves only addition, subtraction, and multiplication of rational numbers and whole number powers of a variable.

The algebraic equation $x^2 - 2 = 0$ has $\sqrt{2}$ and $-\sqrt{2}$ as solutions, and thus these numbers are algebraic numbers as well as irrational numbers.

The real numbers that include all numbers that can be represented as decimals (including repeating and non-repeating decimals) and includes all the numbers mentioned above.

The numbers e and π are real numbers that are called ***transcendental numbers***, because they transcend or go beyond algebraic numbers and are not the solution to algebraic equations.

The ***complex number*** $\sqrt{-1}$ is denoted i. Complex numbers are written in the form $a + bi$, where a and b are real numbers. For example, $-3 + 2i$ is a complex number. The number $2i$ is a complex number that is called a ***pure imaginary*** number.

The diagram shows the relationship among the sets of numbers described in this section.

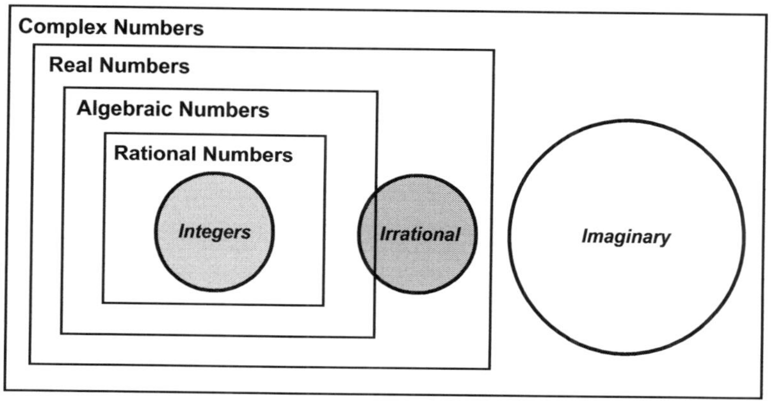

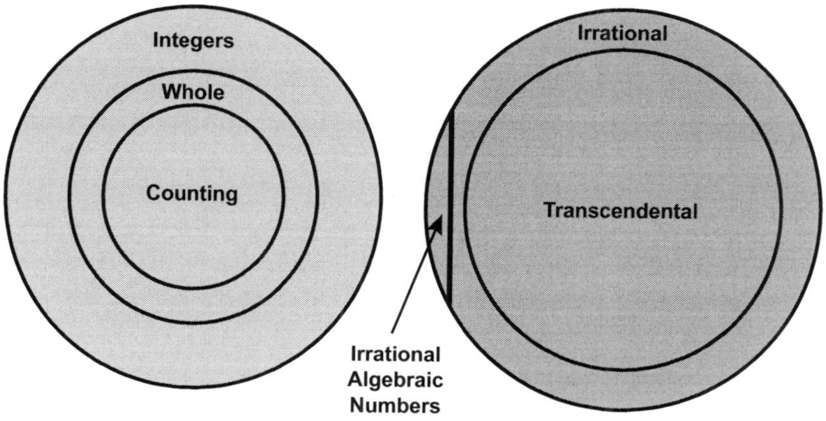

Properties

The ***Distributive Property*** and ***Associative Properties*** that are part of the basis for the Order of Operations apply to all the real numbers, real number systems, and the complex number system.

Addressing Common Errors 1.1

The following table shows some of the most common errors that learners tend to make with the kinds of problems covered in this section.

ERROR: Distributing in a multiplication expression

Incorrect Process	Resolution
$2 \cdot (3 \cdot 7) = 2 \cdot 3 \cdot 2 \cdot 7$ $42 \neq 84$	*Use the Associative Property of Multiplication correctly.*
Correct Process	**Validation**
$2 \cdot (3 \cdot 7) = 2 \cdot 21$	$42 = 42$ ✓

ERROR: Distributing in an addition expression

Incorrect Process	Resolution
$2 + (3 + 7) = 2 + 3 + 2 + 7$ $12 \neq 15$	*Use the Associative Property of Addition correctly.*
Correct Process	**Validation**
$2 + (3 + 7) = 2 + 10$	$12 = 12$ ✓

Preparation Inventory 1.1

I can...	OR ***Here's my question...***
☐ recognize the types of numbers.	☐
☐ give examples of the properties of numbers.	☐
☐ give examples of the types of numbers.	☐
☐ differentiate between the types of numbers.	☐
☐ use the Order of Operations.	☐
☐ use the Distributive Property.	☐

Critical Thinking Questions 1.1

1. What are five examples of different types of numbers?

2. What types of numbers are contained in the complex number system?

3. Which types of numbers have a multiplicative identity?

4. Which types of numbers have an additive identity?

5. Given that π is not an algebraic number, can you think of other numbers that are not algebraic? If so, give an example.

6. What is -0? Is $\sqrt{-0}$ a complex number? Explain your answers.

7. Is $0 - 2$ equivalent to -2? Explain your answer.

Demonstrate Your Understanding 1.1

1. Explain why the solution to the quadratic equation [$x^2 - 5 = 0$] is irrational.

2. Explain why the solution to the quadratic equation [$x^2 = 4$] is rational.

3. Is the solution to the equation [$x - e = \pi$] a rational number? Explain.

4. Is the statement, "Every irrational number is transcendental" true? Explain your answer.

Identify, Describe, & Correct the Errors 1.1

Solve the problem correctly in the second column. In the third column, identify and describe the error you found in the worked solution.

Problem 1	Correct Process	Describe the Error
Is the following number: $\sqrt{36}$ Rational or irrational? Why? **Worked Solution** *(Find the error!)* $\sqrt{36}$ is irrational because it contains a square root symbol.		

Problem 2	Correct Process	Describe the Error
Is the following number: $2 + 0i$ a real number? Why? **Worked Solution** *(Find the error!)* $2 + 0i$ is **not a real number** because it contains i for "imaginary."		

Problem 3	Correct Process	Describe the Error
Is the following number: $\frac{3.4}{1.7}$ rational or irrational? Why? **Worked Solution** *(Find the error!)* $\frac{3.4}{1.7}$ is an **irrational** number because neither the numerator or the denominator is an integer.		

Problem 4

Is the following number: 0

a counting number?

Worked Solution *(Find the error!)*

0 **is** a counting number because it comes before 1 when you count: "...0, 1, 2, 3..."

Correct Process

Describe the Error

Problem 5

Is the following number:

i^2

a pure imaginary number? Why?

Worked Solution *(Find the error!)*

i^2 **is** a pure imaginary number because it contains an i and no real part.

Correct Process

Describe the Error

What Did You Learn About Learning Math? 1.1

1. What role do Critical Thinking Questions play in helping you to learn the material more effectively and efficiently?

2. How do you learn mathematical notation (written mathematical symbols)? For example, how did you learn to represent a complex number?

3. What role do goals play in learning?

Section 1.2 Working with Numbers Represented by Radicals

Why? 1.2

Some numbers, like the diagonal of a square with sides 1 unit long, are represented with the use of radicals. These numbers can be added, subtracted, multiplied, and divided just like any of the numbers that were introduced in Section 1.1. In this section, you will learn how to use these numbers in expressions. Back to our square: The diagonal of the square with sides 1 unit long is $\sqrt{2}$ units. The *sum* of the diagonals of two such squares is $\sqrt{2} + \sqrt{2}$ units. This sum is an *expression* that contains radicals.

What Do You Already Know? 1.2

How do you multiply 5^3 with 5^7?

How do you represent "the cube root of 5," mathematically?

How does "the square root of 4" differ from the expression $\sqrt{4}$?

Goals 1.2

When you complete this activity, you should be able to:

1. Add, subtract, multiply and divide with expressions involving radicals
2. Understand and explain the meaning of radical indices

Building Mathematical Language 1.2

New Terms: radical radicand radical index

Key Concepts 1.2

Rule Table

Operation	Example
Addition	$\sqrt{2}+\sqrt{2}=2\sqrt{2}$
Subtraction	$5\sqrt{2}-2\sqrt{2}=3\sqrt{2}$
Multiplication	$\sqrt{2}\bullet\sqrt{2}=2$
Division	$\frac{2}{\sqrt{2}}=\frac{2\bullet\sqrt{2}}{\sqrt{2}\bullet\sqrt{2}}=\frac{2\bullet\sqrt{2}}{2}=\sqrt{2}$

Exponentiation	$\left(\sqrt{2}\right)^2 = \sqrt{2} \bullet \sqrt{2} = 2$ $\left(\sqrt{2}\right)^4 = \left(\sqrt{2} \bullet \sqrt{2}\right) \bullet \left(\sqrt{2} \bullet \sqrt{2}\right) = 2 \bullet 2 = 4$
Rationalize the denominator (remove all radicals from the denominator)	$\frac{1}{\sqrt{2}} = \frac{\sqrt{2}}{\sqrt{2}} \bullet \frac{1}{\sqrt{2}} = \frac{\sqrt{2}}{2}$

Interesting observation: When we square a square root, the result is the number itself. For example: $\left(\sqrt{3}\right)^2 = 3$

Methodology 1.2

Simplifying Radical Expressions

The methodology (steps for performing a task) for solving the problems in Examples 1 and 2 is shown in the following table. You should complete the Your Turn problem in the same table, following each step in the methodology.

	Example 1	**Example 2**	**Your Turn**
Simplify:	$\sqrt{12} + \sqrt{27}$	$\sqrt{18}+\left(5\sqrt{2}\right)^2-\sqrt{\frac{3}{2}}+\sqrt{12}-\sqrt{8}$	$\sqrt{75}-\sqrt{\frac{3}{2}}+\sqrt{54}-\sqrt{32}+\left(2\sqrt{3}\right)^2$

Steps	**Discussion**
1 Eliminate exponents of radicals	**All exponents need to be eliminated first by using the exponentiation rule for radicals.**

Ex 1 Done (no exponents).

Ex 2 $\sqrt{18}+(25 \bullet 2)-\sqrt{\frac{3}{2}}+\sqrt{12}-\sqrt{8}$

Your Turn

Steps	Discussion
2 Rationalize the denominator	By using the "division of radicals" rule, eliminate any fractional radicands.

Ex 1

Done (no denominators).

Ex 2

$$\sqrt{18}+(25 \bullet 2)-\frac{\sqrt{3}}{\sqrt{2}}+\sqrt{12}-\sqrt{8}=$$

$$\sqrt{18}+50-\frac{\sqrt{3} \bullet \sqrt{2}}{\sqrt{2} \bullet \sqrt{2}}+\sqrt{12}-\sqrt{8}=$$

$$\sqrt{18}+50-\frac{\sqrt{6}}{2}+\sqrt{12}-\sqrt{8}$$

Your Turn

Steps	Discussion
3 Factor each radicand	For each radicand, produce the prime factorization.

Ex 1

$$\sqrt{2^2 \bullet 3} + \sqrt{3^3}$$

Ex 2

$$\sqrt{2 \bullet 3^2} + 50 - \frac{\sqrt{2 \bullet 3}}{2} + \sqrt{2^2 \bullet 3} - \sqrt{2^2 \bullet 2}$$

Your Turn

Steps	Discussion
4 Pull out the perfect roots	Using the multiplication rule, pull out all the perfect squares.

Ex 1

$$2\sqrt{3} + 3\sqrt{3}$$

Ex 2

$$3\sqrt{2}+50-\frac{\sqrt{2 \bullet 3}}{2}+2\sqrt{3}-2\sqrt{2}$$

Your Turn

Steps	Discussion
5 Combine like terms	Using the associative and distributive properties, combine all like terms.

Ex 1

$$(2+3)\sqrt{3}$$
$$5\sqrt{3}$$

Ex 2

$$3\sqrt{2}+50-\frac{\sqrt{2 \cdot 3}}{2}+2\sqrt{3}-2\sqrt{2}$$
$$=50+\left(3\sqrt{2}-2\sqrt{2}\right)-\frac{\sqrt{2 \cdot 3}}{2}+2\sqrt{3}$$
$$=50+\sqrt{2}+2\sqrt{3}-\frac{\sqrt{6}}{2}$$

Your Turn

6 Validate	Estimate the answer of the original expression with a calculator. The results should be nearly the same.

Ex 1

$$\sqrt{12}+\sqrt{27} \stackrel{?}{=} 5\sqrt{3}$$
$$3.4641+5.1962 \stackrel{?}{\approx} 5(1.7321)$$
$$8.6603 \approx 8.6605 \quad \checkmark$$

Ex 2

$$\sqrt{18}+\left(5\sqrt{2}\right)^2-\sqrt{\frac{3}{2}}+\sqrt{12}-\sqrt{8} \stackrel{?}{=}$$
$$50+\sqrt{2}+2\sqrt{3}-\frac{\sqrt{6}}{2}$$
$$4.2425+50-1.2247+3.4641-2.8284 \stackrel{?}{=}$$
$$50+1.4142+3.4641-1.2247$$
$$53.6535 \approx 53.6536 \quad \checkmark$$

Your Turn

Addressing Common Errors 1.2

The following table shows some of the most common errors that learners tend to make with the kinds of problems covered in this section.

ERROR: Incorrectly multiplying cube roots as if they were square roots

Incorrect Process	Resolution
$\sqrt[3]{5} \cdot \sqrt[3]{5} = 5$	*Multiply the numbers under the cube roots.*

Correct Process	Validation
$\sqrt[3]{5} \cdot \sqrt[3]{5} = \sqrt[3]{5 \cdot 5} = \sqrt[3]{25}$	$\sqrt[3]{5} \cdot \sqrt[3]{5} =$ $\approx 1.70998 \cdot 1.70998$ ≈ 2.92403 ✓ $\sqrt[3]{25} =$ ≈ 2.92402 ✓

ERROR: Adding two square roots by adding the numbers under the radical signs

Incorrect Process	Resolution
$\sqrt{2} + \sqrt{8} = \sqrt{10}$	*Write each radical expression in simplest radical form and then add.*

Correct Process	Validation
$\sqrt{2} + \sqrt{8}$ $= \sqrt{2} + \sqrt{2^3}$ $= \sqrt{2} + 2\sqrt{2}$ $= 3\sqrt{2}$	$\sqrt{2} + \sqrt{8}$ $\approx 1.41421 + 2.82843 \approx 4.24264$ $3\sqrt{2} \approx 3 \cdot 1.41421 \approx 4.24263$ ✓ These two approximations are nearly the same.

ERROR: Adding two radical numbers by adding the numbers under the radical sign

Incorrect Process	Resolution
$\sqrt{7} + \sqrt{7} = 7$	*The two radical numbers are the same, so their sum will be twice the radical number.*

Correct Process	Validation
$\sqrt{7} + \sqrt{7} = 2\sqrt{7}$	$\sqrt{7} + \sqrt{7} \approx 2.64575 + 2.64575 \approx 5.2915$ $2 \cdot \sqrt{7} \approx 5.2915$ ✓ These two approximations are nearly the same.

Preparation Inventory 1.2

I can...	OR *Here's my question...*
☐ add radical expressions.	☐
☐ multiply radical expressions.	☐
☐ divide radical expressions.	☐
☐ simplify radical expressions.	☐

Critical Thinking Questions 1.2

1. What does a radical symbol mean?

2. What is the square root of 2?

3. What are the characteristics of a reduced, simplified radical expression?

4. Why do you eliminate exponents and fractional components before you start combining terms?

5. What is the grouping you should use when combining terms?

6. Why does calculator approximation work as a way to validate an answer?

Demonstrate Your Understanding 1.2

Simplify each of the following expressions.

a) $4\sqrt{7} \cdot 2\sqrt{7}$

b) $\sqrt[4]{32}$

c) $\dfrac{\sqrt{6}}{\sqrt{3}}$

d) $\sqrt{3}+4\sqrt{5}-2\sqrt{3}+7\sqrt{3}$

e) $\sqrt{9}-\sqrt{\frac{24}{3}}+\left(\sqrt{2}\right)^4+2\sqrt{5}-\sqrt{45}$

f) $\sqrt{\frac{2}{3}}+\sqrt{6}$

Create and Solve the Hardest Problem 1.2

Your Hardest Problem	Problem Solution

What makes a problem involving radicals and exponents hard to solve?

Identify, Describe, & Correct the Errors 1.2

Solve the problem correctly in the second column. In the third column, identify and describe the error you found in the worked solution.

Problem 1	Correct Process	Describe the Error
Simplify: $\sqrt{2}+\sqrt{2}$ **Worked Solution** *(Find the error!)* $\sqrt{2}+\sqrt{2}=2\sqrt{4}=2 \bullet 2=4$		

Problem 2	Correct Process	Describe the Error
Simplify: $\sqrt{(-4)^2}$ **Worked Solution** *(Find the error!)* $\sqrt{(-4)^2}=-4$		

Problem 3

Simplify:

$$\sqrt{12}-\sqrt{8}+2\sqrt{32}$$

Worked Solution *(Find the error!)*

$$\sqrt{12}-\sqrt{8}+2\sqrt{32}$$
$$=2\sqrt{3}-\sqrt{8}+2\cdot 2\sqrt{8}$$
$$=2\sqrt{3}-\sqrt{8}+4\sqrt{8}$$
$$=2\sqrt{3}+3\sqrt{8}$$

Correct Process

Describe the Error

Problem 4

Simplify:

$$\sqrt[3]{54}$$

Worked Solution *(Find the error!)*

$$\sqrt[3]{54}=3\sqrt[3]{6}$$

Correct Process

Describe the Error

What Did You Learn About Learning Math? 1.2

1. What did you learn about converting between symbolic representations of expressions?

2. How did you use the methodology when you worked through the Demonstrate Your Understanding section?

3. How did you use awareness of common errors as you were working through the Identify, Describe, and Correct the Errors section?

Section 1.3 Working with Complex Numbers

Why? 1.3

Some equations do not have real number solutions. Introducing a new type of number provides solutions to such equations. These numbers are the complex numbers that you saw in the first section. Complex numbers are used in designing computer chips that are used in smart phones and MP3 players.

What Do You Already Know? 1.3

What is a complex number?

What is a pure imaginary number?

Are all rational numbers also real numbers?

Goals 1.3

When you complete this activity, you should be able to:

1. Add and subtract complex numbers
2. Work with the conjugate of a complex number
3. Multiply and divide expressions involving complex numbers

Building Mathematical Language 1.3

New Terms: conjugate

Key Concepts 1.3

The *standard form of a complex number* is $a + bi$, where a and b are real numbers and i is $\sqrt{-1}$. The real number a is called the *real part* of the complex number. The number bi is called the *imaginary part* of the complex number. $\sqrt{-1}$ or the square root of -1 is not a real number since every real number square is positive or zero. Numbers like $2i$ ($= 0 + 2i$) are called *pure imaginary numbers*.

The *addition of complex numbers* is accomplished by adding the real parts of the complex numbers and the imaginary parts of the complex numbers and simplifying the resulting sum.

For example, $(3+4i)+(-5+i)=(3+(-5))+(4i+i)=-2+(4+1)i=-2+5i$.

The ***conjugate*** of a complex number $(a + bi)$ is the number $(a - bi)$.

The product of a complex number and its conjugate is a real number: $(a+bi) \cdot (a-bi)=a^2+b^2$.

Methodology 1.3

Multiplying Complex Numbers

The methodology for solving the problems in Examples 1 and 2 is shown in the following table. Students should complete the Your Turn problem in the same table, following each step in the methodology.

	Example 1	Example 2	Your Turn
Multiply:	$(2 + i)(3 + 5i)$	$(3 + 2i)(4 - 3i)$	$(2 - i)(5 + 3i)$

Steps	Discussion	Ex 1	Ex 2	Your Turn
1 Multiply the real number components	Pull out the real number components and multiply them.	$2 \bullet 3 = 6$	$3 \bullet 4 = 12$	
2 Multiply the imaginary number components	Pull out the imaginary number components and multiply them, remembering that i^2 is –1.	$i \bullet 5i = 1(5) \bullet i^2$ $= 5 \bullet (-1)$ $= -5$	$2i \bullet (-3i) = 2(-3) \bullet i^2$ $= (-6)(-1)$ $= 6$	
3 Multiply the real with the imaginary number components	For each real number component of one number, multiply it by the imaginary component of the other number	$2 \bullet 5i = 10i$ $i \bullet 3 = 3i$	$3(-3i) = -9i$ $2i \bullet 4 = 8i$	
4 Add real numbers	Add the results of Steps 1 and 2 to produce the real number component of the product.	$6 + (-5) = 1$	$12 + 6 = 18$	
5 Add imaginary numbers	Add the results in Step 3 to produce the imaginary component of the product.	$10i + 3i = 13i$	$-9i + 8i = -i$	

Steps	Discussion
6 Present as a complex number	Take the results of Steps 4 and 5 and present them as a complex number.

Ex 1	Ex 2	Your Turn
$1+13i$	$18-i$	

Steps	Discussion
7 Validate through division	Divide the product by one of the original complex numbers, using the technique for dividing complex numbers (illustrated below).

Ex 1

$$\frac{1+13i}{2+i}$$

$$\frac{1+13i}{2+i}\bullet\frac{2-i}{2-i}$$

$$=\frac{2+13+(26i-1i)}{4+1}$$

$$=\frac{15+25i}{5}=3+5i \checkmark$$

Ex 2

$$\frac{18-i}{3+2i}$$

$$\frac{18-i}{3+2i}\bullet\frac{3-2i}{3-2i}$$

$$=\frac{54-2+(-3i-36i)}{9+4}$$

$$=\frac{52-39i}{13}=4-3i \checkmark$$

Your Turn

Methodology 1.3

Dividing Complex Numbers

The technique for solving the problem in Example 1 is shown in the following table. Complete the Your Turn problem in the same table, following each step in the methodology.

	Example 1	Your Turn
Divide:	$\frac{18-i}{3+2i}$	$\frac{5-2i}{2-3i}$

Steps	Discussion
1 Multiply by the conjugate fraction	Multiply the numerator *and* the denominator by the conjugate of the denominator.

Ex 1	Your Turn
$\frac{18-i}{3+2i}\cdot\frac{3-2i}{3-2i}$	

Steps	Discussion
2 Simplify the fraction	Calculate and simplify the complex number.
Ex 1 $\dfrac{\cancel{52}^{4}-\cancel{39}^{3}i}{\cancel{13}^{1}}$	**Your Turn**
3 Present in complex form	Present the number in standard form: (a + b*i*).
Ex 1 $4-3i$	**Your Turn**

Addressing Common Errors 1.3

The following table shows some of the most common errors that learners tend to make with the kinds of problems covered in this section.

ERROR: Incorrectly multiplying complex numbers

Incorrect Process	Resolution
$(2+4i)\bullet(3-2i)=6-8i$	*Use the methodology for multiplying complex numbers.*

Correct Process	Validation
$(2+4i)\bullet(3-2i)$ $=6-8i^2+12i-4i$ $=6+8+8i=14+8i$	$\dfrac{14+8i}{2+4i}=\dfrac{14+8i}{2+4i}\bullet\dfrac{2-4i}{2-4i}$ $=\dfrac{28-32i^2-56i+16i}{4+16}=\dfrac{28+32-40i}{20}=\dfrac{60-40i}{20}=3-2i$

ERROR: Incorrectly dividing complex numbers by dividing real parts and complex parts

Incorrect Process	Resolution
$\dfrac{3-4i}{5-2i}=\dfrac{3}{5}+2i$	*Use the methodology for dividing complex numbers.*

Correct Process	Validation
$\dfrac{3-4i}{5-2i}=\dfrac{3-4i}{5-2i}\bullet\dfrac{5+2i}{5+2i}$ $=\dfrac{15-8i^2+6i-20i}{25+4}=\dfrac{15+8-14i}{29}=\dfrac{23}{29}-\dfrac{14}{29}i$	$\left(\dfrac{23}{29}-\dfrac{14}{29}i\right)\bullet(5-2i)$ $=\dfrac{115}{29}+\dfrac{28}{29}i^2-\dfrac{70}{29}i-\dfrac{46}{29}i=\dfrac{87}{29}-\dfrac{116}{29}i=3-4i$

ERROR: **Multiplying only the denominator by the complex conjugate**

Incorrect Process

$$\frac{5-3i}{2+5i}=\frac{5-3i}{(2+5i)(2-5i)}=\frac{5-3i}{4+25}=\frac{5}{29}-\frac{3}{29}i$$

Resolution

Use the methodology for dividing complex numbers.

Correct Process

$$\frac{5-3i}{2+5i}=\frac{5-3i}{2+5i}\bullet\frac{2-5i}{2-5i}$$

$$=\frac{10+15i^2-25i-6i}{4+25}=\frac{10-15-31i}{29}$$

$$=\frac{-5-31i}{29}=\frac{-5}{29}-\frac{31}{29}i$$

Validation

$$\left(\frac{-5}{29}-\frac{31}{29}i\right)\bullet(2+5i)$$

$$=\frac{-10}{29}-\frac{155}{29}i^2+\frac{-25}{29}i-\frac{62}{29}i$$

$$=\frac{145}{29}-\frac{87}{29}i=5-3i$$

Preparation Inventory 1.3

I can...	OR *Here's my question...*
☐ add and subtract complex numbers.	☐
☐ multiply complex numbers.	☐
☐ create the conjugate of a complex number.	☐
☐ divide complex numbers.	☐

Critical Thinking Questions 1.3

1. How do complex numbers differ from real numbers?

2. What happens when you multiply two imaginary numbers together?

3. What happens when you multiply a complex number by its conjugate?

4. Why does the technique of dividing complex numbers use the multiplication of the conjugate of the denominator?

5. How is the process of finding the product of $(2 + 3x)(4 - 2x)$ similar to the process of finding the product of $(2 + 3i)(4 - 2i)$? How are these processes different?

6. Why is 3 a complex number? What is its complex conjugate?

7. Why does $x^2 = -1$ have both i and $-i$ as solutions?

8. How would you find the value of i^{22}?

Demonstrate Your Understanding 1.3

Perform the given operation for each of the following.

1. Subtract: $(-2 + 3i) - (5 + 2i)$	2. Multiply $(2 + 5i)$ by its conjugate
3. Multiply: $(-3 + 4i)(-1 - 2i)$	4. Divide: $\dfrac{-4+3i}{2-i}$

Identify, Describe, & Correct the Errors 1.3

Solve the problem correctly in the second column. In the third column, identify and describe the error you found in the worked solution.

Problem 1

Subtract: $(2 + 5i) - (3 - 4i)$

Worked Solution *(Find the error!)*

$(2+5i)-(3-4i)$

$2+5i-3-4i$

$2-3+5i-4i$

$-1+i$

Correct Process

Describe the Error

Problem 2

Multiply: $(3 - 2i)(4 - i)$

Worked Solution *(Find the error!)*

$(3-2i)(4-i)$

$(3)(4)+(-2i)(-i)+(3)(-i)+(-2i)(4)$

$12+2i^2-3i-8i$

$12-9i$

Correct Process

Describe the Error

Problem 3

Multiply: $(-2-5i)(-2+5i)$

Worked Solution *(Find the error!)*

$(-2-5i)(-2+5i)$

$(-2)(-2)+(5)(5i)+(-2)(5i)+(-5i)(-2)$

$4+25i-10i+10i$

$4+25i$

Correct Process

Describe the Error

Problem 4

Divide: $\dfrac{2+3i}{-1+2i}$

Worked Solution *(Find the error!)*

$\dfrac{2+3i}{-1+2i}$

$\dfrac{2+3i}{-1+2i}\bullet\dfrac{-1-2i}{-1-2i}$

$\dfrac{(2)(-1)+(2)(-2i)+(3i)(-1)+(3i)(-2i)}{1-4}$

$\dfrac{-2-4i-3i+6}{-3}$

$\dfrac{4-7i}{-3}$

$\dfrac{-4+7i}{3}$

Correct Process

Describe the Error

What Did You Learn About Learning Math? 1.3

1. What are the differences between the skill of adding real numbers and adding complex numbers? How did you learn this difference?

2. What did you see as the most difficult part of learning how to divide complex numbers?

3. What do you find interesting about the definition of the pure imaginary number i?

Section 1.4 Interpreting Word Problems

Why? 1.4

Throughout life you will encounter problems that require logic and quantitative reasoning. This section will develop your ability to understand and conceptualize such problems with the use of symbols. For example, you may need to know how to compute your payment for work done at a grocery or retail store where your pay is dependent upon the hours you work, whether you worked overtime, and whether you are paid a commission (a percentage of your sales). Your company may give this information as a formula or you may have to compute the formula yourself. This section will develop your ability to do that.

What Do You Already Know? 1.4

What symbols are used for addition, subtraction, multiplication and division?

What words or phrases do you associate with addition? With subtraction?

What words or phrases do you associate with multiplication? With division?

What are three words that can be ambiguous in quantitative problems?

What do you call a letter that is used to represent a number?

Goals 1.4

When you complete this activity, you should be able to:

1. Understand some of the ambiguities in translating English statements into clear, precise, and accurate mathematical statements
2. Identify variables, constants, and assumptions when working with quantitative problems
3. Demonstrate an improved ability to solve quantitative problems

Building Mathematical Language 1.4

New Terms: constant variable

Key Concepts 1.4

Consider the terms *precision*, *accuracy*, and *clarity*. The following table offers examples of English and mathematical statements that demonstrate differing degrees of precision, accuracy, and clarity.

It is important to note that precision and accuracy are often treated the same in English; we tend to use the words interchangeably. They have different meanings in mathematics, however. An inaccurate answer in mathematics is an error. A mathematical solution may be very precise and still be inaccurate. For example: $16 \div 3.8 \approx 4.220528$. The correct answer, to the same number of decimal places (one way to measure the degree of precision) is: 4.210526.

	English		Mathematics	
	Less	*More*	*Less*	*More*
Precision	Sarah and her best friend went to a movie.	Sarah and Jane went to see the latest blockbuster.	$\frac{1}{3} \approx 0.3$	$\frac{1}{3} \approx 0.3333$
Accuracy	I didn't hear my alarm go off.	I heard my alarm go off, but hit the snooze button and fell back asleep.	$\pi \approx 3.15$	$\pi \approx 3.14$
Clarity	He feels bad, coughing and stuff like that, you know—sick.	He does not feel well, has a cough, and other symptoms of illness.	$\left[\left(9 \times \sqrt{4}\right) + \left(\frac{3}{2}\right)\right] - \left(1^2\right)$	18.5

In both English and mathematics, the more precise a statement, the more likely it is that someone else can interpret it accurately. One of the most common ways that inaccuracies are introduced in mathematics is through a lack of neatness. The more neatly you write as you perform calculations or when you present your work, the less likely it is that you will misread your writing (mistaking "2" for "z" for example) or have your work misread by others.

Clarity is a kind of neatness of thinking or process. The clearer your understanding, the more accurate your work will be. In the process of communication, clarity means that what you say or write is easily understood by others. Consider how frustrating it is to follow unclear directions when you're trying to get somewhere. The clearer the directions, the better your chances of arriving at the accurate destination. (Precision helps here too! Being told to "turn left after a few blocks" isn't very helpful.) In all mathematics courses, as well as other courses which use mathematics, you will be asked to "show your work." This really means *to make clear the reasoning you used to arrive at your answer.* Your reasoning and its clarity are as important as the accuracy of your answer.

In this course, we value precision and clarity, as they help to eliminate ambiguity and produce accuracy in all our work.

When we are confronted with word problems that must be solved mathematically, we must translate the meaning of the problem into mathematical notation. This does **not** mean assigning a mathematical symbol or number to individual words (*transliteration*); doing this is guaranteed to lead to wrong answers. The act of translating requires that you first *understand* the meaning of the problem in English. Then you *interpret* that meaning, selecting and ordering mathematical symbols and numbers that represent the same meaning. Only then can you work to solve the problem.

In translating word problems into precise mathematical language, we want to ***eliminate ambiguity*** completely. This means that we must write mathematical statements that precisely, accurately, and clearly reflect the meaning of the English language description of the problem situation. Inaccuracies can be introduced into mathematical statements by not adhering to conventions agreed upon by mathematicians. For example, 20% of 75 is 15, because the convention **in this particular situation** is "*of* means multiplication." However, the phrase "5 of 8" may mean several different things. Just by adding one word here or there, you can drastically change the meaning of the statement. Consider the following examples, where the meaning for the word ***of*** differs:

Statement	Context	Mathematical meaning
5 chances **of 8** chances	What fraction of the time will you win a lottery if you hold 5 of 8 winning tickets?	5/8 of the time you will win the lottery
5 teams **of 8** players	What is the total number of players?	$5 \times 8 = 40$ players
5 percent **of 8** dollars	How much is the discount on purchasing an $8 item?	$5\% \times \$8 = 0.05 \times 8 = \$\,0.40$ or 40¢
5 slices out **of 8** slices	How much of a pizza pie has been eaten?	5/8 of the pizza pie
choose **5** elements out **of 8**	How many different 5 member committees can be chosen from 8 possible members?	${}_8C_5$ possible committees

That the meaning of words (such as *of*) varies by context is why *transliteration* (substituting mathematical symbols for words) causes errors.

When you translate statements such as those in the first column of the table above into mathematical language, you must use mathematical symbols that communicate the ***meaning*** of the English statement, not the individual words. This meaning is based on your understanding and interpretation of the statement. When there is ambiguity in an English language presentation of a situation, you can do one of two things. You can choose to interpret the statement based on inference (a tentative conclusion based on clues) or you can ask for more information.

Limitation/Caution: Mathematical conventions do not exhaustively cover all English language usage. You cannot be sure that the person who wrote a problem statement in English used or even knew any of the conventions established by mathematicians. This means that when ambiguity exists, you may not correctly understand and interpret the context or meaning intended by the writer.

The methodologies in this book have been created by experts as a way to help others learn to achieve clarity in mathematical reasoning that leads to accuracy in worked-out solutions. Your use of these methodologies can help you increase the clarity of your reasoning and work, therefore helping you reduce misinterpretations and inaccuracies.

Limitation/Caution: A methodology is simply a model of what experts feel are the critical steps in mathematical reasoning that are needed to solve a given type of problem.

Methodology 1.4

Interpreting Word Problems

Most quantitative problems in the real world are presented in sentences and phrases. The meaning of these sentences and phrases can be translated into mathematical expressions and equations. This process of translation requires a clear understanding of the problem and a fluency with the language in which the problem is stated (English for this book) and the conventions and symbols of mathematics.

Limitation/Caution: All human languages tend to be imprecise and there are often ambiguities in written and spoken language that lead to multiple possible interpretations.

Example 1

Solve:

> The number of mobile phones in the United States has increased enormously since the approval of the Advanced Mobile Phone System by the FCC in 1982 when there were fewer than 50,000 mobile phone users. By the year 2000, there were 110 million mobile phone users in the United States. Use has steadily increased since then, and the number of users reached 270 million in 2008. In 2008, more than three-quarters of the population of the United States had a mobile phone. Predict the number of mobile phones in 2012.

Your Turn

Solve:

> A Health Maintenance Organization (HMO) suggests that a woman's recommended weight can be computed. For each inch over 5 feet tall, 5 pounds should be added to 100 pounds. Some women are, of course, 4 feet 11 inches or shorter, but this recommended weight is fairly reliable for 90% of United States women. What does this HMO's formula recommend for the weight of a woman 6 feet 1 inches tall?

Steps	Discussion
1 Read and state the problem	**Read the given problem description carefully and give a clear problem statement based on your interpretation of the problem description.**

Ex 1

Predict the number of phones there will be in the U.S. in 2012, based on the information in the problem.

Your Turn

Steps	Discussion
2 Inventory the information	**Make three lists:** 1. **All information *given* in the problem description** 2. **What information you feel is *missing*** 3. **What you need to *figure out***

Ex 1

In 1982 there were less than 50,000 mobile phones

In 2000 there were 110 million mobile phones

In 2008 there were 270 million mobile phones

Three quarters of the United States population used mobile phones in 2008

Your Turn

Steps	Discussion
3 Eliminate irrelevant information	**Carefully examine the information to see which items can be eliminated because they are not pertinent to solving the problem.**

Ex 1

In 1982 there were less than 50,000 mobile phones

In 2000 there were 110 million mobile phones

In 2008 there were 270 million mobile phones

~~Three quarters of the United States population used mobile phones in 2008~~

Your Turn

Steps	Discussion
4 Identify variables, constants and assumptions	Identify and define all variables, constants, and assumptions that are required.

Ex 1

Variables and Constants
y = the number of years after 2000 n = the number of phones in millions after y years r = rate of increase in the number of phones per year b = the beginning number of phones in millions (110)
Assumptions
The increase in number of phones from 2000 to 2008 is linear; the period from 2000 to 2008 is likely to be a better predictor of the change from 2008 to 2012.

Your Turn

5 Determine the form for the solution	How does the writer of the problem description want you to present the answer?

Ex 1

The number of mobile phones in 2012.

Your Turn

Steps	Discussion
6 Simplify the problem	**Identify logical subdivisions of the problem that match up with processes/solutions you are already familiar with.**

Ex 1

1) Determine the rate at which the number of phones increases each year.
2) Determine the model for the number of phones, a certain number of years after 2000.

Your Turn

Steps	Discussion
7 Model each sub-problem	**Precisely translate each sub-problem into mathematical expressions, using mathematical conventions.**

Ex 1

1) Determine the rate at which the number of phones increases each year from the year 2000 to the year 2008.

   110 million to 270 million mobile phones in 8 years

   $$r = \frac{(270 \text{ million phones}) - (110 \text{ million phones})}{8 \text{ years}}$$

   $$r = \frac{160 \text{ million phones}}{8 \text{ years}}$$

2) Determine the linear model for the number of phones after a number of years

   $n = b + r \times y$ (in millions of phones).

Your Turn

Steps	Discussion
8 Integrate the sub-problems	Using mathematical reasoning, sequentially solve each of the sub-problems leading to the final desired solution. Remember to observe both clarity and neatness in your process and presentation!

Ex 1

1) $r = \dfrac{160 \text{ million phones}}{8 \text{ years}} = \mathbf{20\ million\ phones\ per\ year}$

2) $n = b + r \times y$ (in millions of phones)

$n = 110 + 20 \times y$

In year 2012, $y = \mathbf{12}$

$n = 110 + 20 \times 12 = 110 + 240 =$ **350 (in millions)**

Your Turn

9 Present the solution	Present the solution in the required form (as stated in Step 5).

Ex 1

There will be 350 million mobile phones in the U.S. in 2012

Your Turn

Steps	Discussion
10 Validate the solution	**Validate the solution using the validation techniques presented in Section 2.4.**

Ex 1

Check the model:

In year 2000, $y = 0$ and $n = 110 + 20 \times 0 = 110$ ✓

In year 2008, $y = 8$ and $n = 110 + 20 \times 8 = 110 + 160 = 270$ ✓

Check the number of mobile phones in 2012:

$$n = 110 + 20 \times y$$

$$350 \stackrel{?}{=} 110 + 20 \times 12$$

$$350 \stackrel{?}{=} 110 + 240$$

$$350 = 350 \checkmark$$

Your Turn

Note that this is only one method of solving the problem; there are other, equally valid approaches.

Model 1.4

In 2002, a mathematics teacher purchased her home in Fort Wayne, Indiana for \$120,000. The value of her home fell at the same rate each year, and in 2010, the value of her home was \$104,000. If she retires in 2012, what will her home be worth?

Step 1 Read and restate the problem

What will the value of her home be in 2012?

Step 2 Inventory the information

Her home is in Fort Wayne, Indiana.

The value of her home was \$120,000 in 2002.

The value of her home was \$104,000 in 2010.

Step 3 Eliminate irrelevant information

~~Her home is in Fort Wayne, Indiana.~~

The value of her home was \$120,000 in 2002.

The value of her home was \$104,000 in 2010.

Step 4 Identify variables, constants, and assumptions

Variables and Constants
v = value of the home in dollars y years after 2002
y = the number of years since 2002
r = the rate at which the value of the home decreased each year (in dollars per year)
b = the value of the home in dollars in 2002
Assumptions
The value of the home decreases at the same rate after 2010.

Step 5 Determine the form for the solution

State what the value of her home would be in 2012.

Step 6 Simplify the problem

1) How much does the value of the home decrease each year?
2) How much is her home worth after a number of years?

Step 7 Model each sub-problem

1) How much does the value of the home decrease each year?

$$r = \frac{\$120000 - \$104000}{8 \text{ years}}$$

2) How much is her home worth after a number of years?

$$v = b - r \times y$$

Step 8 Integrate the sub-problems

1) $r = \dfrac{\$120000 - \$104000}{8 \text{ years}} = \dfrac{\$16000}{8 \text{ years}} = \mathbf{2000\dfrac{\$}{yr}}$

2) In 2012, she will have owned the home for 10 years.

$$v = b - r \times y$$
$$v = 120000 - 2000 \times 10$$
$$v = 120000 - 20000$$
$$v = \mathbf{100000}$$

Step 9 Present the solution

Her home will be worth $100,000 in the year 2012.

Step 10 Validate the solution

Check the value of her home at the beginning when $y = 0$ and at the end when $y = 8$:

$$v = 120000 - 2000 \times 0 = 120000 \checkmark$$
$$v = 120000 - 2000 \times 8 = 120000 - 160000 = 104000 \checkmark$$

Check the computation of the value of the home.

$$100000 + 20000 = 120000 \checkmark$$

Addressing Common Errors 1.4

The following table shows some of the most common errors that learners tend to make with the kinds of problems covered in this section.

ERROR: Misunderstanding ambiguous words

Incorrect Process

Interpret the sentence:
"There are six times as many girls as boys."

Think:

6 times the number of girls is the number of boys

Let g = the number of girls

Let b = the number of boys

$$6 \cdot g = b$$

Resolution

Read the sentence carefully to understand that there are more girls than boys.

Correct Process

Think:

6 times the number of boys is the number of girls

Let g = the number of girls

Let b = the number of boys

$$6 \cdot b = g$$

Validation

If there are 5 boys there will be 30 girls. Hence, there are six times as many girls as boys. ✓

ERROR: Incorrectly identifying variables

Incorrect Process

The driving time from Boston to New York is 8 hours shorter than the driving time from Chicago to New York.

Think:

Let d = the difference

Let t = the driving time

$$d = 8t$$

Resolution

Read carefully to understand that the difference is a difference between two times. The shorter time is between Boston and New York, because Boston is closer to New York than Chicago is.

Correct Process

Think:

Let t = the time it takes to drive from Boston to New York

Let T = the time it takes to drive from Chicago to New York

$T - t$ = the positive difference between the times it takes to drive to New York from Chicago and to New York from Boston.

$$T - t = 8$$

Validation

The differences in the driving times will be a positive number if the lesser time is subtracted from the greater time. That difference should be 8 hours. ✓

ERROR: Using confusing variable names

Incorrect Process

Interpret the following sentence:

Find the volume of a box whose sides have lengths 12 inches, 3 feet, and 2 meters.

Let f = the length of the first side

Let s = the length of the second side

Let t = the length of the third side

$$V = f \cdot s \cdot t$$

Resolution

Read the sentence carefully to understand that the measures are not appropriate variable labels. You will see how to convert measures in Section 1.7.

Correct Process

Let l_1 = the length of side 1, in feet

Let l_2 = the length of side 2, in feet

Let l_3 = the length of side 3, in feet

$V = l_1 \cdot l_2 \cdot l_3$

Validation

The measures are all converted to the same measure. The volume is the product of three lengths. ✓

Preparation Inventory 1.4

I can...	OR *Here's my question...*
☐ convert English statements to clear mathematical statements.	☐
☐ read and identify variables, constants, and assumptions.	☐
☐ determine the form of a solution to a word problem.	☐
☐ present a solution to a word problem.	☐

Critical Thinking Questions 1.4

1. When inventorying information, what are five potential types of information in the problem description?

2. How do you determine that you have completely inventoried a problem description?

3. How do you identify relevant information?

4. If there is information that you do not have that is required to solve the problem, what are your choices to complete the problem solution?

5. How do you determine the specifications for the final answer?

6. What do you find to be the most difficult part of the methodology for interpreting word problems? Why?

7. What does it mean to validate a word problem?

Demonstrate Your Understanding 1.4

Interpret and solve each of the following word problems.

1. Your bank assesses a $5.00 monthly fee for having a checking account and $0.20 for each check you cash. At the end of the year, you reorder your checks for $10.00 for 100 checks, which is the annual usage of checks. What is your monthly checking account cost?

2. In preparing to go to the movie theater, your mom has collected cereal boxes that offer a movie promotion. Six boxes can be converted to a free movie pass, while three boxes get you a food certificate for $3.00. You have a total of 15 boxes saved up, and you want to go to the matinee that costs $5.50. How should you optimize the use of the tickets, given that you have three people going to the theater? And how much money do you need in order to make sure that everyone has a $3.00 budget for food? The movie starts at 3:30 and is likely to sell out since it has gotten rave reviews.

3. In planning the party for Grandpa's 80th birthday you are going to rent out the luxury suite at the local country club. A promotion says that you must have a minimum of 40 people and that each person will be charged $35. This includes full buffet for every person, the dedicated room, at-the-table preparation of cherries flambé for dessert (1 dessert for every 5 people), valet parking for every guest (value of $8) and free admission (general admission is $8). You plan on negotiating the rate down to $25/person since the country club is undergoing some renovation and tends to be far less busy during the Saturday of Thanksgiving week. You want to make it affordable for all the grandchildren, so you will only charge them $5 per person. Given that 32 grandchildren and friends plan to come to the party, what would be the cost for each of Grandpa's children, given that there are six siblings and each will be bringing their spouse?

Create and Solve the Hardest Problem 1.4

Your Hardest Problem

Problem Solution

What makes a word problem hard to solve?

Identify, Describe, & Correct the Errors 1.4

Identify the error you find in each of the following worked solutions and describe the error made. Solve the problem correctly in the space provided. You are welcome to solve the problem first, if you wish.

Problem 1

In 2005, the revenue for a community college baseball team was $13,000 (the sum of gifts from its booster club and ticket sales from home games). In 2010, the team began the season with $10,000 in booster club funds. Tickets for home games are sold at $4.00 each. Find the revenue for the club if 5,000 home game tickets are sold.

Worked Solution *(Find the error!)*

Read and state the problem	*Inventory the information*	*Eliminate irrelevant information*
What is the revenue if 10000 tickets are sold?	Revenue in 2005 was $13000 In 2010, the booster club gave $5000 In 2005, 10000 tickets were sold Tickets cost $4.00	~~Revenue in 2005 was $13000~~ In 2005, the booster club gave $5000 In 2010, 10000 tickets were sold Tickets cost $4.00
Identify the variables, constants, and assumptions	*Determine the form of the solution*	*Simplify the problem*
Variables and Constants Rt = the revenue in dollars for the number of tickets sold R = total revenue t = number of tickets sold 4 = ticket cost in dollars $5000 = booster club gift 10000 = tickets sold **Assumptions:** None	How much of the revenue comes from ticket sales? What is the revenue in dollars if 10000 tickets are sold?	1) The revenue from tickets sales is the price of each ticket times the number of tickets sold. 2) The total revenue is the booster gifts plus the revenue from ticket sales.
Model each sub-problem	*Integrate the sub-problems*	*Present the solution*
1) The revenue from tickets sales is the price of each ticket times t tickets: $Rt = 4 \times t$ 2) The total revenue is the booster gifts plus the revenue from ticket sales: $R = 5000 + Rt$	1) $Rt = 4 \times t = 4 \times 10000 = \mathbf{40000}$ 2) $R = 5000 + Rt$ $R = 5000 + 40000$ $R = \mathbf{45000}$	**The revenue for the club was $45,000.**

Describe the Error

Correct Process

Problem 2

In 2008, a local newspaper reported that a nearby river is polluted with 285 parts per million (ppm) of a toxic substance. An ecology instructor at your community college at that time estimated that the pollution could be reduced by 15 ppm each year in his book, *50 Parts Per Million: The Increasing Threat of Toxic Runoff*. According to the instructor's estimate, what should the pollution be in 2014?

Worked Solution *(Find the error!)*

Read and state the problem

Given a starting pollution level in a given year and an increase each year, what is the pollution level in 2014?

Inventory the information

Increasing toxic threat

Each year the pollution increases by 50 ppm

There were 285 ppm in 2008

There are 6 years from 2008 to 2014

Eliminate irrelevant information

~~Increasing toxic threat~~

Each year the pollution increases by 50 ppm

There were 285 ppm in 2008

There are 6 years from 2008 to 2014

Identify the variables, constants, and assumptions

Variables and Constants

P = the amount of pollution in ppm

y = the number of years since 2008

285 ppm = amount of pollution in 2008

50 ppm = rate at which pollution increases each year

Assumptions: None

Determine the form of the solution

State the pollution in ppm in 2014.

Simplify the problem

1) The pollution each year
2) The total pollution in 2014

Model each sub-problem

1) $P = 285 + 50 \times y$
2) The pollution increased for the years from 2008 to 2014

Integrate the sub-problems

1) $P = 285 + 50 \times y$
2) The pollution increased for 6 years (2008 to 2014) so $y = \mathbf{6}$

$P = 285 + 50 \times y$

$P = 285 + 50 \times 6 = \mathbf{585}$

Present the solution

The pollution will increase to 585 ppm in 2014.

Validate the solution

Check the ppm for 2008 ($y = 0$): $P = 285 + 50 \times y = 285 + (50 \times 0) = 285$ ✓

Describe the Error

Correct Process

What Did You Learn About Learning Math? 1.4

1. How do you think about solving a word problem?

2. What part of the methodology did you find most difficult to understand?

3. What did you learn about reading carefully?

4. What other learning skills did you use in this section?

Section 1.5 Simplifying Algebraic Expressions

Why? 1.5

Part of mastering the language of mathematics involves writing algebra using the fewest symbols possible. In the case of algebraic expressions, that means eliminating parentheses, reducing the number of operations to a minimum, and reducing duplicate instances of variables, so that, if you substitute particular values for the variables, you will have the fewest possible number of steps to go through to evaluate the expression.

What Do You Already Know? 1.5

How does the Commutative Property differ from the Associative Property?

When do you use the Distributive Property?

Give an example where the Order of Operations must be used.

Goals 1.5

When you complete this activity, you should be able to:

1. Change an expression to an equivalent expression
2. Simplify an expression
3. Recognize when an expression is fully simplified

Building Mathematical Language 1.5

New Terms: binary operator like terms unary operator

Key Concepts 1.5

Like terms are terms that have the same variables with the same exponents.

A ***unary operator*** operates on one number. Taking the negative of a number (–) is a unary operator. Taking the negative of 3 is –3.

A ***binary operator*** requires two numbers. Subtraction (–) is a binary operator. For example, 5 subtract 3 is (5 – 3).

Two expressions are ***equivalent*** if substituting any values for the variables in one expression yields the same result as substituting the same values for the same variables in the other expression

A simplified algebraic expression: the sequence of terms begins with the highest power term on the left and has no parentheses, no double signs, and all fractions have been reduced.

Technique 1.5
Removing Parentheses

Removing the parentheses is a technique for rewriting an expression or equation.

Limitation: If there are more than two factors involved with parentheses, errors can occur.

Use the Distributive Property to substitute for parts of an expression that have parentheses and validate the signs when removing the parentheses.

Example 1: Remove the parentheses:

$3(a-2b)+4(2a-b)-5$

$3a-6b+8a-4b-5$

Here $3(a-2b)$ is equivalent to $3a-6b$ and $4(2a-b)$ is equivalent to $8a-4b$ by the Distributive Property. We replace $3(a-2b)$ and $4(2a-b)$ with their equivalent expressions.

Example 2: Remove the parentheses

$2-3x(2-5x-4y)+7$

$2-(3x(2-5x-4y))+7$

$2-(6x-15x^2-12xy)+7$

$2-6x+15x^2+12xy+7$

The expression $3x(2-5x-4xy)$ is replaced by the equivalent expression $(6x-15x^2-12xy)$ and we validate the signs when removing the parentheses. In this process, $-3x$ was ***distributed over*** $(2-5x-4y)$. Note that this expression has **not** yet been fully simplified.

Methodology 1.5

Simplifying Algebraic Expressions

The methodology for solving the problem in Example 1 is shown in the following table. Students should complete Example 2 in the same table, following each step in the methodology.

Note: Algebraic expressions come in many forms, some with a few terms, and some with many terms. In working with algebraic terms, experts like to simplify the expression into its easiest interpretable form. The specification for such forms is to have only one of each type of term.

	Example 1	Example 2	Your Turn
Simplify:	$5a - 3a(a^2 + 2)$	$2(x - 4) - 3x^2 - (x - 4)$	$-3x + 2(5 - x) - x(5 + x) - 7$

Steps	Discussion
1 Remove parentheses	Distribute through the expression so as to remove all parentheses in the expression.

Ex 1	Ex 2	Your Turn
$5a - 3a^3 - 6a$	$2x - 8 - 3x^2 - x + 4$	

Steps	Discussion
2 Check signs	Review all coefficients for accuracy in signs, carefully checking for negatives and double negatives.

Ex 1	Ex 2	Your Turn
Checked.	Checked.	

Steps	Discussion
3 Identify the number of terms for simplified answer	Look through every term to inventory how many like terms you will have and produce that number of groupings.

Ex 1	Ex 2	Your Turn
$-3a^3$ $5a, \ -6a$	$2x, -x$ $-3x^2$ $-8, 4$	

Steps	Discussion
4 Set up buckets	Set up the buckets with the variables from Step 3 using parentheses and addition signs.

Ex 1	Ex 2	Your Turn
$(\quad)a^3 + (\quad)a$	$(\quad)x^2 + (\quad)x + (\quad)$ constant	

Steps	Discussion
5 Rearrange the original expression	**Rearrange the original expression by grouping coefficients within the appropriate parentheses.**

Ex 1

$(-3a^3) + (5a - 6a)$

Ex 2

$(-3x^2) + (2x - x) + (-8 + 4)$

Your Turn

Steps	Discussion
6 Validate all terms used	**Make sure that all terms in the previous expression are accounted for by lightly crossing out, term by term, each term that has been incorporated.**

Ex 1

$\not{5}\not{a} - \not{3}\not{a}^{\not{3}} - \not{6}\not{a}$

$(\not{-}\not{3}\not{a}^{\not{3}}) + (\not{5}\not{a} \not{-}\not{6}a)$

Ex 2

$\not{2}\not{x} - \not{8} - \not{3}\not{x}^{\not{2}} - \not{x} + \not{4}$

$(\not{-}\not{3}\not{x}^{2}) + (\not{2}\not{x} - \not{x}) + (-\not{8} + \not{4})$

Your Turn

Steps	Discussion
7 Combine the terms	**Convention says that each variable has a single coefficient and that the computations needed to remove parentheses should be carried out.**

Ex 1

$-3a^3 + 5a - 6a$

$-3a^3 + (-a)$

$-3a^3 - a$

Ex 2

$-3x^2 + (2x - x) + (-8 + 4)$

$-3x^2 + (x) + (-4)$

$-3x^2 + x - 4$

Your Turn

Steps	Discussion
8 Validate by substitution	**Pick a number other than 0,1, and –1 to substitute for the variable in both the original and simplified expressions. If both values are equivalent, you have validated your work.**

Ex 1

Let $x = 2$	*And*
$5a - 3a(a^2 + 2)$	$-3a^3 - a$
$5(2) - 3 \bullet 2(2^2 + 2)$	$-3(2^3) - 2$
$10 - 6(4 + 2)$	$-3(8) - 2$
$10 - 6(6)$	$-24 - 2$
$10 - 36$	-26 ✓
-26	

Ex 2

Let $x = 2$	*And*
$2(x - 4) - 3x^2 - (x - 4)$	$-3x^2 + x - 4$
$2(2 - 4) - 3 \bullet 2^2 - (2 - 4)$	$-3(2^2) + 2 - 4$
$2(-2) - 3 \bullet 4 - (-2)$	$-3(4) + 2 - 4$
$-4 - 12 + 2$	$-12 + 2 - 4$
-14	-14 ✓

Steps	Discussion
8 Validate by substitution (continued)	Pick a number other than 0,1, and –1 to substitute for the variable in both the original and simplified expressions. If both values are equivalent, you have validated your work.

Your Turn

Addressing Common Errors 1.5

The following table shows some of the most common errors that learners tend to make with the kinds of problems covered in this section.

ERROR: Incorrectly distributing a negative sign

Incorrect Process	Resolution
$-(x-3)=-x-3$	*Correctly distribute the negative sign realizing the negative sign represents –1.*

Correct Process	Validation
$-(x-3)=-x+3$	Substitute 5 for x in both expressions. $-(5-3)=-2$ and $-5+3=-2$

ERROR: Double operation symbols

Incorrect Process	Resolution
$-(x-3)=-x--3$	*Use symbols correctly.*

Correct Process	Validation
$-(x-3)=-x-(-3)=-x+3$	Substitute 7 for x in both expressions. $-(\mathbf{7}-3)=-2$ and $-\mathbf{7}+3=-2$

Preparation Inventory 1.5

I can...	OR *Here's my question...*
☐ recognize like terms.	☐
☐ add like terms.	☐
☐ subtract like terms.	☐
☐ use the Order of Operations.	☐

Critical Thinking Questions 1.5

1. What are equivalent expressions?

2. What changes can you make to an expression to yield an equivalent expression?

3. Why do you remove parentheses?

4. How do you remove parentheses?

5. How do you ensure that the signs are correct?

6. How do you ensure that all the terms are used?

7. Why shouldn't you use zero in validating that terms are equivalent?

8. How might you use graphing or the table feature of a calculator to validate that your simplified expression in one variable is equivalent to the original expression?

Demonstrate Your Understanding 1.5

Simplify each of the following expressions.

1. $2x(5 - x^2) + x^3$

2. $a^2 - 4a(a + 3) - 6$

3. $12 - 3(a - 5) - 7a$

4. $-3x(5-x)+3x-3(x+2)$

5. $4x^2-3x-8+7x(x-2)+5$

6. $-(x-2)+3(2-x)+7x^3-x+2x(x+3)$

Create and Solve the Hardest Problem 1.5

Your Hardest Problem
Problem Solution

What makes a problem where you must simplify an algebraic expression a hard problem to solve?

Identify, Describe, & Correct the Errors 1.5

Identify the error you find in each of the following worked solutions and describe the error made. Solve the problem correctly in the space provided.

Problem 1

Simplify:

$$5-2(x-1)$$

Worked Solution *(Find the error!)*

$5-2(x-1)$

$5-2x-1$

$4-2x$

Correct Process

Describe the Error

Problem 2

Simplify and validate:

$$2x-3x+5x^2+4x-5x+7$$

Worked Solution *(Find the error!)*

$2x-3x+5x^2+4x-5x+7$

$(5x^2)+(2x-3x-5x)+(7)$

$5x^2-6x+7$

Validate:

Let $x=0$.

$2x-3x+5x^2+4x-5x+7$

$2\bullet 0-3\bullet 0+5\bullet 0^2+4\bullet 0+7$

$0-0+0+0+7$

7

And

$5x^2-6x+7$

$5\bullet 0^2-6\bullet 0+7$

$0-0+7$

7 ✓

Correct Process

Describe the Error

Problem 3

Simplify:

$$-3x^2-2x(3-x)+7$$

Worked Solution *(Find the error!)*

$$-3x^2-2x(3-x)+7$$
$$-3x^2-6x-2x^2+7$$
$$-5x^2-6x+7$$

Correct Process

Describe the Error

What Did You Learn About Learning Math? 1.5

1. What did you learn about working with algebraic expressions?

2. How did you use the methodologies and common errors in working through the Demonstrate Your Understanding Problems?

3. What methods will you use to remind yourself of important concepts dealing with simplifying algebraic expressions?

4. What learning skills did you use in this section?

Section 1.6 Exponents and Expanding Expressions

Why? 1.6

Expressions have many equivalent forms. You can glean different information from different but equivalent forms of an expression. For example, the cost of manufacturing x carrot cakes is the sum of the cost of the ingredients and the cost of labor. This cost may be represented with the expression $i \cdot x + l \cdot x$ (where i is the cost of ingredients per cake and l is the cost of labor per cake). The equivalent expression, $(i + l) \cdot x$ gives the information that the sum of the cost of ingredients and labor for one cake times the number of cakes is also the manufacturing cost for x carrot cakes. In this section, you will explore a variety of ways of creating equivalent forms of an expression.

What Do You Already Know? 1.6

What is a complex number?

Which of the following are polynomials?

a. $x^5 - 3x^2 + 2$ b. $\sqrt{x} - 5$ c. $2x - 5$ d. $\frac{1}{x^2} + 5x - 5$ e. 3

What is an exponent?

For the polynomial $x^2 - 3x^5 - 5x + 2$, why is its degree 5 and not 2?

Write an expression in p and s.

Goals 1.6

When you complete this activity, you should be able to:

1. Use the Laws of Exponents
2. Expand powers of binomials, trinomials, and general polynomials
3. Use the basic Laws of Logarithms

Building Mathematical Language 1.6

New Terms: degree of a polynomial expanding a polynomial logarithm

Key Concepts 1.6

The ***Properties of Exponents*** are important when operating with expressions.

Limitation/Caution: It is important when working with the Laws of Exponents to remember that the properties do not apply to addition and subtraction.

Properties of Exponents

Exponent Rule	Symbolic Representation	Numeric Example	Algebraic Example
	Observations		
Zero Exponent Rule	$x^0 = 1$	$29^0 = 1$	$p^0 = 1$
	Any number raised to an exponent of zero = 1. This is because a zero exponent shows no factors of the base. Do not confuse this with a base raised to an exponent of 1, such as 6^1 (which is equal to 6).		
Negative Exponent Rule	$x^{-m} = \frac{1}{x^m}$	$7^{-3} = \frac{1}{7^3}$ and $\frac{1}{7^{-3}} = 7^3$	$y^{-3} = \frac{1}{y^3}$ and $\frac{1}{y^{-3}} = y^3$
	Any number raised to a negative exponent is defined to be the reciprocal of the number to the power that is the opposite of that exponent, sign-wise. The negative sign of the exponent indicates the idea of inverse location, not opposite value.		
Product Rule	$x^m x^n = x^{m+n}$	$3^1 \cdot 3^4 =$ $(3)(3 \cdot 3 \cdot 3 \cdot 3) =$ $3^{1+4} = 3^5$	$x^3 \cdot x^4 =$ $xxx \cdot xxxx =$ $x^{3+4} = x^7$
	When multiplying, make sure the bases are the same number (or variable) before adding exponents. Think: multiplication is linked to addition.		
Power Rule	$\left(x^m\right)^n = x^{m \cdot n}$	$\left(7^4\right)^3 =$ $\left(7^4\right)\left(7^4\right)\left(7^4\right) =$ $7^{4 \cdot 3} = 7^{12}$	$\left(n^3\right)^5 =$ $\left(n^3\right)\left(n^3\right)\left(n^3\right)\left(n^3\right)\left(n^3\right) =$ $n^{3 \cdot 5} = n^{15}$
	This rule is often called the "power to power" rule. Multiply the exponents; keep the base unchanged.		
Quotient Rule	$\frac{x^m}{x^n} = x^{m-n}$	$\frac{10^5}{10^2} =$ $= \frac{10 \cdot 10 \cdot 10 \cdot \cancel{10} \cdot \cancel{10}}{\cancel{10} \cdot \cancel{10}}$ $= 10^{5-2} = 10^3$	$\frac{p^4}{p^9} = \frac{\cancel{p}\cancel{p}\cancel{p}\cancel{p}}{\cancel{p}\cancel{p}\cancel{p}\cancel{p}ppppp}$ $= p^{4-9} = p^{-5} = \frac{1}{p^5}$
	When the bases are the same, division is performed by subtracting exponents. Division is repeated subtraction; make the mental connection linking division with subtraction.		

Exponent Rule	Symbolic Representation	Numeric Example	Algebraic Example
	Observations		
"Product to Power" Rule	$(ab)^n = a^n \cdot b^n$	$(3 \cdot 2)^4 =$ $= 3^4 \cdot 2^4$ Validate: $6^4 = 1296$ and $3^4 \cdot 2^4 = 81 \times 16 = 1296$	$(5 \cdot y)^3 =$ $= 5^3 \cdot y^3 = 5^3 y^3$ $= 125y^3$
	The product to power rule is used when different bases are raised to the same exponent. It is not the same as the distributive property—no addition is present—but it does distribute the exponent over multiplication. Use the commutative property to rearrange factors.		
"Quotient to Power" Rule	$\left(\frac{a}{b}\right)^n = \frac{a^n}{b^n}$	$\left(\frac{5}{4}\right)^2 = \frac{5^2}{4^2} = \frac{25}{16}$	$\left(\frac{2}{y}\right)^5 = \frac{2^5}{y^5} = \frac{32}{y^5}$
	This rule allows fractions to be raised to powers. Notice it is not like the quotient rule because the bases in the numerator and denominator are different numbers.		
Combined Rules	$\left(\frac{ax}{by}\right)^m = \frac{a^m x^m}{b^m y^m}$	$\frac{(3 \cdot 2^5)^2}{(3 \cdot 2^3)^4} =$ $\frac{3^2 \cdot 2^{10}}{3^4 \cdot 2^{12}} =$ $\frac{1}{3^2 \cdot 2^2} = \frac{1}{36}$	$\left(\frac{2x}{3y}\right)^2 = \frac{2^2 x^2}{3^2 y^2} = \frac{4x^2}{9y^2}$
	Combined rules show a logical extension of the quotient to power rule with the product to power rule. Situations where exponent rules are combined occur frequently in algebra.		

The ***logarithm of a number*** to a base is the exponent of the number when the number is written as a power of the base. For example, $\log_2(16)$ (where the base is 2) is the number 4 because $16 = 2^4$ and $\log_3\left(\frac{1}{9}\right) = -2$ (where the base is 3) because $\frac{1}{9} = 3^{-2}$. Clearly, logarithms are related to exponents; some might say that logarithms *are* exponents.

Limitation/Caution: The base of a logarithm must be a positive number. Logarithm function values exist as real numbers only for positive real numbers. In courses dealing in greater depth with complex numbers, the logarithm function is defined for negative numbers and zero.

Logarithmic Rule	Symbolic Representation	Numeric Example	Exponential Connection
	Observations		
Logarithm of 1	$\log_b(1) = 0$	$\log_3(1) = 0$	$3^0 = 1$
	The logarithm of 1 for any base is 0.		
Negative Logarithm	$\log_b(x) < 0$ if $0 < x < 1$	$\log_5\left(\frac{1}{5^2}\right) = -2$	$5^{-2} = \frac{1}{5^2} = \frac{1}{25}$
	The logarithm of any positive number less than 1 is negative.		
Product Rule	$\log_b(xy) =$ $\log_b(x) + \log_b(y)$	$\log_2(8 \cdot 16) =$ $\log_2(8) + \log_2(16)$	$2^3 \cdot 2^4 = 2^{3+4}$
	The logarithm of a product is the sum of the logarithms of each of the factors.		
Quotient Rule	$\log_b\left(\frac{x}{y}\right) =$ $\log_b(x) - \log_b(y)$	$\log_2\left(\frac{8}{16}\right) =$ $\log_2(8) - \log_2(16)$	$\frac{2^3}{2^4} = 2^{3-4}$
	The logarithm of a quotient is the difference of the logarithms of each of the factors (with the logarithm of the numerator taken first).		
Power Rule	$\log_b(x^n) =$ $n \cdot \log_b(x)$	$\log_{10}(10^3) =$ $\log_{10}(10 \cdot 10 \cdot 10) =$ $\log_{10}(10) + \log_{10}(10) + \log_{10}(10) =$ $3 \cdot \log_{10}(10)$	$10^3 = 10 \cdot 10 \cdot 10$
	The logarithm of a power of a number is the power times the logarithm of the number.		
Combined Rule	$\log_b\left(\frac{xy}{z}\right) =$ $\log_b(x) + \log_b(y) - \log_b(z)$	$\log_3\left(\frac{9 \cdot 81}{27}\right) =$ $\log_3(9 \cdot 81) - \log_3(27) =$ $\log_3(9) + \log_3(81) - \log_3(27)$	$\frac{3^2 \cdot 3^4}{3^3} = 3^{2+4-3}$
	When dealing with an expression involving powers and rational expressions, decompose the expression over several steps, often starting with the quotient.		

Methodology 1.6

Each Times Every: Multiplying Polynomials

You can transform the product of two polynomials into a sum by expanding the product using the Distributive Property.

One version of the Distributive Property is: $(a + b)c = ac + bc$

If c is a polynomial, say $(x + y + z)$, the Distributive Property still holds and using the Associative Property of Addition we have:

$$\begin{aligned}(a+b)(x+y+z) &= a(x+y+z)+b(x+y+z)\\ &= a\left[(x+y)+z\right]+b\left[(x+y)+z\right]\\ &= a(x+y)+az+b(x+y)+bz\\ &= ax+ay+az+bx+by+bz\end{aligned}$$

Here we can observe that the product is the sum of each term of the first polynomial multiplied by every term of the second polynomial. Hence this technique is called *Each Times Every.*

You may recall a method called FOIL (First, Outer, Inner, Last) for multiplying two binomials. If we use the *Each Times Every* technique to find the product of two binomials, we have:

$$\begin{aligned}(a+b)(x+y) &= a(x+y)+b(x+y)\\ &= ax+ay+bx+by\end{aligned}$$

F O I L

This is obviously the same as the FOIL method, but **only** when we're working with two binomials. The Each Times Every technique will work **regardless** of the number of terms each polynomial has.

Limitation: All terms in the product must be real numbers.

	Example 1	**Your Turn**
Find the product:	$(x+3)(3x^2-5x-7)$	$(2y-1)(4y^2+y-3)$

Steps	Discussion
1 Multiply	**Multiply EACH term in the first expression by EVERY term in the second expression.**

Ex 1

$(x+3)(3x^2-5x-7)=$

$3x^3-5x^2-7x+6x^2-10x-14$

Your Turn

Steps	Discussion
2 Simplify	**Combine like terms.**

Ex 1

$3x^3+x^2-17x-14$

Your Turn

Addressing Common Errors 1.6

The following table shows some of the most common errors that learners tend to make with the kinds of problems covered in this section.

ERROR: Determining the degree of a polynomial by looking at the first term in the polynomial

Incorrect Process	Resolution
The degree of the polynomial $3x^2 - 4x + 5x^4 - 7$ is 2.	*Rewrite the polynomial with the terms in descending order.*
Correct Process	**Validation**
The degree of the polynomial $5x^4 + 3x^2 - 4x - 7$ is 4.	The polynomial is written in descending order and the order of the first term is 4.

ERROR: Not multiplying every term by each term

Incorrect Process	Resolution
$(x^2+4x-6)(2x-3)$ $= 2x^3 - 3x^2 - 12x + 18$	*Be sure to multiply every term of the second polynomial by each term of the first polynomial. There should be as many terms in the first step as the product of the number of terms in the first polynomial and the number of terms in the second polynomial.*
Correct Process	**Validation**
$(x^2+4x-6)(2x-3)$ $= 2x^3 - 3x^2 + 8x^2 - 12x - 12x + 18$ $= 2x^3 + 5x^2 - 24x + 18$	When $x = 2$, $(x^2+4x-6)(2x-3)$ $=\left((2)^2+4(2)-6\right)(2(2)-3)$ $=(6)(1)=6$ And $2x^3+5x^2-24x+18$ $=2(2)^3+5(2)^2-24(2)+18$ $=16+20-48+18=6$ ✓

ERROR: Incorrectly using the Properties of Exponents

Incorrect Process	Resolution
Simplify the expression: $a^2 \bullet a^3 = a^{2 \cdot 3} = a^6$	*Use the Properties of Exponents correctly.*
Correct Process	**Validation**
$a^2 \bullet a^3 = a^{2+3} = a^5$	$a^2 \bullet a^3 = (a \bullet a) \bullet (a \bullet a \bullet a)$ $= a \bullet a \bullet a \bullet a \bullet a = a^5$ ✓

ERROR: **Incorrectly using the Properties of Logarithms**

Incorrect Process

Simplify the following expression:

$\log_3(a) + \log_3(b) = \log_3(a+b)$

Resolution

Use the Properties of Logarithms correctly.

Correct Process

$\log_3(a) + \log_3(b) = \log_3(a \bullet b)$

Validation

The sum of logarithms with the same base is the logarithm of the product of the arguments of the two logarithms. ✓

Preparation Inventory 1.6

I can...	OR *Here's my question...*
☐ multiply numerical expressions involving exponents.	☐
☐ multiply variable expressions involving exponents.	☐
☐ expand the product of two polynomials.	☐
☐ validate multiplication problems involving exponents.	☐

Critical Thinking Questions 1.6

1. How many terms do you expect to occur when using the Each Times Every Technique?

2. What is the preferred method of expanding two multi-term expressions?

3. Why do the multiplication and division of examples in the Properties of Exponents table end up being addition and subtraction of exponents?

4. Why do powers of the examples in the Properties of Exponents table end up being a multiplication problem?

5. How are the Properties of Logarithms related to the Properties of Exponents?

6. Which Properties of Logarithms match up with which Properties of Exponents?

7. Explain why the logarithm of 1, no matter the base, is zero.

Demonstrate Your Understanding 1.6

1. Evaluate the following exponential expressions:

a) $3^4 \cdot 3^5$

Expanded Form	Worked Solution

Validation

b) $3^2 2^3$

Expanded Form	Worked Solution

Validation

c) $\left(2^6\right)^4$

Expanded Form	Worked Solution

Validation

d) $\dfrac{\left(3^{-3}\right)^2}{3^5}$

Expanded Form	Worked Solution

Validation

e) $\left(\dfrac{2^3}{3^2}\right)^2$

Expanded Form	Worked Solution

Validation

2. Simplify the following algebraic expressions:

a) $(x^3)(x^4)(x)$

Expanded Form	Worked Solution

Validation	

b) $\left(2x^5\right)^3$

Expanded Form	Worked Solution

Validation	

c) $\left(\dfrac{3a^5}{6b^0}\right)^2$

Expanded Form	Worked Solution

Validation	

d) $5x^{-3}$

Expanded Form	Worked Solution

Validation

e) $(5x)^{-3}$

Expanded Form	Worked Solution

Validation

f) $\left(3^4 b^{-2}\right) \bullet \left(3^{-5} b^3\right)$

Expanded Form	Worked Solution

Validation

3. Expand:

a) $(2s-3t)(s-5t+6)$

b) $(x+2y)^2$

c) $\log_2(ab^2)$

d) $\log_2\left(\frac{y}{x^3}\right)$

Create and Solve the Hardest Problem 1.6

Your Hardest Problem

Problem Solution

What makes a problem where you must simplify an algebraic expression a hard problem to solve?

Identify, Describe, & Correct the Errors 1.6

Identify the error you find in each of the following worked solutions. Describe the error, solve the problem correctly, and validate your answer in the space provided. You are welcome to solve the problem first, if you wish.

Problem 1

Expand: $\log_3\left(a^4+5\right)$

Worked Solution *(Find the error!)*

$$\begin{aligned}\log_3\left(a^4+5\right) &= \log_3\left(a^4\right)+5\\ &= 4\log_3\left(a\right)+5\end{aligned}$$

Correct Process

Describe the Error

Problem 2

Expand: $\left(x^2-3\right)\left(2x+3y-5\right)$

Worked Solution *(Find the error!)*

$$\left(x^2-3\right)\left(2x+3y-5\right)$$
$$=2x^3+3x^2y-5x^2-6x^2+15$$

Correct Process

Describe the Error

Problem 3

Expand: $\left(x^2\right)^3$

Worked Solution *(Find the error!)*

$$\left(x^2\right)^3=x^5$$

Correct Process

Describe the Error

Problem 4

Expand: $\log_5(2xy)$

Worked Solution *(Find the error!)*

$$\log_5(2xy) = 2\log_5(xy)$$
$$= 2\left(\log_5(x) + \log_5(y)\right)$$
$$= 2\log_5(x) + 2\log_5(y)$$

Correct Process

Describe the Error

What Did You Learn About Learning Math? 1.6

1. What skills did you use in learning how to use Each Times Every to multiply two polynomials?

2. How did you use the methodology tables when working through the Demonstrate Your Understanding problems?

3. What did you observe about the methodology tables?

Section 1.7 Evaluating Formulas

Why? 1.7

Our world abounds with formulas. If you want to order carpet for your bedroom or paint for your walls, you need to know how to use the appropriate area formulas to determine the right amounts to order. Or if you are visiting a country where the currency is the euro and metric units are used, it is helpful to be able to use a simple formula to convert those measurements into ones that are familiar to United States residents: How much are you really paying for gasoline (in the more familiar dollars-per-gallon), if the prices given are in euros-per-liter?

What Do You Already Know? 1.7

What is a formula?

What formulas have you used in the past?

If a formula has three variables, and you want to "evaluate the formula," for how many of the variables do you need to know values?

What is the Order of Operations?

What is a code word or phrase (mnemonic) you can use to help yourself remember the Order of Operations?

Goals 1.7

When you complete this activity, you should be able to:

1. Evaluate formulas and algebraic expressions, given values for some or all of the variables
2. Use the Order of Operations to evaluate formulas and algebraic expressions
3. Understand and explain the role of units in evaluating formulas

Building Mathematical Language 1.7

New Terms: formula unit analysis

Key Concepts 1.7

A ***formula*** defines a quantity that can be calculated by providing specific value(s) for all the variables in the defined expression.

Limitation/Caution: Real world constraints determine the possible values of the variable(s).

Example: The formula for the perimeter of a rectangle in terms of its length and width is $P = 2(L + W)$. You can use the formula to compute the perimeter P when $L = 3$ feet and $W = 5$ feet.

When evaluating a formula, you can check your work by carrying the units of each number through the formula. If your evaluation is correct, then some units should divide out, leaving you with the correct units for the answer. This form of validation is called ***unit analysis*** or ***dimensional analysis***.

Limitation/Caution: This process will not work for a unitless, abstract problem.

Example: The formula for distance in terms of rate and time is: $d = r \bullet t$.

Find the value of d if $r = 30\ \dfrac{\text{miles}}{\text{hour}}$ and $t = 2$ hours.

$d = r \bullet t$

$d = \left(30\ \dfrac{\text{miles}}{\text{hour}}\right)(2\ \text{hours})$

$d = 60\left(\dfrac{\text{miles}}{\text{hour}} \bullet \text{hours}\right) = 60\left(\dfrac{\text{miles}}{\cancel{\text{hour}}} \bullet \dfrac{\cancel{\text{hours}}}{1}\right)$

$d = 60\ \text{miles}$

We can see that the correct unit for distance is *mile*.

Note: Units that are the same type of measure (i.e., length, weight, time, etc) can be related by a fraction equal to 1. For example, minutes and seconds both measure time, and since 60 seconds equals 1 minute, both the fractions $\dfrac{1\,\text{min}}{60\,\text{sec}}$ and $\dfrac{60\,\text{sec}}{1\,\text{min}} = 1$.

Limitation/Caution: Be sure that you are working with the same type of measure, such as length or time, when you create a fraction that is equal to 1.

A Reminder: When dealing with addition or subtraction, we can use the commutative property to reorder the terms in an expression, equation, or formula. Sometimes this is helpful. Consider the following example as a quick review: $3x - x^2 = -x^2 + 3x$. This is true because:

$3x - x^2 = 3x + [-(x)^2]$ *Subtraction is the addition of opposites*

$3x + [-(x)^2] = [-(x)^2] + 3x$ *Use of the commutative property*

$[-(x)^2] + 3x = -x^2 + 3x$ *Removal of parentheses*

Let's try it. Let $x = 4$:

$3x - x^2$	$-x^2 + 3x$
$3(4) - (4)^2 = 12 - 16 = -4$	$-(4)^2 + 3(4) = -16 + 12 = -4$

→ **Be careful!** $-x^2 = -(x)^2$ **NOT** $(-x)^2$**!**

Always follow the Order of Operations (in this case, compute exponents and multiplication before taking the negative).

Methodology 1.7

Using Unit Analysis (Dimensional Analysis) to Convert Units

Simplify:	Example 1	Example 2	Your Turn
	Find the distance traveled if a car is driving 50 miles per hour for 90 minutes.	Find the speed of a car in kilometers per hour, if it is driving at 1 mile per minute.	Find the volume of a suitcase that is 1.5 feet long, 15 inches wide, and 6 inches thick.

Steps	Discussion
1 Identify the units in the problem	**Make a list of all relevant units mentioned in the problem. Identify which units measure the same things.**

Ex 1

miles, hours, minutes:

hours and minutes both measure time

Ex 2

kilometers, hours, miles, minutes:

Km and miles both measure distance; hours and minutes both measure time

Your Turn

Steps	Discussion
2 Substitute given values for the variables	**Include units as you make the substitutions in the formula.**

Ex 1

$$d = r \cdot t$$

$$d = \frac{50 \text{ miles}}{1 \text{ hour}} \cdot 90 \text{ minutes}$$

Ex 2

$$r = \frac{d}{t}$$

$$r = \frac{1 \text{ mile}}{1 \text{ minute}}$$

Your Turn

Steps	Discussion
3 Simplify units	**Divide out any common units that appear in both the numerator and denominator of multiplied fractions.**

Ex 1

no units to simplify

Ex 2

no units to simplify

Your Turn

Steps	Discussion
4 Change other units	Multiply by a fraction equal to 1 to make units that measure the same thing match. (See Key Concepts).

Ex 1

$$d = \frac{50 \text{ miles}}{1 \text{ hour}} \bullet 90 \text{ min} \bullet \frac{1 \text{ hour}}{60 \text{ min}}$$

Ex 2

$$r = \frac{1 \text{ miles}}{1 \text{ minute}} \bullet \frac{60 \text{ min}}{1 \text{ hour}} \bullet \frac{1.61 \text{ km}}{1 \text{ mile}}$$

Your Turn

5 Simplify units again	Divide out any common units that appear in both the numerator and denominator of multiplied fractions.

Ex 1

$$d = \frac{50 \text{ miles}}{1 \cancel{\text{hour}}} \bullet 90 \cancel{\text{min}} \bullet \frac{1 \cancel{\text{hour}}}{60 \cancel{\text{min}}}$$

Ex 2

$$r = \frac{1 \cancel{\text{miles}}}{1 \cancel{\text{minute}}} \bullet \frac{60 \cancel{\text{min}}}{1 \text{ hour}} \bullet \frac{1.61 \text{ km}}{1 \cancel{\text{mile}}}$$

Your Turn

6 Perform all operations	Keep the remaining units as you carry out the operations.

Ex 1

$d = 75$ miles

Ex 2

$$r = \frac{196.6 \text{ km}}{\text{hour}}$$

Your Turn

7 Validate your results	Confirm that the remaining units are appropriate for the problem.

Ex 1

Yes, the distance is in miles.

Ex 2

Yes, the speed is in kilometers per hour.

Your Turn

Addressing Common Errors 1.7

The following table shows some of the most common errors that learners tend to make with the kinds of problems covered in this section.

ERROR: Substituting values for the wrong variables

Incorrect Process

Example: Find the length of a rectangle whose area is 80 ft² and width is 5 feet.

$$A = 80 \bullet 5 = 400$$

Resolution

Think about the formula you're using and carefully identify the variables.

Correct Process

$$A = l \bullet w$$

$$A = 80\text{ ft}^2,\ w = 5\text{ feet, find } l$$

$$80\text{ ft}^2 = l \bullet (5\text{ feet})$$

$$l = \frac{80\text{ ft}^2}{5\text{ feet}} = 16\text{ feet}$$

Length = 16 feet

Validation

Substitute 16 feet for l and calculate area:

$$16\text{ feet} \bullet 5\text{ feet} = 80\text{ ft}^2$$

ERROR: Not multiplying every term by each term

Incorrect Process

The area of a trapezoid is $A = \frac{1}{2}h(m+n)$

Find the area if the height, h, is 6 cm, and the bases are m =10 cm and n = 8 cm.

$$A = \frac{1}{2}(6\text{ cm})(10\text{ cm} + 8\text{ cm})$$

$$A = 3\text{ cm}(5\text{ cm} + 4\text{ cm})$$

$$A = 3\text{ cm}(9\text{ cm})$$

$$A = 27\text{ cm}^2$$

Resolution

Use the correct order of operations: Perform the addition inside the parentheses before multiplying. When multiplying, work from left to right.

Correct Process

$$A = \frac{1}{2}(6\text{ cm})(10\text{ cm} + 8\text{ cm})$$

$$A = \frac{1}{2}(6\text{ cm})(18\text{ cm})$$

$$A = (3\text{ cm})(18\text{ cm})$$

$$A = 54\text{ cm}^2$$

Validation

Compute using the distributive property:

$$A = \frac{1}{2}(6\text{ cm})(10\text{ cm} + 8\text{ cm})$$

$$A = (3\text{ cm})(10\text{ cm} + 8\text{ cm})$$

$$A = (3\text{ cm})(10\text{ cm}) + (3\text{ cm})(8\text{ cm})$$

$$A = 30\text{ cm}^2 + 24\text{ cm}^2 = 54\text{ cm}^2$$

ERROR: **Not including units**

Incorrect Process

Find the distance traveled if a car travels for 90 minutes at 60 miles per hour.

$$d = rt$$
$$d = 90 \bullet 60$$
$$d = 6300$$

Resolution

Include units and make sure they match.

Correct Process

$d = rt$

$r = \dfrac{60 \text{ miles}}{1 \text{ hour}}, \quad t = 90 \text{ minutes}$

conversion between hour and minutes: $\dfrac{1 \text{ hour}}{60 \text{ minutes}}$

$d = \left(\dfrac{60 \text{ miles}}{1 \text{ hour}}\right)(90 \text{ minutes})\left(\dfrac{1 \text{ hour}}{60 \text{ minutes}}\right)$

$d = \left(\dfrac{\cancel{60} \text{ miles}}{1 \cancel{\text{hour}}}\right)(90 \cancel{\text{minutes}})\left(\dfrac{1 \cancel{\text{hour}}}{\cancel{60} \cancel{\text{minutes}}}\right) = 90 \text{ miles}$

Validation

Divide distance (d) by time (t) to verify the rate (r) originally given:

$90 \text{ miles} = r \bullet (90 \text{ minutes})$

$r = \dfrac{90 \text{ miles}}{90 \text{ minutes}} = \dfrac{1 \text{ mile}}{1 \text{ minute}}$

convert this rate to **miles per hour** (the units originally given):

$\left(\dfrac{1 \text{ mile}}{1 \text{ minute}}\right)\left(\dfrac{60 \text{ minutes}}{1 \text{ hour}}\right)$

$= \dfrac{60 \text{ miles}}{1 \text{ hour}} = 60 \dfrac{\text{miles}}{\text{hour}}$ ✓

Preparation Inventory 1.7

I can...	OR *Here's my question...*
☐ correctly multiply a fraction by an integer.	☐
☐ substitute correctly in a formula.	☐
☐ apply the order of operations correctly.	☐
☐ use unit analysis correctly in a formula.	☐

Critical Thinking Questions 1.7

1. In what order do you perform the steps when you use the Order of Operations?

2. Which step in the Order of Operations do you find the most difficult to carry out? Why?

3. What is the purpose of the Order of Operations?

4. How is the fraction bar like parentheses?

5. When evaluating an expression and substituting in specific values for variables, how do you guarantee that the values are substituted in correctly?

6. How does a formula differ from an expression?

7. Why are the units required when evaluating a formula?

Demonstrate Your Understanding 1.7

1. Evaluate the following expressions for the given value(s).

   a) $3x^2 - 5x + 4$ for $x = -2$.

   b) $5(r^2 - 3s) - r + 3$ for $r = -2$ and $s = 4$.

2. The formula for the volume of a cylinder in terms of its radius and height is given by the formula $V = \pi r^2 h$. Find the volume of a cylindrical oil tank with a radius of 7 meters and a height of 20 meters.

3. The formula for the intensity of light, I, in terms of the distance, d, from the source of the light, is given by the equation $I = \frac{k}{d^2}$.

   Find the intensity of light, I, when $d = 10$ meters and the constant $k = 350$ lumens • meters^2.

Create and Solve the Hardest Problem 1.7

Your Hardest Problem
Problem Solution

What makes a problem using or involving formulas a hard problem to solve?

Identify, Describe, & Correct the Errors 1.7

Identify the error you find in each of the following worked solutions and describe the error made. Solve the problem correctly in the space provided. You are welcome to solve the problem first, if you wish.

Problem 1

Evaluate: $-x^2 + 4x - 5$ for $x = -3$

Worked Solution *(Find the error!)*

$-3^2 - 12 - 5$

$9 - 12 - 5$

$-3 - 5$

-8

Correct Process

Describe the Error

Problem 2

Evaluate $3 + 3xy - \dfrac{2x}{x+y}$

for $x = -2$ and $y = 4$.

Worked Solution *(Find the error!)*

$3 + xy - \dfrac{2x}{x+y}$

$3 + (-2)(4) - \dfrac{2(-2)}{(-2)+(4)}$

$3 + (-2)(4) - \dfrac{2\cancel{(-2)}}{\cancel{(-2)}+(4)}$

$3 - 8 - \dfrac{2}{4}$

$3 - 8 - \left(\dfrac{1}{2}\right)$

$-\dfrac{11}{2}$

Correct Process

Describe the Error

Problem 3

Evaluate:

$$3x - 2p + 3\sqrt{s} - \frac{2}{s^2 - 5s}$$

for $s = 4$, $p = 5$ and $x = -2$.

Worked Solution *(Find the error!)*

$$3x - 2p + 3\sqrt{s} - \frac{2}{s^2 - 5s}$$

$$3(-2) - 2(5) + 3\sqrt{(4)} - \frac{2}{(4)^2 - 5(4)}$$

$$3(-2) - 2(5) + 3 \bullet 2 - \frac{2}{-4}$$

$$3(-4) \bullet 5 + 6 + \frac{1}{2}$$

$$-54 + \frac{1}{2}$$

$$-\frac{108}{2} + \frac{1}{2}$$

$$-\frac{107}{2}$$

Correct Process

Describe the Error

Problem 4

The formula for the height, *h*, of a ball dropped from a 200-ft tower after *t* seconds is

$$h = 200 \text{ ft} - a \bullet (t \text{ sec})^2$$

where *t* is measured in seconds and

$$a = 16 \frac{\text{ft}}{\text{sec}^2}.$$

Find the height of the ball after 3 seconds.

Worked Solution *(Find the error!)*

$$h = 200 \text{ ft} - a \bullet (t\,\text{sec})^2$$

$$h = 200 \text{ ft} - \left(16 \frac{\text{ft}}{\text{sec}^2}\right)(3\,\text{sec})^2$$

$$h = 200 \text{ ft} - 48 \left(\frac{\text{ft}}{\cancel{\text{sec}^2}} \bullet \cancel{\text{sec}^2}\right)$$

$$h = 200 \text{ ft} - 48 \text{ ft}$$

$$h = 152 \text{ feet}$$

Correct Process

Describe the Error

What Did You Learn About Learning Math? 1.7

1. What did you learn about correctly using formulas?

2. How did you use the methodology and common errors to help you as you worked through the Demonstrate Your Understanding Problems?

3. What methods will you use to remind yourself of important concepts dealing with evaluating and using formulas in other courses or situations?

4. What learning skills did you use in this section?

Concept Map — Chapter 1

Terminology from the chapter
check the terms used in your map

__binary operator

__complex numbers

__conjugate

__constant

__degree of a polynomial

__expanding a polynomial

__formula

__identities

__inverses

__irrational numbers

__like terms

__logarithm

__number systems

__pure imaginary numbers

__radical

__radical index

__radicand

__rational numbers

__unary operator

__unit analysis

__variable

Your instructor will specify how many terms from the chapter you should use in your concept map.

CHAPTER 2

This chapter develops methods for transforming an equation to an equivalent form using the properties of equations and substitution. You will learn to use these methods to solve linear equations and systems of linear equations. The final section of this chapter explains need for and the importance of validating both your work and how well you have learned new material.

Section 2.1 Equations: Equivalence and Substitution

Why? 2.1

Equations arise when solving mathematical, scientific, and business problems. Such equations can be solved using the principle of substitution. For example, $A=P(1+r)^t$ could be solved using substitution to determine that the amount of money in a savings account that starts off with \$1000 grows to almost \$4300 after 25 years if the interest rate is 5%.

What Do You Already Know? 2.1

Why are the following expressions equivalent or not equivalent? $2(x+5)-3$ and $2x+7$

If $5x-15$ is substituted for $5(x-3)$ in the expression $5(x-3)+2x+6$, is the resulting expression equivalent to the original expression?

What is the difference between an expression and an equation?

How would you validate that $x = 2$ is a solution to the equation $2x-5=-1$?

Goals 2.1

When you complete this activity, you should be able to:

1. Use properties of real numbers to produce equivalent equations
2. Use substitution to produce equivalent equations

Building Mathematical Language 2.1

New Terms: equality (=), equality properties, substitution principle, equivalent equations, solution set

Key Concepts 2.1

Two numbers are equal if, and only if, the difference between the numbers is zero. That is, two numbers ***a*** and ***b*** are equal if, and only if, $\boldsymbol{a - b = 0}$.

An equation is a mathematical statement, using "=" notation, to represent a mathematical relationship. Equations have two values: true or false. The equal sign is part of a statement that can be true or false. Example: $2 + 4 = 6$ is a true statement and $2 + 4 = 8$ is a false statement.

A solution to an equation is a number(s) that makes the equation true. There may be no solutions or there may be an infinite number of solutions to an equation.

The solution set of an equation is the set of all the solutions to the equation. The solution set may contain no elements, one element, or many elements. Throughout this book, we will expand on the algebraic toolbox with techniques and methodologies for finding solution sets for a variety of equation types.

Limitation: The solution set may be empty.

Equivalent equations exist when two equations have the same non-empty solution set. Equivalent equations are used to assure us that when rewriting an equation, at each step, the solution set of the new equation is the same as the solution set of the previous equation. Thus, the final equation will have the same solution set as the original equation.

Limitation: When creating equivalent equations, you must be sure to apply the algebraic properties of equality correctly.

The ***Substitution Principle*** is one of the most common algebraic tools used in solving equations. It states that, for any expression in an equation, you can substitute an equivalent expression to obtain an equivalent equation.

Limitation: The Substitution Principle only works for truly equivalent expressions.

Example: Starting with the equation $2 + 3 = 5$, you can use the Substitution Principle to substitute 5 for $2 + 3$ to obtain the equivalent equation $5 = 5$.

The ***Equality Properties*** can be proved from the *Axioms of Equality* and the *Axioms of the Real Number System* as shown in the following table. These properties are used to create equations that are equivalent to an original equation.

Limitation: These properties can be used only if we assume that the two sides of an equation are equal; that is, if we assume the original equation is true.

Ways to Produce Equivalent Equations Using the Equality Properties

Property	Claim	Justification	Limitation	Example
Addition	If $a = b$, then $a + c = b + c$	If $a=b$, then $a-b=0$ $a-b+1 \cdot (0)=0$ $a-b+1 \cdot (c-c)=0$ $a-b+c-c=0$ $a+c-b-c=0$ $a+c-(b+c)=0$ $a+c=b+c$	none	$5 = 2x + 7$ gives $5 + 9 = 2x + 7 + 9$ or $14 = 2x + 16$
Subtraction	If $a = b$, then $a - c = b - c$	If $a=b$, then $a+(-c)=b+(-c)$ $a-c=b-c$	none	$5 = 2x + 7$ gives $5 - 4 = 2x + 7 - 4$ or $1 = 2x + 3$

Property	Claim	Justification	Limitation	Example
Multiplication	If $a = b$ then $a \bullet c = b \bullet c$	If $a=b$, then $a-b=0$ $c \bullet 0=0$ $c \bullet (a-b)=0$ $a \bullet c-b \bullet c=0$ $a \bullet c=b \bullet c$	If $c = 0$, then $c(a-b) = 0$, whether or not $a = b$. Thus, the equations might not be equivalent.	$5 = 2x + 7$ gives $3 \bullet 5 = 3(2x + 7)$ or $15 = 6x + 21$
Division	If $a = b$ and $c \neq 0$, then $\frac{a}{c}=\frac{b}{c}$	If $c\neq 0$, then $\frac{1}{c}\neq 0$ If $a=b$, $a-b=0$ $\frac{1}{c}(a-b)=0$ $\frac{a}{c}-\frac{b}{c}=0$ $\frac{a}{c}=\frac{b}{c}$	If $c = 0$, then $\frac{1}{c}$ is undefined.	$5 = 2x + 7$ gives $\frac{5}{3}=\frac{(2x+7)}{3}$ or $\frac{5}{3}=\frac{2}{3}x+\frac{7}{3}$

Addressing Common Errors 2.1

The following table shows some of the most common errors that learners tend to make with the kinds of problems covered in this section.

ERROR: Incorrect use of the Addition Property of Equality

Incorrect Process	Resolution
$5 = 2x + 7$ $5 + 7 = 2x$	*Add the opposite of the constant (7) to both sides.*

Correct Process	Validation
$5=2x+7$ $5+(-7)=2x+7+(-7)$ $-2=2x$	Find the solution to the resulting equation and substitute it into the original equation, checking if it is a solution. $-2=2x$ $-1=x$ $5\overset{?}{=}2(-1)+7$ $5\overset{?}{=}-2+7$ $5=5$ ✓

ERROR: Incorrect use of the Multiplication Property of Equality

Incorrect Process

$2x = 6$

$2x - 2 = 6 - 2$

$x = 4$

Resolution

Multiply by the reciprocal of the coefficient of x $\left(\frac{1}{2}\right)$ *or divide by the coefficient (2).*

Correct Process

$2x = 6$

$\frac{2x}{2} = \frac{6}{2}$

$x = 3$

Validation

Find the solution to the resulting equation and substitute it into the original equation.

$2(3) \stackrel{?}{=} 6$

$6 = 6$ ✓

ERROR: Ignoring a sum of zero

Incorrect Process

$5x + 14 = 4$

Subtract 4 from both sides

$5x + 10$

$x = 2$

Resolution

Create an equivalent equation by isolating the variable term (subtract 14 from both sides).

Correct Process

$5x + 14 = 4$

$5x + 14 - 14 = 4 - 14$

$5x = -10$

$x = -2$

Validation

Check that $x = -2$ is a solution to the equation.

$5(-2) + 14 \stackrel{?}{=} 4$

$-10 + 14 \stackrel{?}{=} 4$

$4 = 4$ ✓

Preparation Inventory 2.1

I can...	OR *Here's my question...*
☐ produce equivalent equations using the Addition Property of Equality.	☐
☐ produce equivalent equations using the Multiplication Property of Equality.	☐
☐ produce equivalent equations using substitution.	☐

Critical Thinking Questions 2.1

1. What are at least three properties you can use to create an equation that is equivalent to another equation?

2. What property or properties of the real number system is/are used in rearranging the terms on one side of an equation?

3. How could you represent that two expressions are equivalent?

4. If you are given two equations, each of which has more than one solution, and each also having a solution in common, are they equivalent?

5. What are the easiest ways to create equivalent equations?

6. If you rewrite an algebraic equation multiple times, how do you ensure that the solution you come up with is the solution to the original equation?

7. Why is it that when you multiply an equation by zero, the resulting equation is not equivalent to the original equation?

8. How is the meaning of the word *equivalent* the same in the phrases *equivalent equations* and *equivalent expressions*?

9. How does the meaning of the word *equivalent* differ when used in the phrases *equivalent equations* and *equivalent expressions*?

Demonstrate Your Understanding 2.1

1. Are [$x^2 = 4$] and [$3x = 6$] equivalent equations?

a) circle one: YES NO

b) Demonstrate why your answer is correct.

2. Are $[2x + 5 = 7]$ and $[4x - 2 + 5x = 2 - 3x + 8]$ equivalent equations?

a) circle one: YES NO

b) Demonstrate why your answer is correct.

3. Are $[2(x - 3) = 10]$ and $[x - 3 = 5]$ equivalent equations?

a) circle one YES NO

b) Demonstrate why your answer is correct.

4. Write an equivalent equation for the given equation by using the suggested process. Do not solve the equation.

a) Given equation: $x + 5 = 2(x - 3)$
Process: *Substitution on the right side*

Equivalent Equation:

b) Given equation: $3x = 12 - 6x$
Process: *Divide both sides by 3*

Equivalent Equation:

c) Given equation: $5 - 4x + 3 = 7$

Process: *Substitution on the left side*

Equivalent Equation:

d) Given equation: $x - 5 = 2(x - 3)$

Process: *Add 5 to both sides*

Equivalent Equation:

e) Given equation: $\frac{3}{2} = 7x - 5$

Process: *Multiply both sides by a number*

Equivalent Equation:

f) Given equation: $\frac{4}{x-2} = 2$

Process: *Multiply both sides by (x – 2)*

Equivalent Equation:

g) In problem f, what is the constraint on the multiplier $(x - 2)$?

Create and Solve the Hardest Problem 2.1

Your Hardest Problem

Problem Solution

What makes a problem using or involving equivalence and substitution a hard problem to solve?

Identify, Describe, & Correct the Errors 2.1

Solve the problem correctly in the second column. In the third column, identify and describe the error you found in the worked solution.

Problem 1	Correct Process	Describe the Error
Find an equivalent equation for the following by using **subtraction**: $x + 5 = 7$		
Worked Solution *(Find the error!)*		
$x + 5 = 7$ is equivalent to $\boldsymbol{x + 5 - 5 = 7}$ **(after subtracting 5)**		

Problem 2

Find an equivalent equation for the following by using **division**:

$$2 + 4x = 2x - 6$$

Worked Solution *(Find the error!)*

$$2 + 4x = 2x - 6$$

is equivalent to

$$\mathbf{1 + 4x = x - 6}$$

(after dividing by 2)

Correct Process

Describe the Error

Problem 3

Find an equivalent equation for the following by using **multiplication**:

$$\frac{2}{x} + (x - 3) = \frac{2}{x}$$

Worked Solution *(Find the error!)*

$$\frac{2}{x} + (x - 3) = \frac{2}{x}$$

is equivalent to

$$\mathbf{2 + (x - 3) = 2}$$

(after multiplying by x)

Correct Process

Describe the Error

What Did You Learn About Learning Math? 2.1

1. How did you use the examples of common errors in demonstrating your understanding?

2. How does learning key concepts and terms increase your ability to produce equivalent equations?

3. Why is it valuable to identify and correct errors that other students have made?

Section 2.2 Solving Basic Equations

Why? 2.2

You can solve some equations that arise in the real world by isolating a variable. You can use this method to solve the equation

$$400+\left(1\frac{1}{2}\right)(10)x=460$$

to determine the number of hours of overtime you need to work to earn $460 per week, at an hourly rate of $10, with overtime being time and a half.

What Do You Already Know? 2.2

Give two examples of pairs of equivalent equations.

Use the Addition Property of Equality to create an equation that is equivalent to $6+3x=5$.

Use the Multiplication Property of Equality to create an equation that is equivalent to $-3(x-5)=6$.

Use the Distributive Property and the Substitution Principle to create an equation that is equivalent to $5x+2(3-x)=10$.

Goals 2.2

When you complete this activity, you should be able to:

1. Solve a linear equation using the Isolate the Variable Methodology
2. Solve a literal equation
3. Use the Solving Linear Equations Methodology to find the solution to a linear equation involving fractions and parentheses

Building Mathematical Language 2.2

New Terms: isolate a variable literal equation

Key Concepts 2.2

Some equations have *no solutions* and some equations have an *infinite number of solutions*. These occur when the variable can be totally eliminated from the equation. The resulting equation will state that two numbers are equal. If the two numbers are different, then the equation is never true and there are no solutions to the original equation. If the two numbers are the same, the equation is always true and there are an infinite number of solutions.

A ***literal equation*** is an equation with more than one variable. Engineers, scientists, and government policy makers use these equations because they provide information about relationships between real world quantities such as population and the square feet of retail space required.

Limitation: When dealing with literal equations, you must take care to understand the constraints on the variables in the equations. Assuming that literal equations are true for all values of their variables can cause safety issues in energy plants, on planes, and in factories

Methodology 2.2

Isolating a Variable

Isolating a variable in an equation with no fractions or parentheses, and where the variable appears only to the first power, is useful in solving an equation.

Limitation/Caution: If a term has the specified variable in the denominator, then problems can occur (we will see this later when we work with rational equations).

Example 1	Example 2	Your Turn
Isolate x: $6x - 5 = 7 + 2x$	Isolate x: $6x + 5 - x = 3x + 9$	Isolate a: $2a + 4a = 7a + 2 - 3a$

Steps	Discussion
1 Rewrite the equation	**Using the Equality Properties, rewrite the equation so that all the terms of your chosen variable are on one side of the equation and all the other terms on the other side of the equation.**

Ex 1	Ex 2	Your Turn
$6x - 2x = 7 + 5$	$6x - x - 3x = 9 - 5$	

Steps	Discussion
2 Simplify the expressions	**Simplify the expressions on both sides of the equation.**

Ex 1	Ex 2	Your Turn
$4x = 12$	$5x - 3x = 4$ $2x = 4$	

Steps	Discussion
3 Rewrite the equation	**Use the Distributive Property to rewrite the equation so that the chosen variable is a factor.**

Ex 1	Ex 2	Your Turn
Done	Done	

Steps	Discussion
4 Divide by the coefficient	**Divide both sides of the equation by the coefficient of the chosen variable.**

Ex 1	Ex 2	Your Turn
$4x = 12$ $x = 3$	$2x = 4$ $x = 2$	

Steps	Discussion
5 Validate	Substitute the value of the isolated variable in the equation to verify that the chosen variable has been isolated correctly.

Ex 1

$$6x - 5 = 7 + 2x$$
$$6(3) - 5 \stackrel{?}{=} 7 + 2(3)$$
$$18 - 5 \stackrel{?}{=} 7 + 6$$
$$13 = 13 \checkmark$$

Ex 2

$$6x + 5 - x = 3x + 9$$
$$6(2) + 5 - 2 \stackrel{?}{=} 3(2) - 9$$
$$12 + 5 - 2 \stackrel{?}{=} 6 + 9$$
$$17 - 2 \stackrel{?}{=} 15$$
$$15 = 15 \checkmark$$

Your Turn

Model 1: Isolating a Variable in a Literal Equation 2.2

Isolate y^2: $3y^2 - 4ax = 7ax - 5 + ay^2$

Step 1 Rewrite the equation $3y^2 - ay^2 = 7ax - 5 + 4ax$

Step 2 Simplify $3y^2 - ay^2 = 11ax - 5$

Step 3 Rewrite with chosen variable as a factor $y^2(3 - a) = 11ax - 5$

Step 4 Divide by the coefficient

$$\frac{y^2(3-a)}{(3-a)} = \frac{11ax - 5}{(3-a)}$$

$$y^2 = \frac{11ax - 5}{3 - a}$$

Step 5 Validate

$$3y^2 - 4ax = 7ax - 5 + ay^2$$

$$3\left(\frac{11ax-5}{3-a}\right) - 4ax = 7ax - 5 + a\left(\frac{11ax-5}{3-a}\right)$$

$$\frac{33ax-15}{3-a} - 4ax\left(\frac{3-a}{3-a}\right) = 7ax\left(\frac{3-a}{3-a}\right) - 5\left(\frac{3-a}{3-a}\right) + \frac{11a^2x-5a}{3-a}$$

$$\frac{33ax-15-12ax+4a^2x}{3-a} = \frac{21ax-7a^2x-15+5a+11a^2x-5a}{3-a}$$

$$\frac{4a^2x+21ax-15}{3-a} = \frac{4a^2x+21ax-15}{3-a} \quad \checkmark$$

Methodology 2.2

Clearing Fractions

Equations without denominators (fractions) are easier to work with.

Limitation: Some denominators with variables can cause problems.

	Example 1	**Your Turn**
Clear the fractions:	$\frac{x}{3} + \frac{2x}{5} = 4$	$\frac{3a}{4} + \frac{a^2}{5} = 3$

Steps	**Discussion**
1 Determine LCD	**Determine the LCD of all the denominators of the equation.**
Ex 1: $\frac{x}{3} + \frac{2x}{5} = 4$ The LCD is 3 • 5	**Your Turn**
2 Multiply and simplify	**Multiply both sides of the equation by the LCD of the denominators in the equation and simplify.**
Ex 1: $15\left(\frac{x}{3} + \frac{2x}{5}\right) = 15(4)$ $15\left(\frac{x}{3}\right) + 15\left(\frac{2x}{5}\right) = 60$ $\frac{15x}{3} + \frac{30x}{5} = 60$ $5x + 6x = 60$	**Your Turn**

Methodology 2.2

Solving a Linear Equation

A linear equation in one variable defines possible values for that variable, and can be used to find these values. Linear equations that state a relationship between two or more variables can be solved to determine the relationship one variable has to the other variable(s).

Limitation/Caution: Although some relationships defined by data appear linear for a range of values of the variable(s), assuming that a relationship is linear can lead to errors.

Solve the linear equation for the designated variable:

Example 1	Example 2	Your Turn
for x: $2(x-3)=4x-2$	for a: $2a-\frac{5}{3}a-2=3(2-a)-\frac{1}{2}$	for t: $6-3(t-5)=2t+11$

Steps	Discussion
1 Choose the variable to solve for	**Note the variables and constants and select which variable to solve for.**

Ex 1	Ex 2	Your Turn
The variable is x.	The variable is a.	

Steps	Discussion
2 Clear parentheses	**All variable occurrences should be outside parentheses; carry out any operations required to clear parentheses.**

Ex 1	Ex 2	Your Turn
$2x-6=4x-2$	$2a-\frac{5}{3}a-2=6-3a-\frac{1}{2}$	

Steps	Discussion
3 Clear fractions	**Multiply both sides of the equation by the LCD of the denominators and simplify (*see the Methodology for Clearing Fractions earlier in this section*).**

Ex 1	Ex 2
No fractions	LCD is 6 $6\left(2a-\frac{5}{3}a-2\right)=6\left(6-3a-\frac{1}{2}\right)$ $12a-10a-12=36-18a-3$

Your Turn

Steps	Discussion
4 Simplify	**Simplify the expressions on both sides of the equation, combining like terms whenever possible.**

Ex 1

Simplified

Ex 2

$12a - 10a - 12 = 33 - 18a$

Your Turn

Steps	Discussion
5 Isolate the variable	**Rewrite so that all instances of the chosen variable are on one side, simplify, and rewrite the equation so that the chosen variable is a factor. Finally, divide by any coefficient (*see the Methodology for Isolating a Variable earlier in this section*).**

Ex 1

$$2x - 6 + 6 = 4x - 2 + 6$$
$$2x = 4x + 4$$
$$2x - 4x = 4x - 4x + 4$$
$$-2x = 4$$
$$x = -2$$

Ex 2

$$(12a) - (10a) - 12 = 33 - (18a)$$
$$12a - 10a + 18a = 33 + 12$$
$$20a = 45$$
$$a = \frac{45}{20}$$
$$a = \frac{9}{4}$$

Your Turn

Steps	Discussion
6 Validate	**Substitute the solution into the original equation to check that both sides of the equation are the same.**

Ex 1

$$2(x - 3) = 4x - 2$$
$$2(-2 - 3) \stackrel{?}{=} 4(-2) - 2$$
$$2(-5) \stackrel{?}{=} -8 - 2$$
$$-10 = -10 \quad \checkmark$$

Ex 2

$$2a - \frac{5}{3}a - 2 = 3(2 - a) - \frac{1}{2}$$
$$2\left(\frac{9}{4}\right) - \frac{5}{3}\left(\frac{9}{4}\right) - 2 \stackrel{?}{=} 3\left(2 - \left(\frac{9}{4}\right)\right) - \frac{1}{2}$$
$$\frac{9}{2} - \frac{15}{4} - 2 \stackrel{?}{=} 6 - \frac{27}{4} - \frac{1}{2}$$
$$\frac{18}{4} - \frac{15}{4} - \frac{8}{4} \stackrel{?}{=} \frac{24}{4} - \frac{27}{4} - \frac{2}{4}$$
$$-\frac{5}{4} = -\frac{5}{4} \quad \checkmark$$

(Or validate with a calculator.)

Steps	Discussion

Your Turn

Model 2: Solving a Literal Equation for a Given Variable 2.2

Solve the following literal equation for r:

$$\frac{7}{3} - 2rw = \pi r$$

Step 1 Choose the variable to solve for	The variables are r and w. The symbol π is a constant. We are to solve for r.
Step 2 Clear parentheses	There are no parentheses.
Step 3 Clear fractions	$\frac{7}{3} - 2rw = \pi r$ LCD is 3 $7 - 6rw = 3\pi r$
Step 4 Simplify	There is nothing to simplify.
Step 5 Isolate the variable	$7 - (6rw) = (3\pi r)$ $7 = 3\pi r + 6rw$ $7 = (3\pi + 6w)r$ $\frac{7}{(3\pi + 6w)} = r$ $\frac{7}{3(\pi + 2w)} = r$ or $r = \frac{7}{3(\pi + 2w)}$

Step 6 Validate

$$\frac{7}{3} - 2rw = \pi r$$

$$\frac{7}{3} - 2\left(\frac{7}{3(\pi + 2w)}\right) \cdot w = \pi\left(\frac{7}{3(\pi + 2w)}\right)$$

$$\frac{7}{3} - \frac{14w}{3(\pi + 2w)} = \frac{7\pi}{3(\pi + 2w)}$$

$$\frac{7(\pi + 2w) - 14w}{3(\pi + 2w)} = \frac{7\pi}{3(\pi + 2w)}$$

$$\frac{7\pi + 14w - 14w}{3(\pi + 2w)} = \frac{7\pi}{3(\pi + 2w)}$$

$$\frac{7\pi}{3(\pi + 2w)} = \frac{7\pi}{3(\pi + 2w)} \quad \checkmark$$

Addressing Common Errors 2.2

The following table shows some of the most common errors that learners tend to make with the kinds of problems covered in this section.

ERROR: Incorrect use of the Distributive Property to remove parentheses

Incorrect Process	Resolution
$3x - (2x + 4) = -2$ $3x - 2x + 4 = -2$ $x + 4 = -2$ $x = -6$	*Remove the parentheses, correctly using the Distributive Property.*
Correct Process	**Validation**
$3x - (2x + 4) = -2$ $3x - 2x - 4 = -2$ $x - 4 = -2$ $x = 2$	$3x - (2x + 4) = -2$ $3(2) - (2(2) + 4) \stackrel{?}{=} -2$ $6 - (4 + 4) \stackrel{?}{=} -2$ $6 - 8 \stackrel{?}{=} -2$ $-2 = -2 \checkmark$

ERROR: Incorrect use of the Addition Property of Equality

Incorrect Process	Resolution
$3x + 7 = 11$ $3x = 18$ $x = 6$	*Correct use of the Addition Property of Equality by subtracting 7 from both sides.*

ERROR: Incorrect use of the Addition Property of Equality (con't)

Correct Process

$$3x + 7 = 11$$
$$3x = 4$$
$$x = \frac{4}{3}$$

Validation

$$3x + 7 = 11$$
$$3 \cdot \left(\frac{4}{3}\right) + 7 \stackrel{?}{=} 11$$
$$4 + 7 \stackrel{?}{=} 11$$
$$11 = 11 \checkmark$$

ERROR: Incorrect use of the Multiplication Property of Equality

Incorrect Process

$$3x = 5$$
$$x = 15$$

Resolution

Use the Multiplication Property of Equality correctly by multiplying both sides of the equation by $\frac{1}{3}$

Correct Process

$$3x = 5$$
$$\frac{1}{3} \cdot x = \frac{1}{3} \cdot 5$$
$$x = \frac{5}{3}$$

Validation

$$3x = 5$$
$$3 \cdot \left(\frac{5}{3}\right) \stackrel{?}{=} 5$$
$$5 = 5 \checkmark$$

ERROR: Incorrect use of Order of Operations

Incorrect Process

$$4 + 2x \cdot 3 = 18$$
$$6 \cdot 3x = 18$$
$$18x = 18$$
$$x = 1$$

Resolution

Use the Order of Operations correctly.

Correct Process

$$4 + 2x \cdot 3 = 18$$
$$4 + 6x = 18$$
$$6x = 14$$
$$x = \frac{14}{6} = \frac{7}{3}$$

Validation

$$4 + 2x \cdot 3 = 18$$
$$4 + 2 \cdot \left(\frac{7}{3}\right) \cdot 3 \stackrel{?}{=} 18$$
$$4 + 14 \stackrel{?}{=} 18$$
$$18 = 18 \checkmark$$

ERROR: **Ignoring a sum of zero**

Incorrect Process

$3x + 13 = 7$

$3x + 6$

$x = 2$

Resolution

Leave the 0 on the right side of the equation.

Correct Process

$3x + 13 = 7$

$3x + 6 = 0$

$3x = -6$

$x = -2$

Validation

$3x + 13 = 7$

$3(-2) + 13 \stackrel{?}{=} 7$

$-6 + 13 \stackrel{?}{=} 7$

$7 = 7$ ✓

ERROR: **Adding unlike terms**

Incorrect Process

$4 + 2x = 18$

$6x = 18$

$x = 3$

Resolution

Use the Addition Property of Equality by adding (–4) to both sides.

Correct Process

$4 + 2x = 18$

$(-4) + 4 + 2x = (-4) + 18$

$0 + 2x = 14$

$2x = 14$

$x = 7$

Validation

$4 + 2x = 18$

$4 + 2(7) \stackrel{?}{=} 18$

$4 + 14 \stackrel{?}{=} 18$

$18 = 18$ ✓

Preparation Inventory 2.2

I can...	OR *Here's my question...*
☐ isolate a variable in an equation.	☐
☐ remove parentheses from an equation.	☐
☐ clear fractions in an equation.	☐
☐ solve a linear equation.	☐
☐ solve a literal equation for a given variable.	☐

Critical Thinking Questions 2.2

1. How do you determine which variable to solve for?

2. How do you clear the fractions in an equation?

3. Why eliminate parentheses when solving a linear equation?

4. What are your strategies for isolating the variable term without making a mistake?

5. When isolating the variable, why do you divide both sides of the equation by the coefficient of the variable?

6. How do you make sure that you have isolated the chosen variable correctly?

7. How is isolating a variable different from solving a linear equation?

8. What is the difference between a literal equation and a linear equation?

9. How do the solutions to $2x = 4x - 2x$ and $2x = 4 + 2x$ differ?

Demonstrate Your Understanding 2.2

Solve each literal equation for the indicated variable.

1. Solve for a: $2a + 3 = 7 - 3(2 - 5a)$

2. Solve for x: $5(x - 3) + 4 = -7 + 3x$

3. Solve for x: $-2(x - 5) + \frac{3}{4} = 4x - \frac{5x}{3}$

4. Solve for r: $5 - 3a + 4 = 3 - 2(a + r) - 7$

5. Solve for y: $2(x - 3) = 5 - 3x + y + 5x$

6. Solve for t: $2x - 5t = 3(t - 4) + 7$

Create and Solve the Hardest Problem 2.2

Your Hardest Problem

Problem Solution

What makes a problem with basic equations a hard problem to solve?

Identify, Describe, & Correct the Errors 2.2

Solve the problem correctly in the second column. In the third column, identify and describe the error you found in the worked solution.

Problem 1

Solve for x:

$$x + 5 = 3(4 - 2x)$$

Worked Solution *(Find the error!)*

$$x + 5 = 3(4 - 2x)$$
$$x + 5 = 12 - 2x$$
$$3x = 7$$
$$x = \frac{7}{3}$$

Correct Process

Describe the Error

Problem 2

Solve for x:

$$2x - 5 + x = 6x - 4$$

Worked Solution *(Find the error!)*

$$2x - 5 + x = 6x - 4$$
$$2x + x - 5 = 6x - 4$$
$$3x - 5 = 6x - 4$$
$$3x - 6x = -4 - 5$$
$$-3x = -9$$
$$x = 3$$

Correct Process

Describe the Error

Problem 3

Solve for t:

$-3^2 + 4s + 5t = 6s - 5s - 4(s + 3 - t)$

Worked Solution *(Find the error!)*

$$-3^2 + 4s + 5t = 6s - 5s - 4(s + 3 - t)$$
$$9 + 4s + 5t = 6s - 5s - 4s - 12 + 4t$$
$$9 + 4s + 5t = -3s - 12 + 4t$$
$$4s + 3s = -12 + 4t - 9 - 5t$$
$$7s = -12 - 9 + 4t - 5t$$
$$7s = -21 - t$$
$$s = -3 - \frac{t}{7}$$

Correct Process

Describe the Error

Problem 4

Solve for y:

$$\frac{3}{y} - 4 = 3$$

Worked Solution *(Find the error!)*

$$\frac{3}{y} - 4 = 3$$
$$3 - 4y = 3$$
$$-4y = 0$$
$$y = 0$$

Correct Process

Describe the Error

Problem 5

Solve for x:

$$4x + \frac{2}{3}x = 7$$

Worked Solution *(Find the error!)*

$$4x + \frac{2}{3}x = 7$$

$$12x + 2x = 7$$

$$14x = 7$$

$$x = \frac{1}{2}$$

Correct Process

Describe the Error

What Did You Learn About Learning Math? 2.2

1. What prior knowledge did you find useful for learning the material in this section?

2. How did you learn to distinguish equivalence for expressions from equivalence for equations?

3. What are the key questions to ask when solving a linear equation?

Section 2.3 Solving Systems of Linear Equations

Why? 2.3

There are many times when you need to compare two things, to find out when they are equal or when one exceeds the other. For example, say you want to rent a car for vacation, and you find that the Ace company charges \$35 per day plus \$.04 per mile, but the Best company charges \$25 per day plus \$.09 per mile. By solving a system of linear equations you can find out the number of miles that make the charges the same, or for how many miles one company is cheaper than the other before you decide where to rent your car.

What Do You Already Know? 2.3

What is the Addition Property of Equality?

How would you use the Addition Property of Equality to change $2x + y = 5$ to an equivalent equation that is solved for y?

How would you use substitution to find the value of y if $y = 2x - 5$ and $x = 1$?

Goals 2.3

When you complete this activity, you should be able to:

1. Use substitution to solve a system of linear equations
2. Use elimination to solve a system of linear equations

Building Mathematical Language 2.3

New Terms:

- elimination method
- substitution method
- linear equation
- system of linear equations
- solution of a system

Key Concepts 2.3

Types of Solutions to Systems of Two Linear Equations in Two Variables

Systems of two linear equations in two variables can have one solution, no solution, or an infinite number of solutions. The same types of solutions exist for any system of linear equations.

Limitation: There are other types of systems of equations that are non-linear.

Methodology 2.3

Solving a System of Linear Equations in Two Variables by Substitution

The purpose of this methodology, which we refer to as the ***substitution method***, is to solve a system of two equations in two variables by using the Substitution Principle twice.

Limitation/Caution: When you solve one of the equations for one variable in terms of the other, you should not substitute the result back into that same equation.

	Example 1	**Your Turn**
Solve the system:	$\begin{cases} x+2y=-3 \\ 3x-5y=13 \end{cases}$	$\begin{cases} -3x+4y=17 \\ x+5y=7 \end{cases}$

Steps	Discussion
1 Select a variable and an equation	**Select a variable for the first substitution and an equation to use to determine an expression for the selected variable.**
Ex 1: $\begin{cases} x+2y=-3 \\ 3x-5y=13 \end{cases}$ Select x and the first equation (we choose this equation because x has a coefficient of 1 and can easily be solved for).	**Your Turn**
2 Solve one of the equations	**Solve the selected literal equation for the selected variable.**
Ex 1: $x+2y=-3$ $x=-3-2y$	**Your Turn**
3 Substitute expression	**Substitute the expression for the variable solved for in Step 2 into the other equation.**
Ex 1: $3x-5y=13$ $3(-3-2y)-5y=13$	**Your Turn**
4 Solve for first variable	**Solve the resulting equation in one variable for that variable.**
Ex 1: $-9-6y-5y=13$ $-6y-5y=13+9$ $-11y=22$ $y=-2$	**Your Turn**

Steps	Discussion
5 Substitute first variable	**Substitute the value obtained in Step 4 in the equation you created in Step 2.**

Ex 1

$$x = -3 - 2y$$
$$x = -3 - 2(-2)$$

Your Turn

6 Solve for second variable	**Solve the equation in Step 5 for the other variable.**

Ex 1

$$x = -3 + 4$$
$$x = 1$$

Your Turn

7 Present solution	**Write the solution for each variable.**

Ex 1

(1, –2) or $x = 1$ and $y = -2$

Your Turn

8 Validate	**Validate your work by substituting the values for the variables back into both original equations.**

Ex 1

$$x + 2y = -3 \qquad 3x - 5y = 13$$
$$1 + 2(-2) \stackrel{?}{=} -3 \quad \textbf{and} \quad 3(1) - 5(-2) \stackrel{?}{=} 13$$
$$1 - 4 \stackrel{?}{=} -3 \qquad 3 + 10 \stackrel{?}{=} 13$$
$$-3 = -3\ \checkmark \qquad 13 = 13\ \checkmark$$

Your Turn

Methodology 2.3

Solving a System of Linear Equations in Two Variables by Elimination

The purpose of this methodology, which we refer to as the ***elimination method***, is to solve a system of two equations in two variables by using the Addition Property of Equality to create a third equation in one variable that can be solved.

Limitation/Caution: A system of linear equations may have no solution or an infinite number of solutions. "An infinite number of solutions" (or "infinitely many solutions") for a system of equations is not the same as "all real numbers."

	Example 1	**Your Turn**
Solve the system:	$\begin{cases} x+2y=-3 \\ 3x-5y=13 \end{cases}$	$\begin{cases} 2x+3y=2 \\ 5x-y=22 \end{cases}$

Steps	Discussion
1 Select the variable	**Select the variable to eliminate between the two equations.**
Ex 1: $\begin{cases} x+2y=-3 \\ 3x-5y=13 \end{cases}$ Eliminate y	**Your Turn**
2 Multiply each equation by an appropriate number	**Multiply the two equations by numbers such that the coefficients of the selected variable differ only by a sign.**
Ex 1: $\begin{cases} 5(x+2y=-3) \\ 2(3x-5y=13) \end{cases}$ $\begin{cases} 5x+10y=-15 \\ 6x-10y=26 \end{cases}$	**Your Turn**
3 Add the two equations	**Add the two equations to eliminate the selected variable.**
Ex 1: $11x=11$	**Your Turn**
4 Solve the equation	**Solve the equation from Step 3 for the variable.**
Ex 1: $x=1$	**Your Turn**

Steps	Discussion
5 Substitute first variable	Substitute the value of the variable you solved for in Step 4 into either of the original equations.

Ex 1

$$x + 2y = -3$$
$$(1) + 2y = -3$$

Your Turn

Steps	Discussion
6 Solve for second variable	Solve the equation you obtained in Step 5 for the remaining variable.

Ex 1

$$2y = -3 - 1$$
$$2y = -4$$
$$y = -2$$

Your Turn

Steps	Discussion
7 Present solution	Present the solution for each variable.

Ex 1

$(1, -2)$ or $x = 1$ and $y = -2$

Your Turn

Steps	Discussion
8 Validate	Validate your work by substituting the values for the variables back into both original equations.

Ex 1

$$x + 2y = -3 \qquad 3x - 5y = 13$$
$$1 + 2(-2) \stackrel{?}{=} -3 \quad \textbf{and} \quad 3(1) - 5(-2) \stackrel{?}{=} 13$$
$$1 - 4 \stackrel{?}{=} -3 \qquad 3 + 10 \stackrel{?}{=} 13$$
$$-3 = -3 \checkmark \qquad 13 = 13 \checkmark$$

Your Turn

Addressing Common Errors 2.3

The following table shows some of the most common errors that learners tend to make with the kinds of problems covered in this section.

ERROR: Finding the value of only one variable

Incorrect Process

Solve the system $\begin{cases} 2x+y=9 \\ x-y=6 \end{cases}$

Add the equations to eliminate the variable y:

$3x = 15$

$x = 5$

The solution is $x = 5$.

Resolution

After finding the value of the first variable, substitute that value into one of the equations to find the value of the other variable.

Correct Process

Solve the system $\begin{cases} 2x+y=9 \\ x-y=6 \end{cases}$

Add the equations to eliminate the variable y:

$3x = 15$

$x = 5$

Substitute $x = 5$ into the first equation:

$2(5) + y = 9$

$10 + y = 9$

$y = -1$

The solution is $(5, -1)$.

Validation

$2(5)+(-1)\stackrel{?}{=}9$ and $5-(-1)\stackrel{?}{=}6$

$10-1\stackrel{?}{=}9$ $\quad$ $5+1\stackrel{?}{=}6$

$9=9$ ✓ $\quad$ $6=6$ ✓

ERROR: Eliminating a variable without making the signs opposite

Incorrect Process

Solve the system $\begin{Bmatrix} 3x-2y=0 \\ x-y=1 \end{Bmatrix}$

Multiply the second equation by 3

$$\begin{Bmatrix} \cancel{3x}-2y=0 \\ \cancel{3x}-3y=3 \end{Bmatrix}$$

$$-5y=3$$

$$y=-\frac{3}{5}$$

Find the value of x by eliminating every instance of y (multiply the second equation by 2)

$$\begin{Bmatrix} 3x-\cancel{2y}=0 \\ 2x-\cancel{2y}=2 \end{Bmatrix}$$

$$5x=2$$

$$x=\frac{2}{5}$$

The solution is $\left(\frac{2}{5},-\frac{3}{5}\right)$.

Resolution

Ensure that the coefficients of the variable to be eliminated are the same number but opposite signs.

Correct Process

Solve the system $\begin{Bmatrix} 3x-2y=0 \\ x-y=1 \end{Bmatrix}$

Multiply the second equation by –3 and then add.

$$\begin{Bmatrix} \cancel{3x}-2y=0 \\ \cancel{-3x}+3y=-3 \end{Bmatrix}$$

$$y=-3$$

Substitute $y=-3$ into the second equation to find x

$$x-(-3)=1$$

$$x+3=1$$

$$x=-2$$

The solution is $(-2,-3)$.

Validation

$$3(-2)-2(-3)\stackrel{?}{=}0$$

$$-6+6\stackrel{?}{=}0$$

$$0=0\ \checkmark$$

and

$$(-2)-(-3)\stackrel{?}{=}1$$

$$-2+3\stackrel{?}{=}1$$

$$1=1\ \checkmark$$

Preparation Inventory 2.3

I can...	OR *Here's my question...*
☐ use the substitution principle to create equivalent equations.	☐
☐ use the addition principle of equality to create equivalent equations.	☐
☐ understand what it means for a particular number to be a "solution" to an equation.	☐

Critical Thinking Questions 2.3

1. What does a solution to a system of linear equations in two variables represent?

2. What is the central idea behind the substitution method?

3. What is the central idea behind the elimination method?

4. Both the substitution and elimination methods can be used to solve any system of linear equations. How do you decide which method would be the better one to use on any given problem?

5. As you are solving a particular system, what happens to alert you to the fact that the system has no solution?

6. As you are solving a particular system, what happens to alert you to the fact that the system has infinitely many solutions?

7. How do you validate the solution to a system of linear equations?

8. After you use the elimination method to find a value for your first variable, can you use elimination rather than substitution to find the value of the second variable? When might you want to do this, if at all?

Demonstrate Your Understanding 2.3

Solve each system of equations, using the indicated method.

1. $\begin{cases} 5x - 3y = 2 \\ 5x - 4y = 1 \end{cases}$ (elimination method)

2. $\begin{cases} 3x - y = 2 \\ -6x + 2y = 1 \end{cases}$ (substitution method)

3. $\begin{cases} -3x + 4y = 23 \\ 5x - 4y = 15 \end{cases}$ (elimination method)

4. $\begin{cases} 2x - y = -\dfrac{7}{4} \\ 1 - 4y = 4x \end{cases}$ (using the method of your choice)

Create and Solve the Hardest Problem 2.3

Your Hardest Problem

Problem Solution

What makes a problem with a system of linear equations a hard problem to solve?

Identify, Describe, & Correct the Errors 2.3

Solve the problem correctly in the second column. In the third column, identify and describe the error you found in the worked solution.

Problem 1

Solve by elimination:

$$\begin{cases} 2x - y = 2 \\ -6x + y = -6 \end{cases}$$

Worked Solution *(Find the error!)*

$$\begin{cases} 2x - y = 2 \\ -6x + y = -6 \end{cases}$$

$-4x = -4$

$x = 1$

The solution is $x = 1$.

Correct Process

Describe the Error

Problem 2

Solve by elimination:

$$\begin{cases} -2x + 4y = 12 \\ -2x + 3y = 2 \end{cases}$$

Worked Solution *(Find the error!)*

$$\begin{cases} -2x + 4y = 12 \\ -2x + 3y = 2 \end{cases}$$

$7x = 14$

$x = 2$

$-2(2) + 4y = 12$

$-4 + 4y = 12$

$4y = 16$

$y = 4$

The solution is (2, 4) or $x = 2$ and $y = 4$.

Correct Process

Describe the Error

Problem 3

Solve by substitution:

$$\begin{cases} 2x - y = 2 \\ -6x + y = -6 \end{cases}$$

Worked Solution *(Find the error!)*

$$\begin{cases} 2x - y = 2 \\ -6x + y = -6 \end{cases}$$

$-y = 2 - 2x$

$y = -2 + 2x$

$2x - (-2 + 2x) = 2$

$2x + 2 - 2x = 2$

$0 = 0$

There are infinitely many solutions.

Correct Process

Describe the Error

Problem 4

Solve by substitution:

$$\begin{cases} x - y = 2 \\ 5x - 5y = -6 \end{cases}$$

Worked Solution *(Find the error!)*

$$\begin{cases} x - y = 2 \\ 5x - 5y = -6 \end{cases}$$

$$\begin{cases} -5(x - y = 2) \\ 5x - 5y = -6 \end{cases}$$

$$\begin{cases} -5x + 5y = -10 \\ 5x - 5y = -6 \end{cases}$$

$0 = -16$

The solution set is all real numbers.

Correct Process

Describe the Error

What Did You Learn About Learning Math? 2.3

1. What did you have to learn about solving systems of linear equations using substitution or elimination that you did not already know from solving linear equations?

2. How did you use visualization when you looked at a system of linear equations in order to help you decide how to solve the system?

3. Why was it important to ask yourself what you already knew about solving systems of linear equations?

Section 2.4 Validation

Why? 2.4

Validation is important to ensure that you have the correct answer or truly understand what you are learning. In the workplace, if you don't check your work, someone else may have to. This can get you fired. If you do not check your understanding of what you have learned, you may make mistakes when applying your new knowledge. This, too, can get you fired. But most importantly, if you truly understand what you have learned, you will find personal pleasure in your new knowledge. You will be encouraged to further your own personal growth by continuing to learn.

What Do You Already Know? 2.4

How do you go about checking that your evaluation of a formula is correct?

What are three methods that you know to check that you have simplified an expression correctly?

What are three ways that you can be sure that you have solved a system of two linear equations in two variables?

How do you know that you understand the difference between an equation and an expression?

Goals 2.4

When you complete this activity, you should be able to:

1. Use five methods of validating that you know what you have learned
2. Use five methods of validating your work

Building Mathematical Language 2.4

New Terms: validation validation of learning validation of work

Key Concepts 2.4

Your ability to use ***validation*** is critical to the quality and value of the work you produce. Producing work of the quality that your supervisors or clients expect, value, and demand requires two things. First, you need to be able to identify what you need to learn, then effectively learn it, and finally know that you have learned it so you can perform your job at its highest level. Then once you are on top of your job you must be able to produce quality problem solutions that others know are correct. If they must check your work, your value decreases significantly. These two major aspects of your work are addressed by this activity. The validation process includes both the ***validation of your learning*** and the ***validation of your work*** so that you are able to convince others that your problem solutions satisfy all interested parties.

Reasons for Underperformance

1. Individuals do not see learning opportunities.	The general behavior of problem solvers is to use existing knowledge in solving a current problem. This often leads to solutions not nearly as powerful as those resulting from applying new knowledge to the problem. Mastering the learning process not only includes both knowing when you know, but also knowing when you don't know, thus triggering the need for new learning. In your professional life and practice, you are expected to validate your use of knowledge (including seeking or creating new knowledge) in solving real world problems.
2. Individuals do not thoroughly validate their work.	Every time someone questions the quality of your work because you didn't validate its quality or solution at the level of other people's expectations, your value and future opportunities for advancement are significantly jeopardized. Whether this is because you don't know how to validate your problem solution, or because you don't validate your work doesn't make any difference.
3. Individuals do not validate others' learning and work.	Not only are you responsible for seeing the opportunities for your own learning, but also for seeing others' learning needs. When you use others' problem solutions, your inability to catch their errors will cause you to propagate errors and will reflect negatively on you.

We can ***validate the quality and level of learning*** (knowledge) with three different scales.

Level of Knowledge (quality)	Clarity Index (adjectives)	Level of Knowledge (nouns: Bloom's Taxonomy)	Breadth of Knowledge (Framework/Connections)
Lowest Level	Muddy	Informational	Highlight a piece of knowledge
	Foggy	Comprehension / Understanding	Connect to prior knowledge
	Hazy	Application to New Context	Compare and contrast to prior knowledge
	Clear	Working Expertise	Differentiate knowledge from context
	Crystal Clear	New Creative Enterprise	Build association with other knowledge
Highest Level	Extension of the knowledge	Evaluation / Validation of its Quality	Refine your "concept map" of the knowledge area

Five Techniques for Validating Your Learning (to verify that you KNOW that you know)

1. Teach someone else the specific knowledge.	Helping a student struggling with the specific piece of learning apply the knowledge to new contexts will help you elevate your clarity index. When you succeed, your knowledge will be crystal clear.
2. Solve the most difficult problem.	The process of creating and solving a complex problem by changing every dimension of the problem to the most difficult it can be will allow you to generalize the knowledge so that you can apply it to any specific context. This is the specification for level three of Bloom's Taxonomy on transferability of the knowledge.
3. Create a question that others can't answer but you can.	The process of working with the boundary conditions of the knowledge — its limitations, its connections, and subtleties of how knowledge can be used, in what context, and with what other knowledge — will consolidate your mastery of the knowledge. This mastery can be measured with the depth of knowledge schema you have built.
4. Correct the errors in other people's solutions.	The process of evaluating the quality of the work and solutions of others by understanding how they are thinking and how you can correct misconceptions and issues advances both their performance and yours. Consultants are handsomely paid because of their ability to dissect your learning or the learning of your organization, correct misconceptions, and advance learning so that the ownership of this knowledge expansion resides in the organization.
5. Write a learning journal entry explaining what you know.	Writers have always found that the process of putting ideas on paper for a generalized audience is the most difficult challenge with knowledge because everyone will be skeptical about this knowledge and will do whatever they can to find fault in the presentation of this knowledge. Thus, writing presenting knowledge takes more effort than presenting knowledge another way.

Validating Your Problem Solution

We will begin to understand the place and importance of validation in the Problem Solving Process by stating an expanded version of this process.

The Steps of the Problem Solving Process

1. Identify and define the problem
2. Identify key issues that must be resolved
3. Assess and obtain the necessary information
4. Identify and test the assumptions that will be used to solve the problem
5. Break apart the problem into manageable components
6. Solve the individual sub-problems by reusing previous solutions/models/knowledge
7. Integrate the sub-problems holistically
8. ***Validate the solution to meet the defined problem***
9. Generalize the solution for reuse in the future
10. Communicate the process of solving the problem to others

Step 8 is critical. If you do not know that your solution meets your own and others' expectations, have you truly solved the problem?

Five Key Techniques for Validating a Solution to a Problem

1. Check the common sense realities of the situation	Does the solution to the problem make sense? For example, can a person really run 50 miles in an hour?
2. Work backwards through your problem solving process	Reversing each step in the problem solution provides a check for your work. A simple example is using multiplication to validate a division problem.
3. Apply the solution to the problem	See if your problem solution satisfies the conditions of the problem.
4. Use a second process for solving the problem	Use a second method to find the solution. The two solutions should be the same.
5. Perform a unit analysis	Unit analysis will ensure that the problem was set up correctly.

Model 1: Common Sense Check 2.4

The solution to a problem should make sense within the context of the problem. For example, if you are to determine the speed a car will be traveling in some situation, it does not make sense to arrive at a solution that indicates that the car would be traveling at 500 mph. Some mistake has occurred. It may be simply a computational error or it may be a more difficult-to-discover conceptual error.

Model 2: Working Backwards 2.4

To validate if the factored form (Section 4.2) of $12x^2 - 23x + 5$ is $(6x - 1)(2x - 5)$, we can work backwards by multiplying the two factors together.

$$(6x - 1)(2x - 5) = 6x(2x) + 6x(-5) + (-1 \bullet 2x) + (-1)(-5) = 12x^2 - 30x - 2x + 5 = 12x^2 - 32x + 5$$

The correct result would be $12x^2 - 23x + 5 = (3x - 5)(4x - 1)$ which you can validate by working backwards.

Model 3: Applying the Solution to the Problem 2.4

The solution to a solved equation can be tested by substituting each solution into the equation and making sure that the solution makes the equation correct. Note: This is only part of the validation; if you didn't come up with all of the solutions, this process won't inform you that you have missed one or more of the solutions in the solution set.

Validate that one of the solutions to the equation $2x^2 - 3x - 5 = 9$ is $x = -2$ by substituting this value for x in the equation:

$$2x^2 - 3x - 5 = 9$$
$$2(-2)^2 - 3(-2) - 5 = 9$$
$$2(4) + 6 - 5 = 9$$
$$8 + 6 - 5 = 9$$
$$9 = 9 \quad \checkmark$$

Model 4: Use a Second Process 2.4

If you solve an equation using substitution, you can then validate it using elimination.

For example, the system of equations $\begin{cases} x+y=5 \\ x-y=3 \end{cases}$ can be solved by substitution and checked through elimination.

Solve by **substitution**	Verify by ***elimination***
$\begin{cases} x+y=5 \\ x-y=3 \end{cases}$	$\begin{cases} x+y=5 \\ x-y=3 \end{cases}$
$\begin{cases} x+y=5 \\ x=3+y \end{cases}$	$2x=8$
	$x=4$
$(3+y)+y=5$	$x+y=5$
$3+2y=5$	$4+y=5$
$2y=2$	$y=1$
$y=1$	$(4,1)$ ✓
$x=3+y=3+1=4$	
$(4,1)$	

Model 5: Unit Analysis 1.4

When solving a word problem that involves units, it is possible to check your result by ensuring that the units match. Errors can occur when converting one type of unit to another. This method is also called *dimensional analysis*.

For example, if converting 231 cm to yards, we might have the following computation.

$$231 \bullet \frac{1}{2.54} \bullet 12 \bullet \frac{1}{3} \approx 364\ yd$$

Using unit analysis, we have $231\ \cancel{cm} \bullet \dfrac{1in}{2.54\cancel{cm}} \bullet \dfrac{12in}{ft} \bullet \dfrac{1yd}{3ft} \approx 364\ \dfrac{in^2 yd}{ft^2}$

The dimensions (units) are not correct. The correct result would be:

$$231\ \cancel{cm} \bullet \frac{1\cancel{in}}{2.54\cancel{cm}} \bullet \frac{1\cancel{ft}}{12\cancel{in}} \bullet \frac{1yd}{3\cancel{ft}} \approx 2.53yd$$

Addressing Common Errors 2.4

The following table shows some of the most common errors that learners tend to make with the kinds of problems covered in this section.

ERROR: Failure to validate your work

Incorrect Process

$$\begin{cases} x+y=2 \\ x-y=4 \end{cases} \rightarrow \begin{cases} x+y=2 \\ 2x=-2 \end{cases} \rightarrow \begin{cases} x=-1 \\ -1+y=2 \end{cases} \rightarrow \begin{cases} x=-1 \\ y=3 \end{cases}$$

Resolution

Validate your work.

Correct Process

$$\begin{cases} x+y=2 \\ x-y=4 \end{cases} \rightarrow \begin{cases} x+y=2 \\ 2x=-2 \end{cases} \rightarrow \begin{cases} x=-1 \\ -1+y=2 \end{cases} \rightarrow \begin{cases} x=-1 \\ y=3 \end{cases}$$

$$\begin{cases} x+y=2 \\ x-y=4 \end{cases} \rightarrow \begin{cases} -1+3\stackrel{?}{=}2 \\ -1-3\stackrel{?}{=}4 \end{cases} \rightarrow \begin{cases} 2=2 \\ -4\neq 4 \end{cases}$$

$$\begin{cases} x+y=2 \\ 2x=6 \end{cases} \rightarrow \begin{cases} x=3 \\ 3+y=2 \end{cases} \rightarrow \begin{cases} x=3 \\ y=-1 \end{cases}$$

$$\begin{cases} x+y=2 \\ x-y=4 \end{cases} \rightarrow \begin{cases} 3+(-1)\stackrel{?}{=}2 \\ 3-(-1)\stackrel{?}{=}4 \end{cases} \rightarrow \begin{cases} 2=2 \\ 4=4 \end{cases}$$

Validation

The work has been validated by substituting the solution back into the original system of equations.

ERROR: Failure to validate your learning

Incorrect Process

Read a section in the book without reflecting on what was learned.

Resolution

Use a reading journal to solidify your work.

Correct Process

List three ways that the learned material could be used.

Validation

Look back at the reading material to check that the uses were correct.

Preparation Inventory 2.4

I can...	OR *Here's my question...*
☐ name five techniques for validating my learning.	☐
☐ apply three different scales for measuring the quality of my learning.	☐
☐ name five techniques for validating my work.	☐

Critical Thinking Questions 2.4

1. Why is validation important to you?

2. Why do managers expect you to use a validation process?

3. Why is validation important to the expanded Problem Solving process?

4. What is the most important part of the validation process?

5. What are five situations in which you can validate with a different method?

6. What do you think is the most important scale for measuring quality of learning?

7. Why is it important to refine your "concept map"?

Demonstrate Your Understanding 2.4

1. Validate that $x = \frac{4}{3}$ is a solution to $-3x^2 + x + 4 = 0$.

2. Validate that the factored form of 2800 is $2^4 \bullet 5^2 \bullet 7$.

3. How would you validate that $\left(2, \frac{1}{2}\right)$ is the solution to the system of equations $\begin{cases} x + 2y = 3 \\ 3x - 2y = 5 \end{cases}$ other than by substitution?

4. Validate that $x = 1$ is the solution to the equation $3x - (2 - 4x) = 1$.

5. Use a technique from this section to validate whether a problem that asks for the speed at which a person is walking could have a solution of 65 mph.

6. Validate that $\frac{x^2 + 4x - 21}{x - 7}$ is equivalent to $x + 3$ for $x \neq 7$.

7. Validate that $2x - 5 = 3$ is equivalent to $2x = 8$.

8. Validate that the expression $5x - 3(2 - 3x) + 7$ is equivalent to $14x + 1$.

What Did You Learn About Learning Math? 2.4

1. Why is validating your learning an important learning skill?

2. When validating your work, why is it important to use prior knowledge?

3. Why is it important to know how to establish the requirements of a solution?

4. What would you advise another student about validation?

Concept Map — Chapter 2

Terminology from the chapter
check the terms used in your map

__elimination method

__equality (=)

__equality properties

__equivalent equations

__isolate a variable

__linear equation

__literal equation

__solution of a system

__solution set

__substitution method

__substitution principle

__system of linear equations

__validation

__validation of learning

__validation of work

Your instructor will specify how many terms from the chapter you should use in your concept map.

CHAPTER 3

This chapter develops methods of solving various types of inequalities. In addition, you will revisit the basic concepts involved needed to graph lines in the coordinate plane including graphing points, intercepts, and slope of a line. In the final two sections, you will learn how to graph linear equations in two variables (lines) and how to solve linear inequalities in two variables.

Section 3.1 Solving and Graphing a Linear Inequality of a Single Variable

Why? 3.1

How many books does a publisher need to sell in order to have their income exceed their expenses? Suppose a textbook publisher sells books to bookstores at an average price of $30 per book; its average printing cost per book is $6; and each month it spends $250,000 for its warehouse, offices, and salaries. How many books must it sell per year? Such questions can be addressed with inequalities.

What Do You Already Know? 3.1

What is a number line?

What does it mean to specify the "solution set" of an equation or inequality?

Explain in general terms, and with a specific example, how to solve an equation in one variable.

Using the symbols $<$,$>$, $>$, and $<$, create four true statements.

Goals 3.1

When you complete this activity, you should be able to:

1. Solve inequalities in one variable
2. Indicate solution sets of inequalities on a number line and in interval notation

Building Mathematical Language 3.1

New Terms:	closed interval	Multiplicative Law of Inequalities
	interval notation	open interval

Key Concepts 3.1

Similarities and Differences Between Solving Equations and Solving Inequalities

Solving inequalities involves exactly the same methodology as solving equations with the exception that multiplying or dividing by a negative number reverses the order of the inequality.

Methodology: Solving a Linear Inequality 3.1

This methodology highlights, when and how to eliminate fractions, when and how to clear parentheses, when to simplify expressions, when and how to address negative coefficients, and how to denote the end points of a solution, both symbolically and graphically. The solution to linear inequalities is usually an infinite number of values.

Limitation/Caution: The need to reverse the direction of the inequality when multiplying or dividing by a negative number can be confusing.

	Example 1	Example 2	Your Turn
Solve for x:	$-3x + 8 \geq 2(3 - x)$	$\frac{1}{2}(x-3) < \frac{1}{3}(4x+2)$	$2(4 - x) > \frac{2}{5}x + 3$

Steps	Discussion
1 Clear the parentheses	Use the Distributive Property in the clearing the parentheses.

Ex 1

$$-3x + 8 \geq 6 - 2x$$

Ex 2

$$\frac{1}{2}x - \frac{1}{2} \cdot 3 < \frac{1}{3} \cdot 4x + \frac{1}{3} \cdot 2$$

Your Turn

Steps	Discussion
2 Eliminate fractions	Multiply the inequality by the LCM (Least Common Multiple) to eliminate fractions within the inequality (Multiplicative Law of Inequalities).

Ex 1

Done.

Ex 2

$$6\left(\frac{x}{2} - \frac{3}{2}\right) < 6\left(\frac{4x}{3} + \frac{2}{3}\right)$$

$$3x - 9 < 8x + 4$$

Your Turn

Steps	Discussion
3 Simplify	Simplify the inequality by combining like terms and dividing by common factors.

Ex 1

Done.

Ex 2

Already simplified.

Your Turn

Steps	Discussion
4 Decide on direction	Think through which direction to isolate the variable.

Ex 1

Move $-2x$ to the left side because adding $2x$ to both sides is easy.

Ex 2

Move x to the right side because $8x - 3x$ is easier to simplify.

Your Turn

Steps	Discussion
5 Isolate the variable in the inequality	Use the Additive Property of Inequalities to create an inequality with all terms of the variable on the same side of the inequality.

Ex 1

$$-3x + 8 \geq 6 - 2x$$

$$-3x + 2x + 8 \geq 6 - 2x + 2x$$

$$-x + 8 \geq 6$$

Ex 2

$$3x - 9 < 8x + 4$$

$$-3x + 3x - 9 < -3x + 8x + 4$$

$$-9 < 5x + 4$$

Your Turn

Steps	Discussion
6 Continue the isolation	Use the Additive Property of Inequalities to create an inequality with the constants on the other side of the inequality.

Ex 1

$$-x + 8 \geq 6$$

$$-x + 8 - 8 \geq 6 - 8$$

$$-x \geq -2$$

Ex 2

$$-9 < 5x + 4$$

$$-4 - 9 < -4 + 5x + 4$$

$$-13 < 5x$$

Your Turn

Steps	Discussion
7 Complete the isolation	**Use the Multiplicative Property of Inequalities to divide each side by the coefficient of the variable. If the coefficient is negative, you must reverse the direction of the inequality.**

Ex 1

Coefficient: –1

$$\frac{-x}{-1} \le \frac{-2}{-1}$$

$$x \le 2$$

Ex 2

Coefficient: 5

$$\frac{-13}{5} < \frac{5x}{5}$$

$$\frac{-13}{5} < x$$

Your Turn

Steps	Discussion
8 Represent the solution set graphically and in interval notation	**Identify the end of the line segment, determine if open or closed, and identify the direction of the solution set.**

Ex 1

Point is 2, closed, and goes to negative infinity

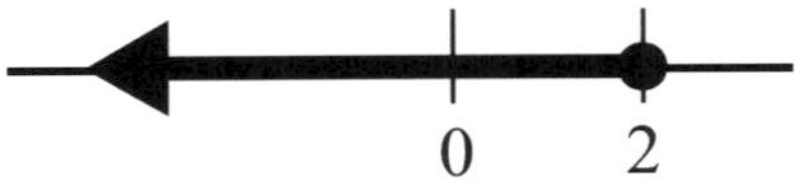

OR

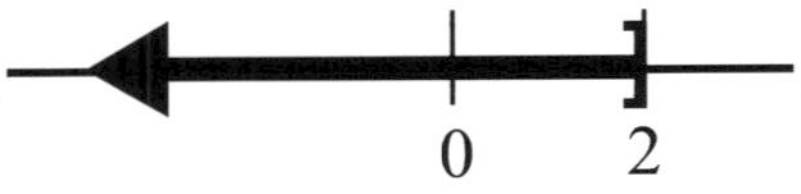

AND $(-\infty,\ 2]$

Ex 2

Point is $-\frac{13}{5}$, open, and goes to positive infinity

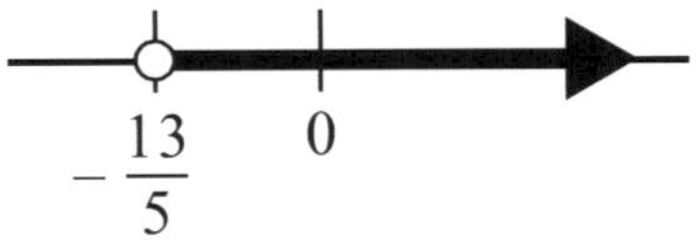

OR

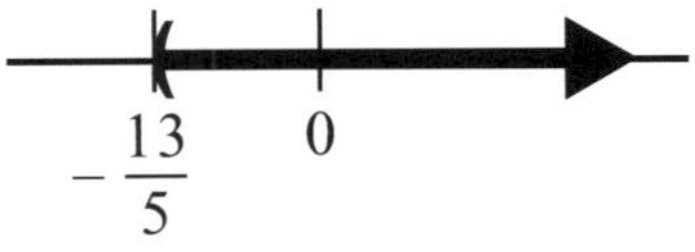

AND $\left(-\frac{13}{5},\ +\infty\right)$

Your Turn

Steps	Discussion
9 Test and validate solution set	Pick two points from the graph (one in the solution set and one not in the solution set; pick points that can be validated easily).

Ex 1

Pick: **0** (in the solution set) and **3** (not in the solution set)

$$-3x+8\geq 2(3-x)$$

$$-3(\mathbf{0})+8 \overset{?}{\geq} 2(3-\mathbf{0})$$

$$0+8 \overset{?}{\geq} 6$$

$$8\geq 6 \quad \textbf{(yes)} \checkmark$$

$$-3(\mathbf{3})+8 \overset{?}{\geq} 2(3-\mathbf{3})$$

$$-9+8 \overset{?}{\geq} 2(0)$$

$$-1\geq 0 \quad \textbf{(no)} \checkmark$$

Ex 2

Pick: **0** (in the solution set) and **–10** (not in the solution set)

$$\frac{1}{2}(x-3)<\frac{1}{3}(4x+2)$$

$$\frac{1}{2}(\mathbf{0}-3) \overset{?}{<} \frac{1}{3}(4\bullet\mathbf{0}+2)$$

$$-\frac{3}{2}<\frac{2}{3} \quad \textbf{(yes)} \checkmark$$

$$\frac{1}{2}(\mathbf{-10}-3) \overset{?}{<} \frac{1}{3}(4(\mathbf{-10})+2)$$

$$-\frac{13}{2} \overset{?}{<} -\frac{38}{3}$$

$$-6.5<-12.66 \quad \textbf{(no)} \checkmark$$

Your Turn

Addressing Common Errors 3.1

The following table shows some of the most common errors that learners tend to make with the kinds of problems covered in this section.

ERROR: Order of the inequality

Incorrect Process	Resolution
$-3x < 6$ $x < -2$	*When multiplying or dividing by a negative number, the order of the inequality reverses.*

Correct Process	Validation
$-3x < 6$ $x > -2$	Choose an x which is greater than −2, such as 5. $-3(5) \overset{?}{<} 6$ $-15 < 6$ ✓

ERROR: Writing an incorrect solution in interval notation

Incorrect Process	Resolution
$3 + 4x > 6x + 9$ $3 - 2x > 9$ $-2x > 6$ $x < -3$ Solution: $[-3, -\infty]$	*Use [ ] where the endpoints are included in the solution, but ( ) where they are not included. Write the ordered pair with the smaller number to the left of the larger number. Note that ∞ and −∞ are not numbers themselves, but are used to indicate the "ends" of the number line (which in fact, never ends).*

Correct Process	Validation
$3 + 4x > 6x + 9$ $3 - 2x > 9$ $-2x > 6$ $x < -3$ Solution: $(-\infty, -3)$	Let $x = -5$ (which is a number < -3) and $x = 0$ (which is a number > -3). $3+4x \overset{?}{>} 6x+9$ $3+4(-5) \overset{?}{>} 6(-5)+9$ $3-20 \overset{?}{>} -30+9$ $-17 > -21$ (yes) $3+4(0) \overset{?}{>} 6(0)+9$ $3 > 9$ (no) ✓

ERROR: Graphing the solution incorrectly on the number line

Incorrect Process

$$5 > 3 + x$$
$$2 > x$$

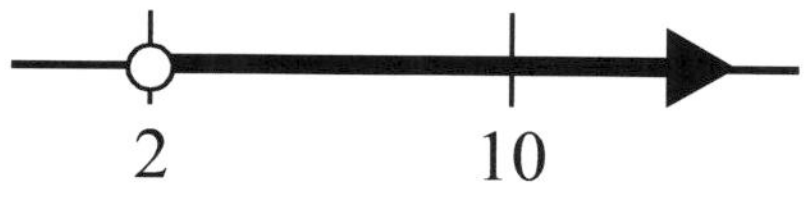

Resolution

Read (or rewrite) the inequality starting with the variable.

Correct Process

$$5 > 3 + x$$
$$2 > x$$
$$x < 2$$

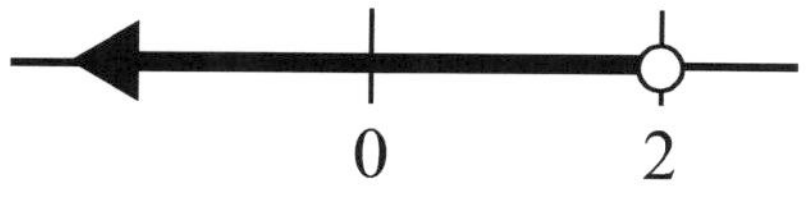

Validation

Choose convenient values for x; $x = 0$ and $x = 3$.

$5 \overset{?}{>} 3 + x$	$5 \overset{?}{>} 3 + x$
$5 \overset{?}{>} 3 + 0$	$5 \overset{?}{>} 3 + 3$
$5 > 3$ (yes)	$5 > 6$ (no) ✓

Preparation Inventory 3.1

I can...	OR *Here's my question...*
☐ understand when to reverse the order of an inequality.	☐
☐ solve a linear inequality in one variable.	☐
☐ graph a solution to an inequality on a number line.	☐
☐ present a solution to an inequality using interval notation.	☐

Critical Thinking Questions 3.1

1. What symbols are used to represent different types of inequalities?

2. How do you represent a solution set on a number line?

3. How do you represent a solution set in interval notation?

4. How do you represent an end point ***not*** in the solution set, in the following ways?
   a. symbolically

   b. graphically

   c. in interval notation

5. What properties do you use when you solve linear inequalities?

6. What are the critical techniques or steps in the methodology for solving linear equations?

7. Which of the techniques or steps from Question 6 can you use when solving linear inequalities?

8. Which steps in the methodology for solving a linear inequality are new, or different, from the steps in the methodology for solving a linear equation?

9. What is the most common error that occurs when dividing by a negative coefficient?

10. What issues can you identify that might make solving linear inequalities difficult for some students?

11. Why is $(2, -\infty)$ an incorrect way of writing a solution in interval notation?

12. How do you determine the order of the numbers when writing a solution in interval notation?

13. How do you validate a linear inequality?

Demonstrate Your Understanding 3.1

Solve each inequality for the indicated variable. Record your answer graphically and in interval notation.

1. Solve for x: $3x - 4 + 5x > 7 - 9x$

2. Solve for a: $3(2a - 4) \geq -3 + 2(3a + 7)$

3. Solve for x: $\frac{2}{3}(5-x)+2<\frac{3}{2}(x-2)$

4. Solve for x: $2x - 4 < 2x + 5$

5. Solve for y: $-2y+\frac{3}{5}(4y-5)\leq 5+2y-11$

Create and Solve the Hardest Problem 3.1

Your Hardest Problem
Problem Solution

What makes a problem involving graphing a linear inequality in one variable a hard problem to solve?

Identify, Describe, & Correct the Errors 3.1

Describe the error you found in the worked solution in the appropriate space. Solve the problem correctly under "Correct Process." You are welcome to solve the problem first and then describe the error afterward.

Problem 1

Solve for x: $\frac{1}{2}(x-5)+2<7$

Worked Solution *(Find the error!)*

$\frac{1}{2}(x-5)+2<7$

$x-5<5$

$x<10$

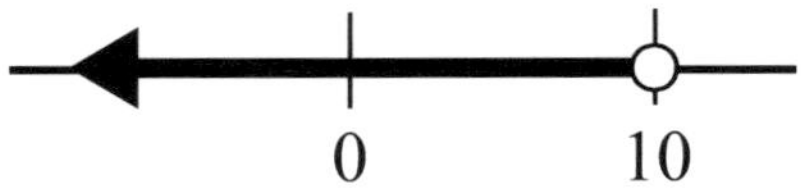

Describe the Error

Correct Process

Problem 2

Solve for x:

$$6 < -3(x-4)$$

Worked Solution *(Find the error!)*

$$6 < -3(x-4)$$
$$6 < -3x+12$$
$$-6 < -3x$$
$$2 > x$$

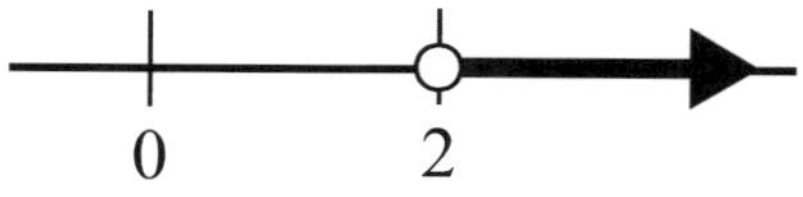

Correct Process

Describe the Error

Problem 3

Solve for x:

$$3x-4 \le 7$$

Worked Solution *(Find the error!)*

$$3x-4 \le 7$$
$$3x \le 11$$
$$x \le \frac{11}{3}$$

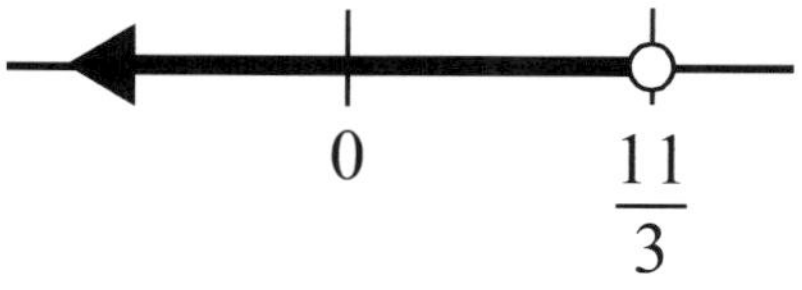

Correct Process

Describe the Error

Problem 4

Solve for s:

$$-2s > 5$$

Worked Solution *(Find the error!)*

$$-2s > 5$$

$$\frac{\cancel{-2}s}{\cancel{-2}} > \frac{5}{-2}$$

$$s > -\frac{5}{2}$$

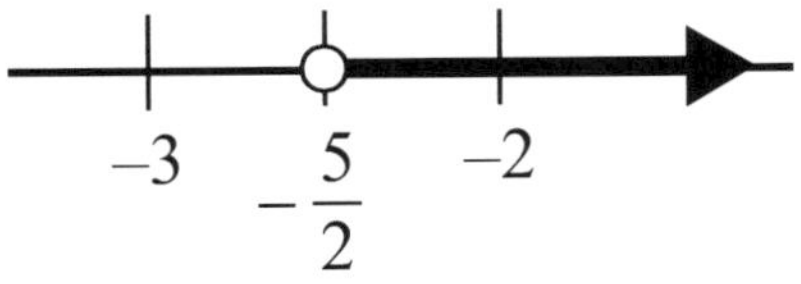

Correct Process

Describe the Error

What Did You Learn About Learning Math? 3.1

1. How does recalling previous knowledge about equations help in learning new knowledge about inequalities?

2. How does comparing previous knowledge with new knowledge strengthen your skill with the new knowledge?

Section 3.2 Solving and Graphing Compound Inequalities of a Single Variable

Why? 3.2

Compound inequalities occur in real-life situations. For example, the measure of obesity is defined in terms of compound inequalities.

Obesity is usually measured using Body Mass Index (BMI). The BMI is the mass of the person in kg divided by the square of the person's height in meters.

BMI $<$ 18.5 is underweight

18.5 $<$ BMI $<$ 25 is normal weight

25.0 $<$ BMI $<$ 30 is overweight

30.0 $<$ BMI $<$ 40 is obese

BMI $>$ 40.0 is severely (or morbidly) obese

Absolute value inequalities occur when you have scientific measurements that are not exact If a measured length, x, is off by no more than 2 cm either way from the exact length R, we write

$$|R \text{ cm} - x \text{ cm}| \leq 2 \text{ cm or just } |R - x| \leq 2$$

What Do You Already Know? 3.2

Give an example of an inequality in one variable.

What does it mean to solve an inequality in one variable?

What does the solution of an inequality in one variable look like?

How do you graph an inequality in one variable?

What does $|x|$ (absolute value) mean?

Give an example of a conjunction.

Goals 3.2

When you complete this activity, you should be able to:

1. Solve a compound inequality in one variable created by connecting two inequalities with "and" or "or"
2. Change an absolute value inequality into a compound inequality
3. Solve an absolute value inequality

Helpful Website 3.2

www.khanacademy.org/math/algebra/solving-linear-equations-and-inequalities/v/absolute-value-inequalities

Building Mathematical Language 3.2

New Terms: absolute value inequality compound inequality three-part inequality

Key Concepts 3.2

The ***absolute value*** of a quantity can be thought of as the distance of that quantity from zero on a number line.

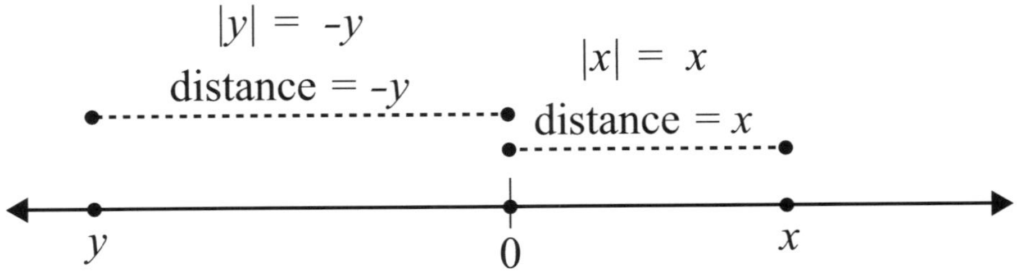

A ***compound inequality*** is two or more inequalities linked together with a conjunction (and, or). For example, [$x < 4$ ***or*** $x > 5$] as well as [$x > 4$ ***and*** $x < 5$] are compound inequalities. Such inequalities occur in situations where conditions are placed upon variables.

Limitation/Caution: In the ***and*** situation both inequalities must be true, whereas in the ***or*** situation only one needs to be true (but both can be true).

A ***three-part inequality*** has two inequality signs that face the same direction. For example, $2 < 3 < 5$ is equivalent to the compound inequality [$2 < 3$ ***and*** $3 < 5$].

Limitation/Caution: The inequality signs need to be facing the same direction. Therefore, $x > 4$ and $x > 5$ cannot be written as $4 < x > 5$.

Methodology: Changing an Absolute Value to a Compound Inequality 3.2

Limitation/Caution: Absolute value inequalities with more than one set of absolute value bars are difficult to understand or to solve.

Limitation/Caution: The solution to some inequalities can be all real numbers or no solution.

$|x| < 2$ *implies* $-2 < x < 2$

because the distance of x from 0 is less than 2, so x must be **between** 2 and –2.

$|x| > 2$ *implies* $x < -2$ ***or*** $x > 2$

because the distance of x from 0 is greater than 2.

Example 1:	Example 2:
Change into a compound inequality:	Change into a compound inequality:
$\lvert y-2\rvert < 3$	$\lvert 3-2x\rvert > 5$
$-3 < y-2 < 3$	$3-2x < -5$ ***or*** $3-2x > 5$

Methodology: Solving Compound Inequalities 3.2

The methodology highlights, how to (1) change compound linear inequalities into simple linear inequalities, (2) solve these linear inequalities, and (3) present a solution set symbolically and graphically, including the appropriate notation for the set's boundaries.

Limitation/Caution: This methodology does not work for non-linear compound inequalities.

	Example 1	**Example 2**	**Your Turn**
Solve for x:	$\|2x+7\| > 5$	$(x-3) < \frac{1}{3}(4x+2) \leq 10$	$\frac{1}{3} \leq 2(3x-1) < 5x$

Steps	**Discussion**
1 Set up and identification	**Set up the problem and identify its type**

Ex 1	Ex 2	Your Turn
$\|2x+7\| > 5$	$(x-3) < \frac{1}{3}(4x+2) \leq 10$	

2 If possible, draw a diagram	**If it is an absolute value inequality, draw a diagram (if possible)**

Ex 1

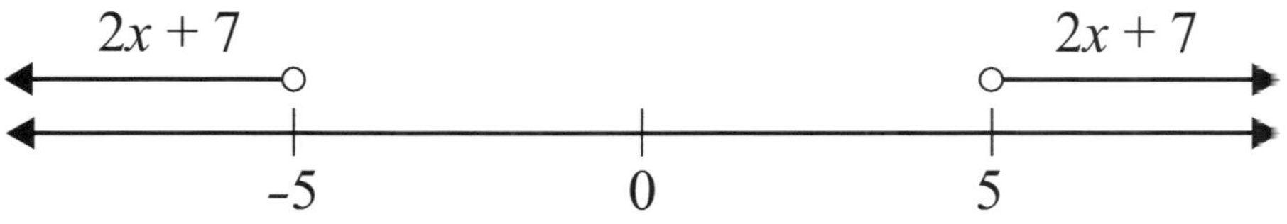

Ex 2

(Not an absolute value inequality; no diagram needed)

Your Turn

3 Separate compound inequality	**If you are presented with a compound inequality, you must solve each inequality separately and then move on to step 4.**

Ex 1 Problem 1:	Ex 1 Problem 2:	Ex 2 Problem 1:	Ex 2 Problem 2:
$2x+7 < -5$	$2x+7 > 5$	$(x-3) < \frac{1}{3}(4x+2)$	$\frac{1}{3}(4x+2) \leq 10$

Your Turn

Steps	Discussion
4 Solve the inequalities	**Use the Methodology for Solving Inequalities in One Variable.**

Ex 1

Problem 1:

$2x < -12$

$x < -6$

Problem 2:

$2x > -2$

$x > -1$

Ex 2

Problem 1:

$(x-3) < \frac{1}{3}(4x) + \frac{1}{3}(2)$

$3(x-3) < 3\left(\frac{4x}{3} + \frac{2}{3}\right)$

$3x - 9 < 4x + 2$

$-9 - 2 < 4x - 3x$

$-11 < x$

Problem 2:

$\frac{1}{3}(4x) + \frac{1}{3}(2) \le 10$

$3\left(\frac{4x}{3} + \frac{2}{3}\right) \le 3 \cdot 10$

$4x + 2 \le 30$

$4x + 2 \le 30$

$4x \le 28$

$x \le 7$

Your Turn

Steps	Discussion
5 Present the compound solution set	**Both solutions must be true. If appropriate, present as a three-part inequality.**

Ex 1

$x < -6$ or $x > -1$

Ex 2

$-11 < x$ and $x \le 7$

$-11 < x \le 7$

Your Turn

Steps	Discussion
6 Describe the solution set	**Identify the end(s) of the line segments, determine if they are open or closed, and identify the direction(s) of the solution sets.**

Ex 1

Points 6 and –1 are open.

Ex 2

Point –11 is open and point 7 is closed.

Your Turn

Steps	Discussion
7 Present the compound solution set graphically	**Determine the end point(s), open or closed, and the interval(s).**

Ex 1

Ex 2

Your Turn

Steps	Discussion
8 Test and validate the solution set	**Pick three or more points, some in the solution set and some not in solution set (pick easy calculation points, if possible).**

Ex 1

We choose –7 and 0 (in the solution set) and –2 (not in the solution set)

$\lvert 2(-7)+7 \rvert \overset{?}{>} 5$	$\lvert 2(0)+7 \rvert \overset{?}{>} 5$	$\lvert 2(-2)+7 \rvert \overset{?}{>} 5$
$\lvert -14+7 \rvert \overset{?}{>} 5$	$\lvert 0+7 \rvert \overset{?}{>} 5$	$\lvert -4+7 \rvert \overset{?}{>} 5$
$\lvert -7 \rvert \overset{?}{>} 5$	$\lvert 7 \rvert \overset{?}{>} 5$	$\lvert 3 \rvert \overset{?}{>} 5$
$7 > 5$ **(yes)** ✓	$7 > 5$ **(yes)** ✓	$3 > 5$ **(no)** ✓

Steps	Discussion
Ex 2	We choose 0 and 7 (in the solution set) and –10 and 20 (not in the solution set) $(x-3)<\frac{1}{3}(4x+2)\le 10$ $(0-3)\overset{?}{<}\frac{1}{3}(4\cdot 0+2)\overset{?}{\le}10$ $-3<\frac{2}{3}\le 10$ **(yes)** ✓ $(7-3)\overset{?}{<}\frac{1}{3}(4\cdot 7+2)\overset{?}{\le}10$ $4\overset{?}{<}\frac{1}{3}(30)\overset{?}{\le}10,\ 4<10\le 10$ **(yes)** ✓ $(-12-3)\overset{?}{<}\frac{1}{3}(4(-12)+2)\overset{?}{\le}10$ $-15<-\frac{46}{3}\le 10$ **(no)** ✓ $(20-3)\overset{?}{<}\frac{1}{3}(4\cdot 19+2)\overset{?}{\le}10$ $17\overset{?}{<}\frac{1}{3}(78)\overset{?}{\le}10$ $17<26\le 10$ **(no)** ✓
Your Turn	

Addressing Common Errors 3.2

The following table shows some of the most common errors that learners tend to make with the kinds of problems covered in this section.

ERROR: Separating compound inequalities incorrectly (wrong conjunction)

Incorrect Process	Resolution
$-2<x<5$ $-2<x$ or $x<5$	*A three-part inequality always means one of the inequalities AND the other.*

Correct Process	Validation
$-2<x<5$ $-2<x$ AND $x<5$	Choose an x value that is between −2 and 5, such as 0: Is it true that $0>-2$ (yes) AND $0<5$ (yes)? Yes. ✓ Choose an x value that is NOT between −2 and 5, such as 10. Is it true that $10>-2$ (yes) AND $10<5$ (no)? No. ✓

ERROR: Using only the positive value of the absolute value inequality

Incorrect Process

$|x+3|>6$

$x+3>6$

$x>3$

Resolution

An absolute value inequality must be separated into two separate inequalities.

Correct Process

$|x + 3| > 6$

The distance of $x + 3$ from 0 is greater than 6.

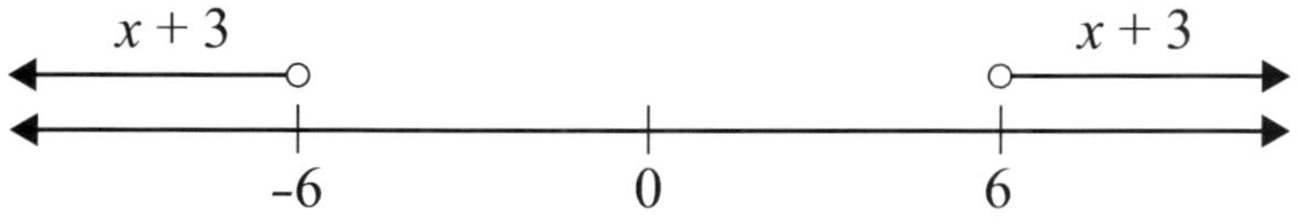

$x+3<-6$ OR $x+3>6$

$x<-9$ OR $x>3$

Validation

Select an x which is less than −9, such as −10 (should be a solution)

Select an x which is greater than 3, such as 4 (should be a solution)

Select an x which is between −9 and 3, such as 0 (should not be a solution)

Substitute the values into the original inequality:

Is $|-10 + 3| \overset{?}{>} 6$ OR $|-10 + 3| \overset{?}{<} -6$?

$7 \overset{?}{>} 6$ OR $7 \overset{?}{<} -6$

(yes) (no) So −10 is a solution.

Is $|4 + 3| \overset{?}{>} 6$ OR $|4 + 3| \overset{?}{<} -6$?

$7 \overset{?}{>} 6$ OR $7 \overset{?}{<} -6$

(yes) (no) So 4 is a solution.

Is $|0 + 3| \overset{?}{>} 6$ OR $|0 + 3| \overset{?}{<} -6$?

$3 \overset{?}{>} 6$ OR $3 \overset{?}{<} -6$

(no) (no) So 0 is not a solution. ✓

ERROR: **Splitting an absolute value inequality into two separate inequalities, but one of them is incorrect**

Incorrect Process

$|x+3|>6$

$x+3>6$ OR $x+3>-6$

$x>3$ OR $x>-9$

Resolution

Draw a number line graph, placing the expression inside the absolute value signs more than 6 units away from 0 on either side. Then translate the diagram into two inequalities.

Correct Process

$|x+3|>6$

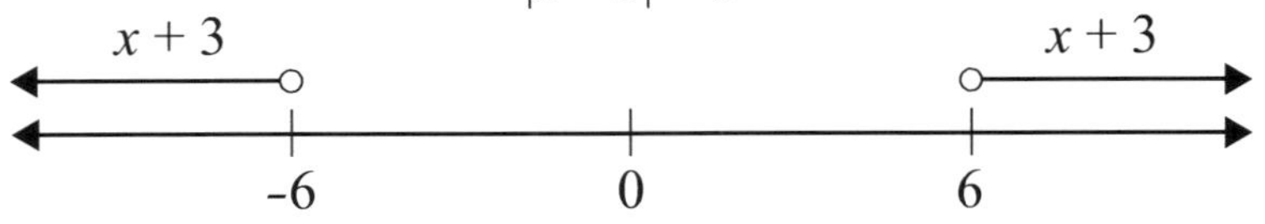

$x+3<-6$ OR $x+3>6$

$x<-9$ OR $x>3$

Validation

Select an x which is less than -9, such as -10 (should be a solution)

Select an x which is greater than 3, such as 4 (should be a solution)

Select an x which is between -9 and 3, such as 0 (should not be a solution)

Substitute the values into the original inequality:

Is $|-10+3| \overset{?}{>} 6$ OR $|-10+3| \overset{?}{<} -6$?

$7 \overset{?}{>} 6$ OR $7 \overset{?}{<} -6$

(yes) (no) So -10 is a solution.

Is $|4+3| \overset{?}{>} 6$ OR $|4+3| \overset{?}{<} -6$?

$7 \overset{?}{>} 6$ OR $7 \overset{?}{<} -6$

(yes) (no) So 4 is a solution.

Is $|0+3| \overset{?}{>} 6$ OR $|0+3| \overset{?}{<} -6$?

$3 \overset{?}{>} 6$ OR $3 \overset{?}{<} -6$

(no) (no) So 0 is not a solution. ✓

Preparation Inventory 3.2

I can...	OR *Here's my question...*
☐ correctly separate a three-part inequality into two separate inequalities.	☐
☐ correctly change an absolute value inequality into separate or a three-part inequality.	☐
☐ draw a number line representation of an absolute value inequality.	☐
☐ validate the solution to a compound inequality.	☐

Critical Thinking Questions 3.2

1. How do you represent an end point not in a solution set symbolically, graphically, and with interval notation?

2. What conjunctions are used when separating compound inequalities into separate inequalities?

3. How do you change an absolute value inequality into two separate inequalities?

4. What is the most common error that occurs when dividing by a negative coefficient while solving an inequality? Why do you think this is error is so common?

5. What are the similarities and differences between solving an inequality and solving a compound inequality?

6. How do you decide on the conjunction when presenting a solution to a compound inequality?

7. Why are $2 > x < 5$ and $2 > x > 5$ incorrect three-part inequalities?

8. What issues can you identify that make solving linear inequalities difficult for some students?

Demonstrate Your Understanding 3.2

1. On the number line, represent the set of numbers for which $x > \frac{9}{2}$ *or* $x < -2$.

2. Solve: $-2 < 2x - 5 \leq 7$

3. Solve and present the solution in interval notation: $2 < \frac{1}{3}(x-2) < 5x$

4. Solve: $-3(x-1) \le \frac{2}{3}(5-x) < \frac{3}{4}$

5. Solve and graph: $|2x-5|>7$

6. Solve and present the solution in interval notation: $|3x-5|\le 2$

7. Discuss the solution sets of these two absolute value inequalities and defend your answer:

   $|3x-5| \le -2$

   $|3x-5| \ge -2$

Create and Solve the Hardest Problem 3.2

Your Hardest Problem

Problem Solution

What makes a problem where you must solve and graph compound inequalities in one variable a hard problem?

Identify, Describe, & Correct the Errors 3.2

Describe the error you found in the worked solution in the appropriate space. Solve the problem correctly under "Correct Process." You are welcome to solve the problem first and then describe the error afterward.

Problem 1

Solve: $2 \le 2 - x < 4$

Worked Solution *(Find the error!)*

$$2 \le 2 - x < 4$$
$$2 - 2 \le -x < 4 - 2$$
$$0 \le -x < 2$$
$$0 \le x < -2$$

Correct Process

Describe the Error

Problem 2

Solve: $-7 < x - 4 < 5$

Worked Solution *(Find the error!)*

$$-7 < x - 4 < 5$$
$$-7 < x - 4 \; and \; x - 4 < 5$$
$$-3 < x \; and \; x < 9$$
$$So, \; x < 9$$

Correct Process

Describe the Error

Problem 3

Solve: $-2x < 5(x-3) < 7$

Worked Solution *(Find the error!)*

$$-2x < 5(x-3) < 7$$
$$-2x < 5(x-3) \quad and \quad 5(x-3) < 7$$
$$-2x < 5x-15 \quad and \quad 5x-15 < 7$$
$$-7x < -15 \quad and \quad 5x < 22$$
$$x < \frac{15}{7} \quad and \quad x < \frac{22}{5}$$
$$x < \frac{15}{7}$$

Correct Process

Describe the Error

Problem 4

Solve: $|2x-3| > 4$

Worked Solution *(Find the error!)*

$$|2x-3| > 4$$
$$(2x-3) < -4 \ and \ (2x-3) > 4$$
$$2x < -1 \ and \ 2x > 7$$
$$x > -\frac{1}{2} \ and \ x > \frac{7}{2}$$
$$So, \ x > \frac{7}{2}$$

Correct Process

Describe the Error

Problem 5

Solve: $-8 < -2x - 4 < 2$

Worked Solution *(Find the error!)*

$-8 < -2x - 4 < 2$

$-8 < -2x - 4 \; and \; -2x - 4 < 2$

$-4 < -2x \; and \; -2x > 6$

$2 > x \; and \; x < -3$

$2 > x < -3$

Correct Process

Describe the Error

What Did You Learn About Learning Math? 3.2

1. How did prior knowledge help you in learning how to solve compound inequalities?

2. How does using patterns help when solving various types of inequalities?

3. How do visual representations (graphs) help clarify inequality relationships?

Section 3.3 Graphing Lines Using Tables

Why? 3.3

A visual representation of a relationship between two quantities that change at a constant rate is important for interpreting real world situations. For example, starting at a shopping mall 2 miles from your home, the distance you are from home, walking at a constant rate of 3 miles per hour, depends on the time you have walked and can be represented as

$$\textit{distance} = 2 - 3 \times \textit{time}$$

where *distance* is in miles, *time* is in hours and 3 means a rate of 3 miles per hour. One visual representation of this situation is a line that provides useful information.

What Do You Already Know? 3.3

How would you find a named street with street map coordinates A3?

What is a linear equation in one variable?

What are some uses of the number line?

What kinds of numerical tables are you familiar with?

Goals 3.3

When you complete this activity, you should be able to:

1. Plot points on a coordinate plane
2. Explain what a constant rate of change means
3. Use a table to graph a linear equation in two variables
4. Use the standard form of a linear equation in two variables to graph a line

Building Mathematical Language 3.3

New Terms: dependent variable, rate of change, x-intercept, independent variable, slope, y-intercept, origin, table of values (T-table), standard form of a linear equation in two variables

Key Concepts 3.3

Coordinate axes are two perpendicular number lines intersecting at their zero values as shown here. The point at the intersection of the two number lines is called the ***origin*** and is marked in this case with a capital O. Together, these axes define the coordinate plane.

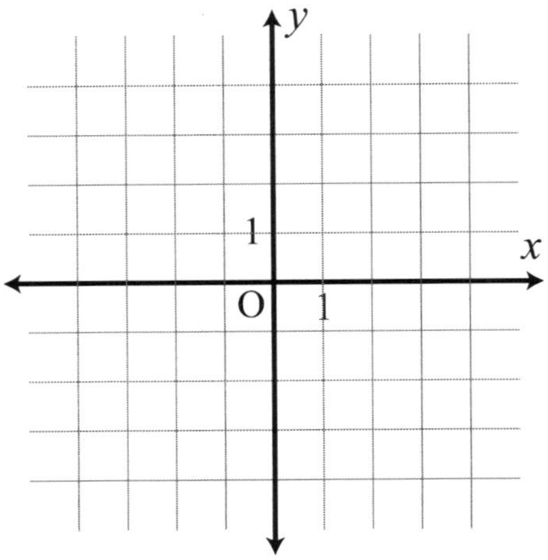

A pair of numbers in parentheses like (2, 3) is called an *ordered pair*. Such ordered pairs can be placed or plotted on the coordinate plane. The first coordinate is assigned to the horizontal axis, the second to the vertical axis. Thus, the point (2, 3) is located 2 to the right on the horizontal axis and 3 up parallel to the vertical axis. The horizontal axis is often called the *x-axis* and the vertical axis the *y-axis*.

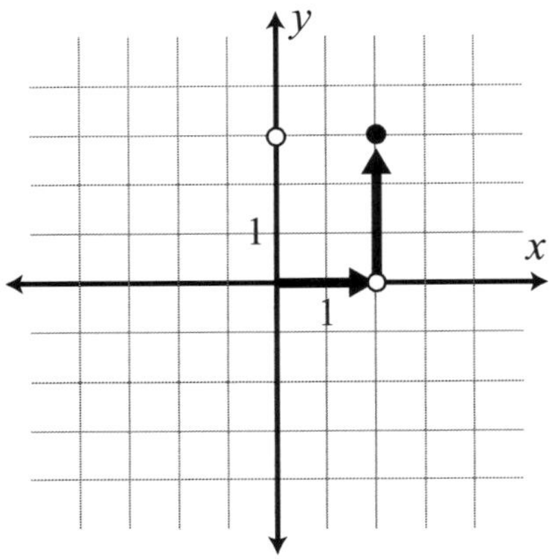

By convention, the first variable of an ordered pair is called the ***independent variable*** and the second variable is called the ***dependent variable*** regardless of the letters used to label the axes. For example, in the coordinate plane, at right, t is the independent variable and d is the dependent variable.

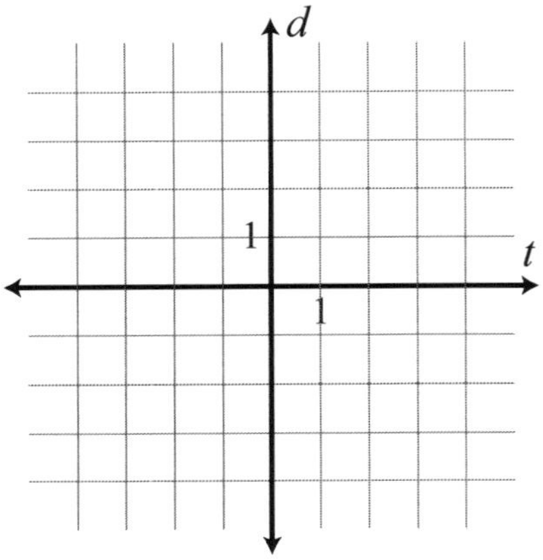

The linear equation $2x - y = 4$ is a ***linear equation in two variables in standard form*** where both variables are linear and on one side of the equation and the constant term is on the other side. Such equations have graphs that are lines. A ***table of values***, called a ***T-table***, showing the value of one quantity when the other quantity is given results in a list of ordered pairs of numbers which are solutions to the equation. For example, in this case a table might look like this (vertically or horizontally):

x	y
−2	−8
−1	−6
0	−4
1	−2
2	0
3	2

x	−2	−1	0	1	2	3
y	−8	−6	−4	−2	0	2

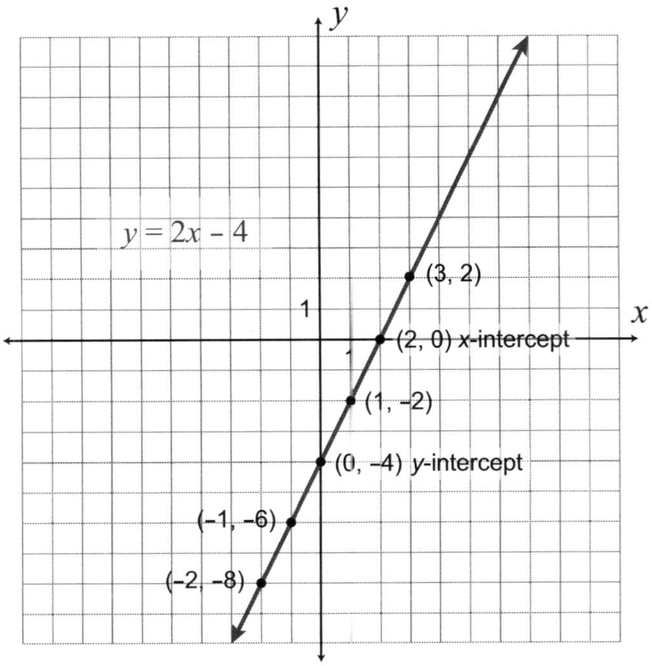

The point (0, –4) on the fourth row of the vertical table (the fourth column of the horizontal table) is called the ***y-intercept*** of the equation because it is where the line crosses the *y*-axis. Similarly, the point (2, 0) on the fifth line of the table is called the ***x-intercept***. The coordinates of such intercepts are easy to determine from the standard form ($2x - y = 4$) by setting *x* equal to zero and *y* equal to zero.

If we solve this equation, $2x - y = 4$, for *y* in terms of *x*, we have $y = 2x - 4$, and it is easier to see that when *x* is 2, *y* will be 0. We will often make this kind of conversion when we are creating a table of values. To enter an equation into a calculator, you must make this kind of conversion.

These ordered pairs can be placed on the coordinate plane one at a time as shown previously. Notice that as the first variable increases by one, the second variable always increases by two.

For relationships like this, we say that there is a ***constant rate of change***. Linear equations in two variables always have such a constant rate of change. An equation with a constant rate of change, when graphed, appears as a line.

In the table for the linear equation

$$2x - y = 4$$

given below,

x	–2	–1	0	1	2	3
y	–8	–6	–4	–2	0	2

this constant rate of change means that between any two points in the table, the rate of change will be the same. For example, the rate of change between (–2, –8) and (–1, –6) is the same as the rate of change between (0, –4) and (3, 2).

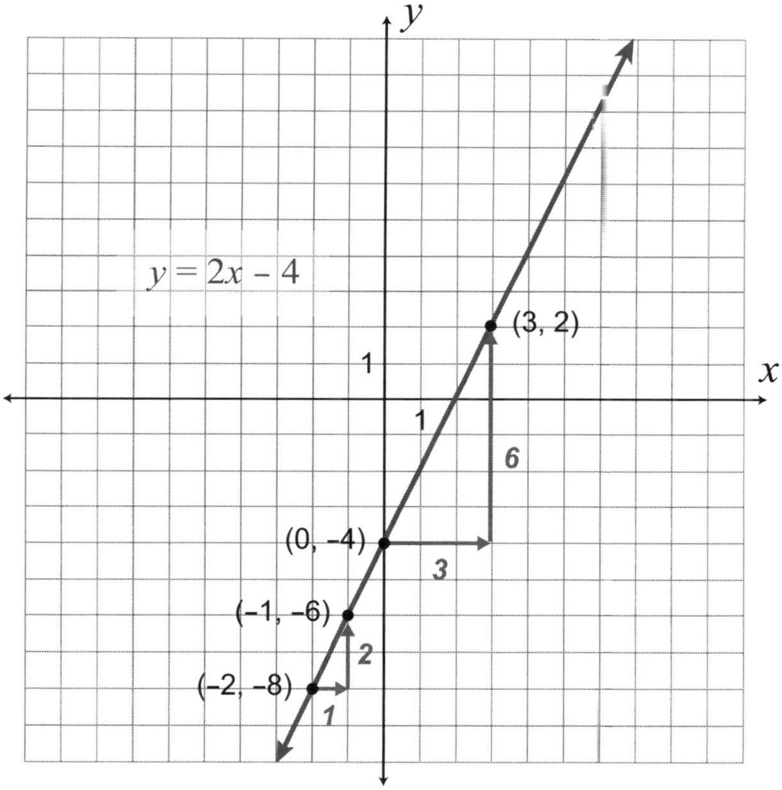

$$\text{rate of change} = \frac{\text{change in horizontal direction}}{\text{change in vertical direction}} = \frac{\text{change in } y}{\text{change in } x}$$

$(-2,-8)$ and $(-1,-6)$

$$\text{rate of change} = \frac{-6-(-8)}{-1-(-2)} = \frac{2}{1} = 2$$

$(0,-4)$ and $(3,2)$

$$\text{rate of change} = \frac{2-(-4)}{3-0} = \frac{6}{3} = 2$$

In both cases, the constant rate of change of 2 means for every change of 1 unit horizontally, there is a 2 unit change vertically.

The rate of change of a line on a coordinate plane representing a linear equation can be found using an algebraic formula. Let two points on the line be designated Point 1, (x_1, y_1), and Point 2, (x_2, y_2). Notice that the source (Point 1 or Point 2) of the coordinates of the points is designated with subscripts.

The ***slope***, m, can be determined algebraically from the formula: $m = \dfrac{\text{vertical change}}{\text{horizontal change}} = \dfrac{\Delta y}{\Delta x}$

$$= \frac{y_2 - y_1}{x_2 - x_1}$$

The following diagram illustrates the formula.

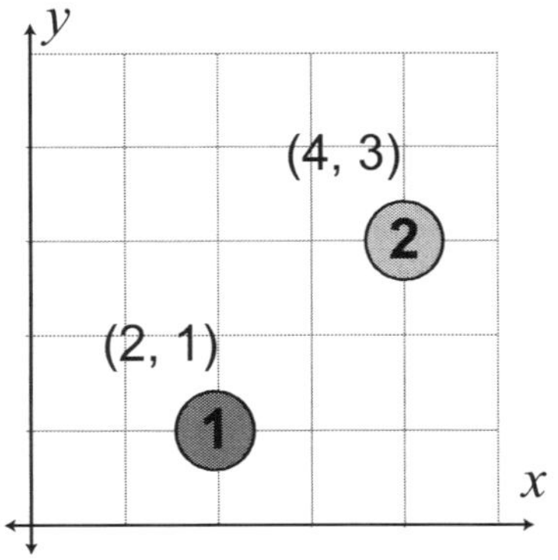

Point 1	**Point 2**
(2, 1)	**(4, 3)**
x coordinate = 2	**x coordinate = 4**
y coordinate = 1	**y coordinate = 3**

$$m = \frac{(y_2) - (y_1)}{(x_2) - (x_1)}$$

$$m = \frac{(\text{y of Point 2}) - (\text{y of Point 1})}{(\text{x of Point 2}) - (\text{x of Point 1})}$$

$$= \frac{(3) - (1)}{(4) - (2)} = \frac{2}{2} = \mathbf{1}$$

Methodology: Graphing an Equation in Two Variables 3.3

If the equation has one variable by itself on the left side, then that variable is the dependent variable and the other variable is the independent variable. For linear equations in two variables, either variable can be the independent variable.

	Example 1	**Example 2**	**Your Turn**
Graph:	$y = 3x - 4$	$5n + 10d = 50$	$2x - y = 3$

Steps	**Discussion**
1 Determine the independent and dependent variables	**If there is no context that specifies otherwise, then the first letter alphabetically is the independent variable. If they have a context, then the context can determine which variable is the independent variable and which variable is the dependent variable.**
Ex 1	When x and y are the variables, by convention x is the independent variable and y is the dependent variable.
Ex 2	Let d be the independent variable and n be the dependent variable
Your Turn	

Steps	Discussion
2 Create a table of values	**The table will have two columns. The first column will have values of the independent variable, and the second column will have the corresponding calculated values of the dependent variable. The table should have at least 3 ordered pairs (two points determine a line and the third is to check it).**

Ex 1

Two points determine a line
and an extra point to check.

Technique 1: Using x-intercept and y-intercepts

x	0	4/3	2
y	–4	0	2

Technique 2: Solving for y in terms of x
(already done).

$$y = 3x - 4$$

x	0	2	4
y	–4	2	8

Ex 2

Two points determine a line
and an extra point to check

Technique 1: Using x-intercept and y-intercepts

d	0	5	3
n	10	0	4

Technique 2: Solving for n in terms of d

$$5n + 10d = 50$$

$$5n = 50 - 10d$$

$$n = \frac{50}{5} - \frac{10d}{5}$$

$$n = 10 - 2d$$

d	1	2	4
n	8	6	2

Your Turn

Steps	Discussion
3 Plot the ordered pairs from the table	**Plot each point using the first coordinate as the horizontal coordinate and the second coordinate as the vertical coordinate.**

Ex 1

Technique 1

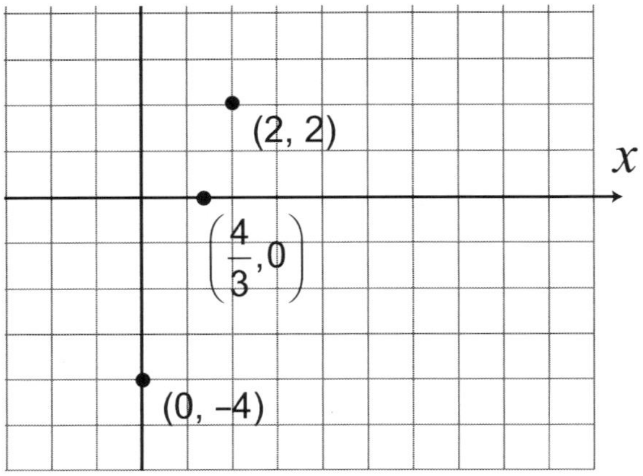

Technique 2

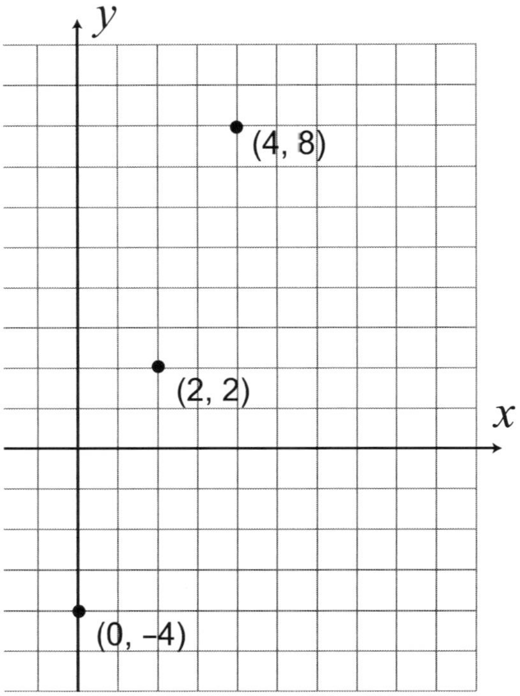

Ex 2

Technique 1

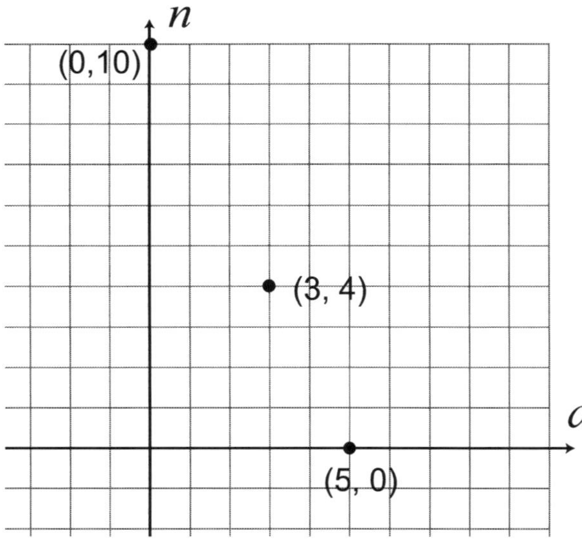

Technique 2

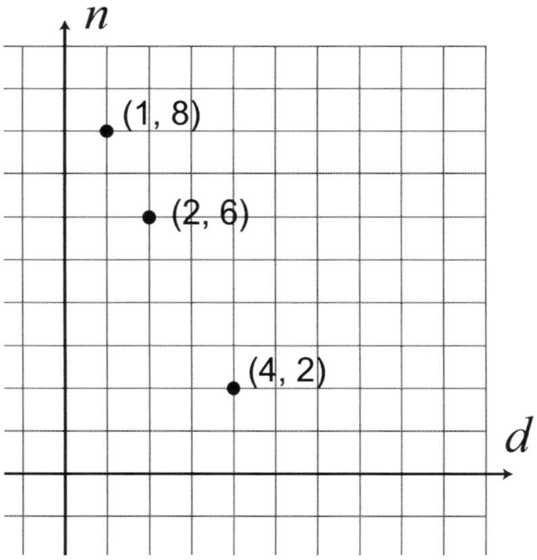

Steps	Discussion

Your Turn

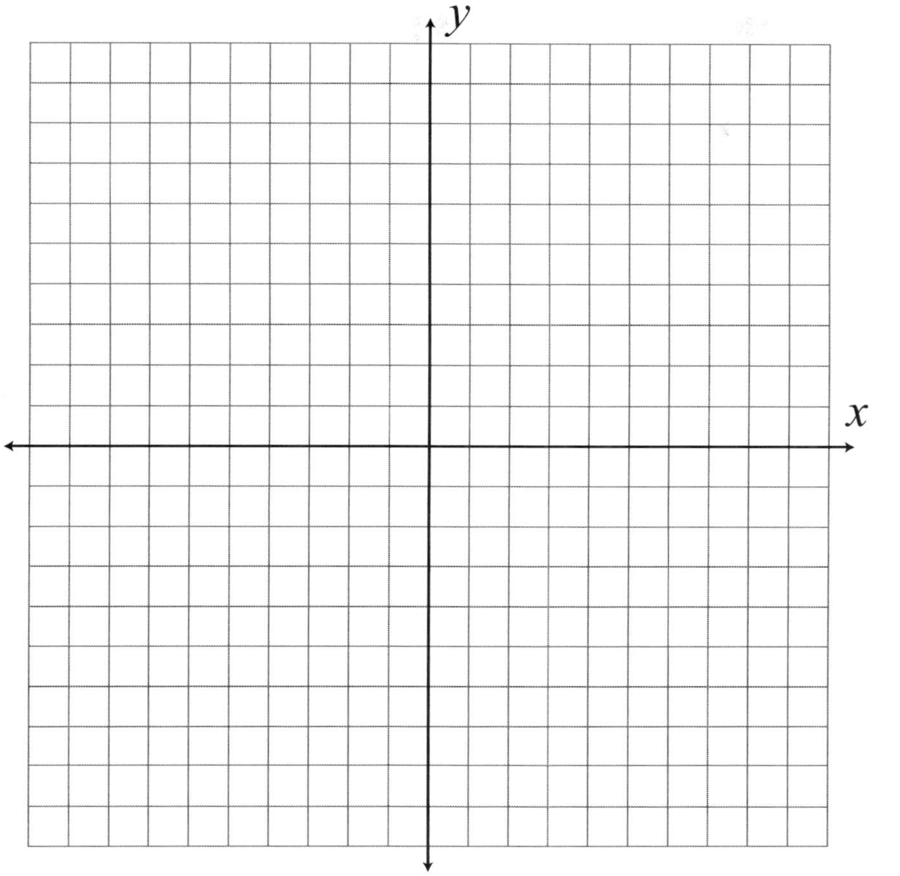

4 Draw a line through the points	The line should pass through all three points. In real world problems, you should consider which values are reasonable for the situation.

Ex 1

Technique 1

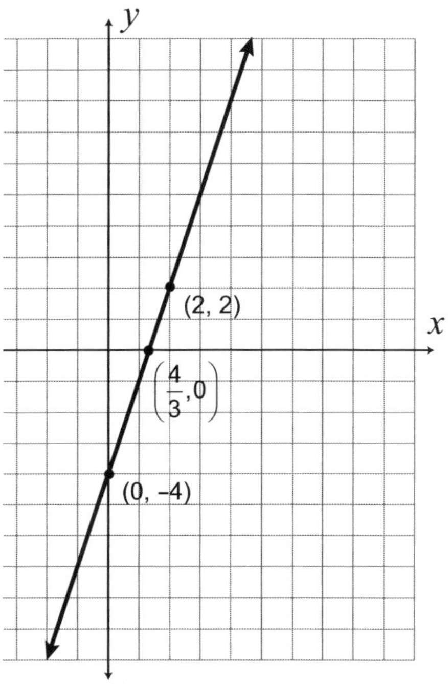

Technique 2

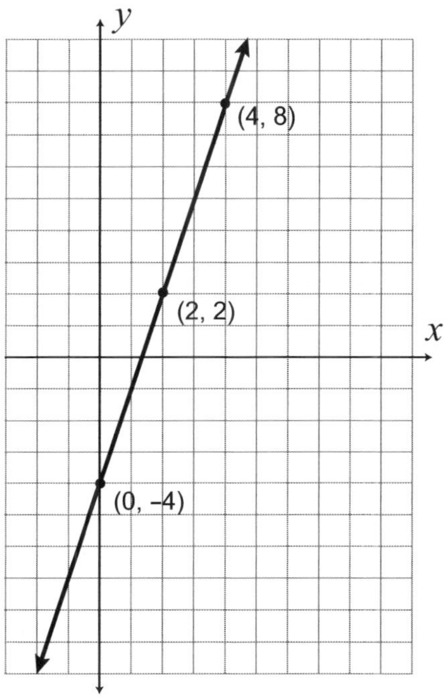

Steps	Discussion

Ex 2

Technique 1

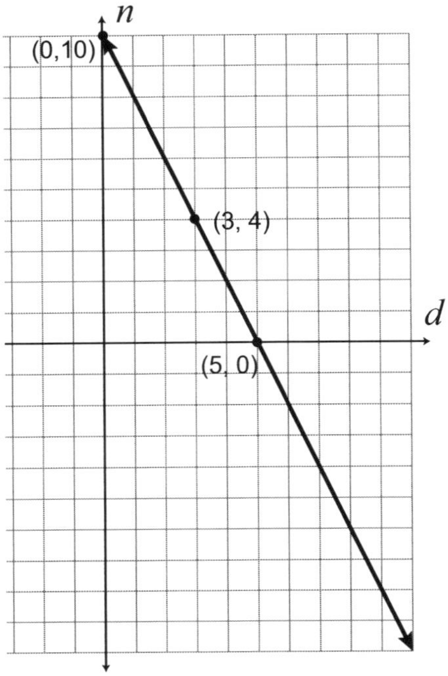

Technique 2

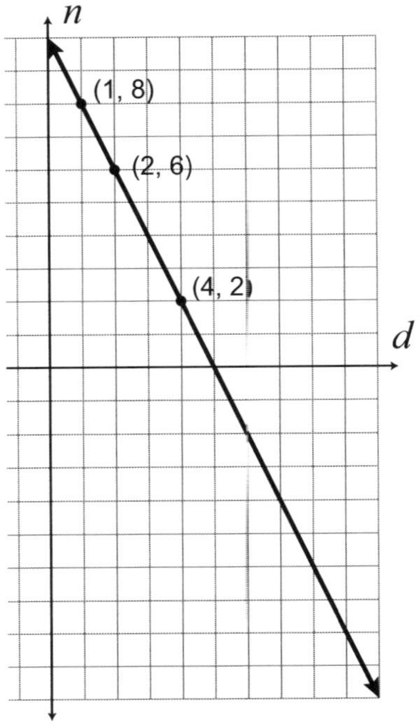

Your Turn

Use your previous graph.

5 Validate the line	Check that a fourth point on the line satisfies the linear equation.

Ex 1

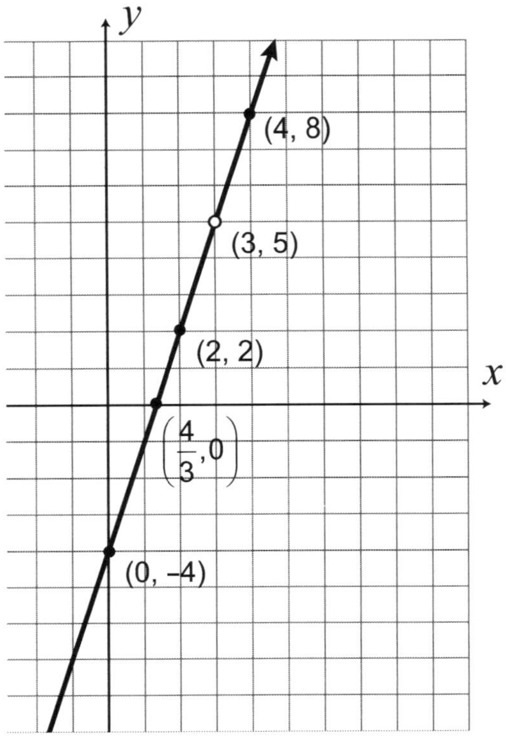

The points from both techniques are represented on the graph. We can see that the point (3, 5) should satisfy the linear equation, as it appears to be on the line.

Substitute those values back into the original equation:

$(3,5)$

$y\overset{?}{=}3x-4$

$5\overset{?}{=}3(3)-4$

$5\overset{?}{=}9-4$

$5=5$ ✓

Steps	Discussion

Ex 2

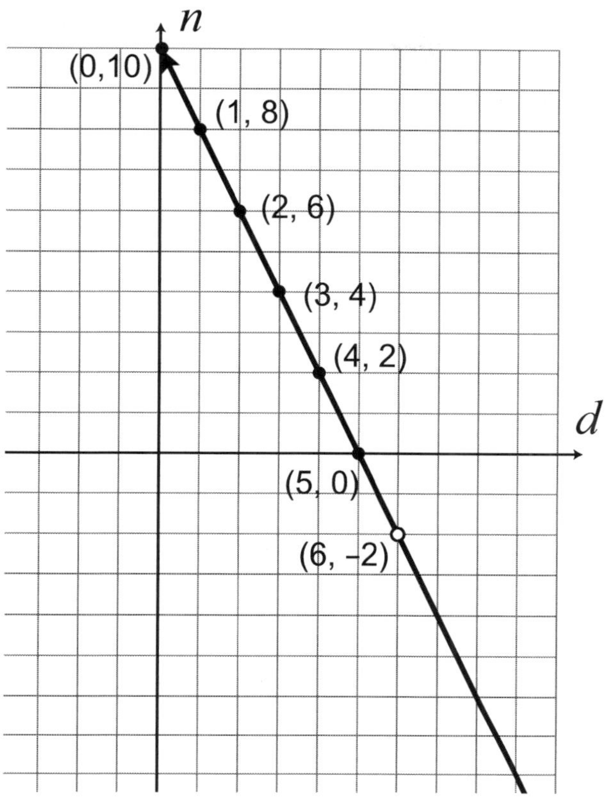

The points from both techniques are represented on the graph. We can see that the point (6, –2) should satisfy the linear equation, as it appears to be on the line.

Substitute those values back into the original equation:

$(6,-2)$

$5n+10d=50$

$5(-2)+10(6) \stackrel{?}{=} 50$

$-10+60 \stackrel{?}{=} 50$

$50=50$ ✓

Your Turn

(You're welcome to use your previous graph)

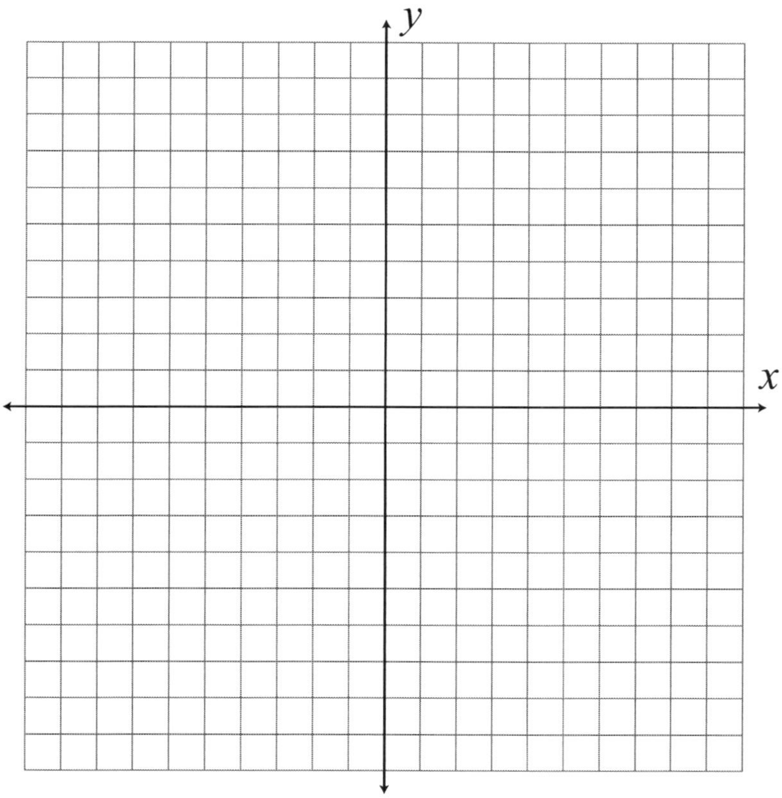

Addressing Common Errors 3.3

The following table shows some of the most common errors that learners tend to make with the kinds of problems covered in this section.

ERROR: Not fully simplifying the equation (leaving *y* with a coefficient of –1)

Incorrect Process

Create a table of values: $3x - y = 4$

$$3x - y = 4$$
$$-y = -3x + 4$$

x	-1	2	4
y	7	-2	-8

Resolution

Correctly solve the equation for y, not −y.

Correct Process

$$3x - y = 4$$
$$-y = -3x + 4$$
$$y = 3x - 4$$

x	-1	2	4
y	-7	2	8

Validation

Is $3x - y = 4$ true when $x = 4$ and $y = 8$?

$$3x - y = 4$$
$$3(4) - 8 \stackrel{?}{=} 4$$
$$12 - 8 \stackrel{?}{=} 4$$
$$4 = 4 \checkmark$$

ERROR: Error in dividing both sides of the equation by 2

Incorrect Process

Create a table of values: $2y = 3x + 4$

$$2y = 3x + 4$$
$$y = 3x + 2$$

x	-1	2	4
y	-1	8	14

Resolution

Divide both sides by 2 correctly.

Correct Process

$$2y = 3x + 4$$
$$y = \frac{3}{2}x + 2$$

x	-1	2	4
y	$\frac{1}{2}$	5	8

Validation

Try $2y = 3x + 4$ for $x = 2$ and $y = 5$

$$2y = 3x + 4$$
$$2(5) \stackrel{?}{=} 3(2) + 4$$
$$10 \stackrel{?}{=} 6 + 4$$
$$10 = 10 \checkmark$$

ERROR: **Plotting the points in a table by reversing the coordinates of the ordered pairs**

Incorrect Process

Plot the points in the table:

x	−2	−1	0	1
y	2	4	6	8

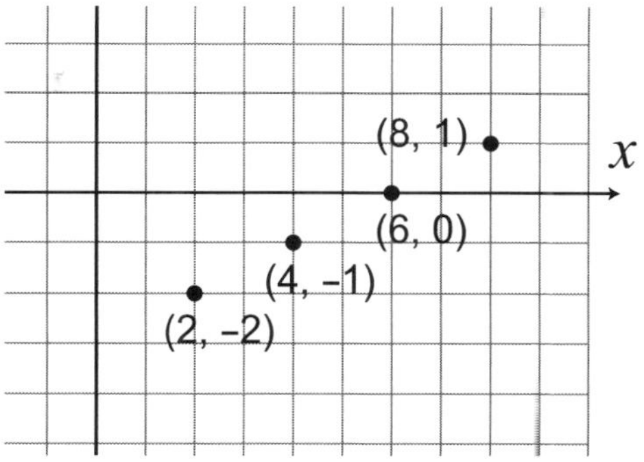

Resolution

Plot the independent variable (x) on the horizontal axis and the dependent (y) on the vertical axis.

Correct Process

Plot the points in the table:

x	−2	−1	0	1
y	2	4	6	8

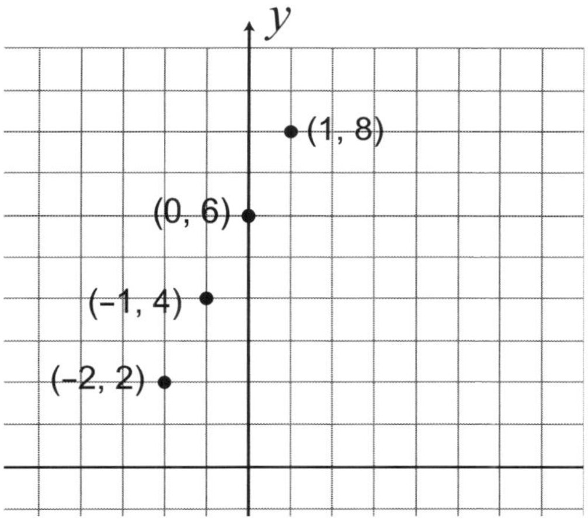

Validation

The point (−1, 4) is to the left of the origin 1 and up 4 in the vertical direction.

Preparation Inventory 3.3

I can...	OR *Here's my question...*
☐ plot points on the coordinate plane.	☐
☐ define *constant rate of change*.	☐
☐ use a table to graph a linear equation in two variables.	☐
☐ use the standard form of a linear equation in two variables to graph a line.	☐

Critical Thinking Questions 3.3

1. How do you remember which coordinate of an ordered pair is graphed with respect to the horizontal axis? The vertical axis?

2. How do you graph a point from a two-column table?

3. What is the simplest strategy for finding an ordered pair solution to a linear equation in two variables?

4. What is the y-coordinate of an ordered pair solution at the x-intercept of a linear equation?

5. What does it mean to "solve for y in terms of x?"

6. Why do you create a table of more than two points when you are graphing a linear equation in two variables?

7. What are the differences between the two given techniques for graphing a linear equation?

8. What would the equation of a vertical or horizontal line be?

Demonstrate Your Understanding 3.3

1. Graph the points (−2, 3), (3, 3), (4, 3) and (6, 3). How would you describe the relationship among these four points?

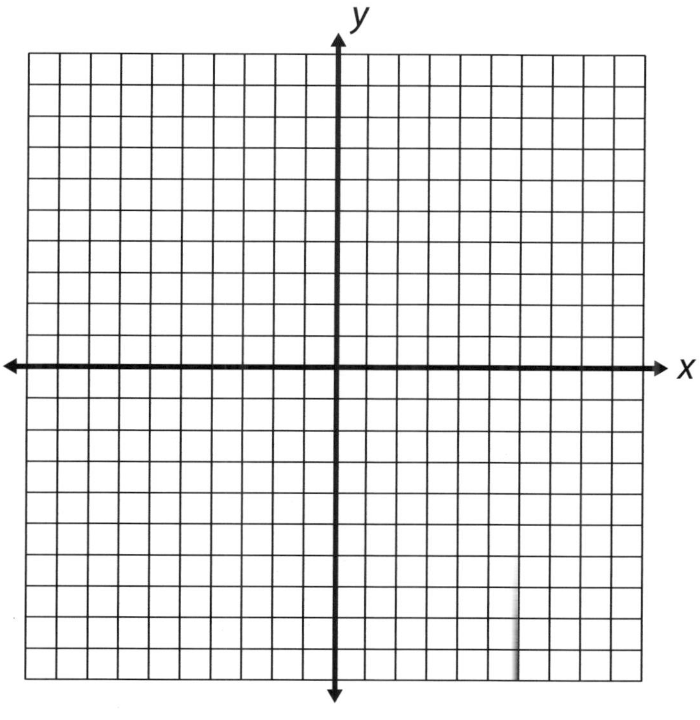

2. Graph the equation $x - 2y = 4$

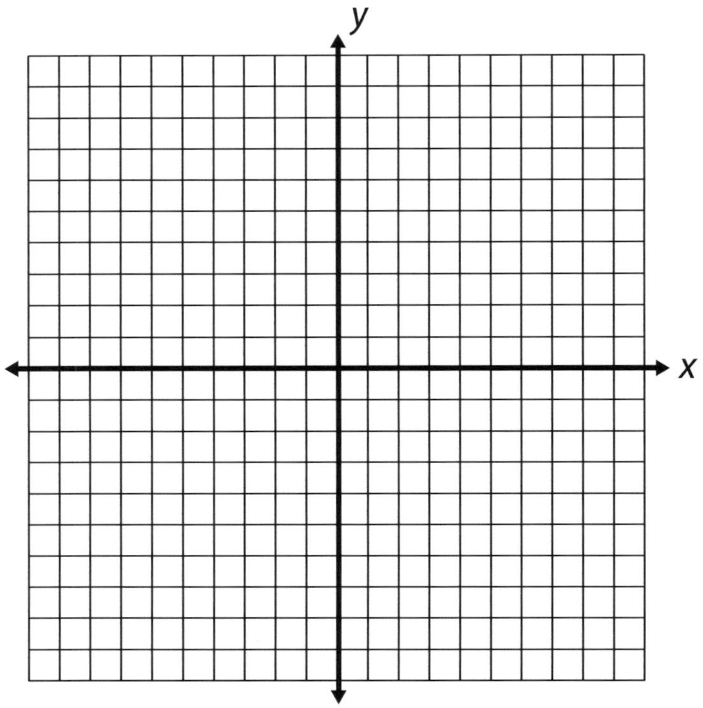

3. Graph the equation $y = -2x + 4$

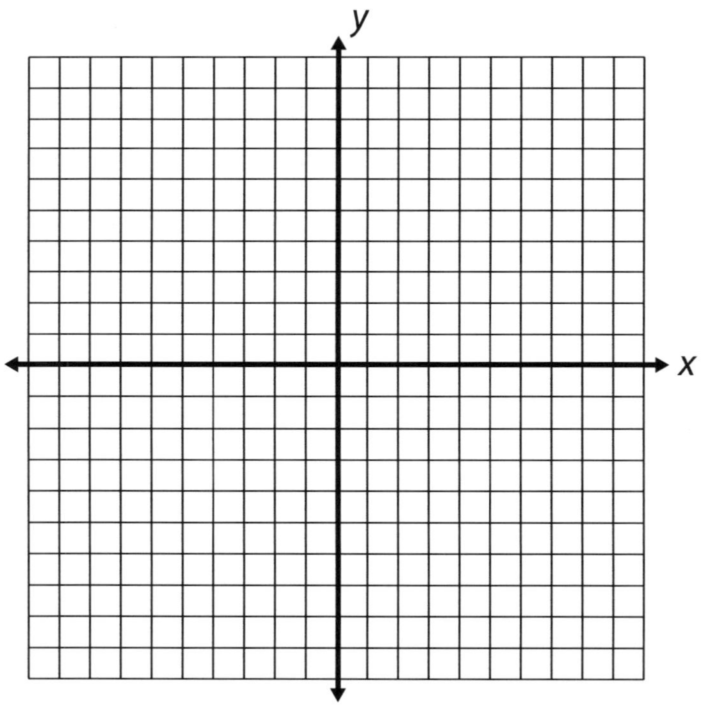

4. Write an equation that is a vertical line through (2, 0).

5. Find the constant rate of change for the linear relationship that is represented by the table below and interpret the meaning of this constant rate of change in terms of the variables *quantity* and *cost*.

quantity	1	3	5	6	7	9	11	12
cost	8	14	20	23	26	32	38	41

Create and Solve the Hardest Problem 3.3

Your Hardest Problem

Problem Solution

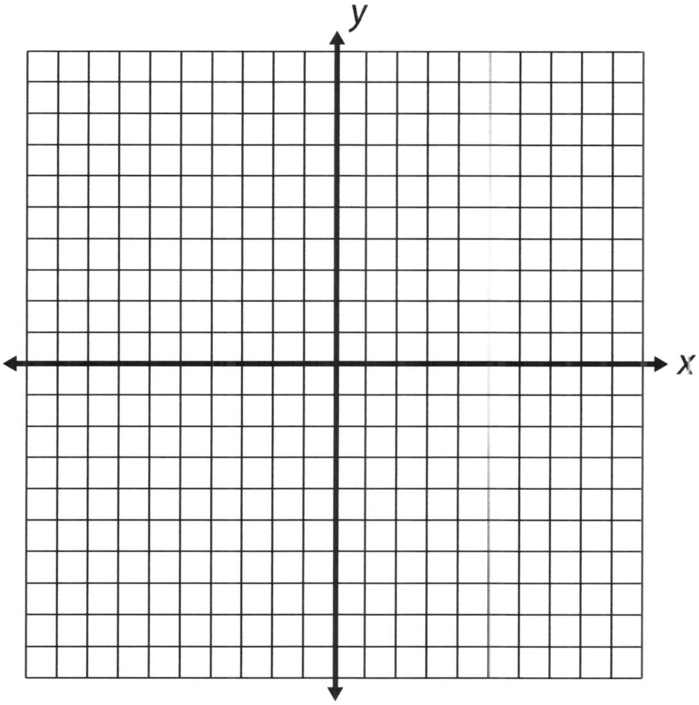

What makes a problem where you graph lines by using tables a hard problem to solve?

Identify, Describe, & Correct the Errors 3.3

Describe the error you found in the worked solution in the appropriate space. Solve the problem correctly under "Correct Process." You are welcome to solve the problem first and then describe the error afterward.

Problem 1

Plot the point (3,–2) on the coordinate plane.

Worked Solution *(Find the error!)*

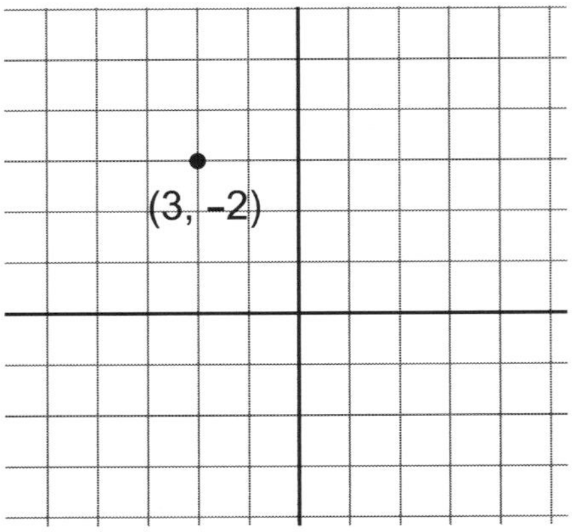

Correct Process

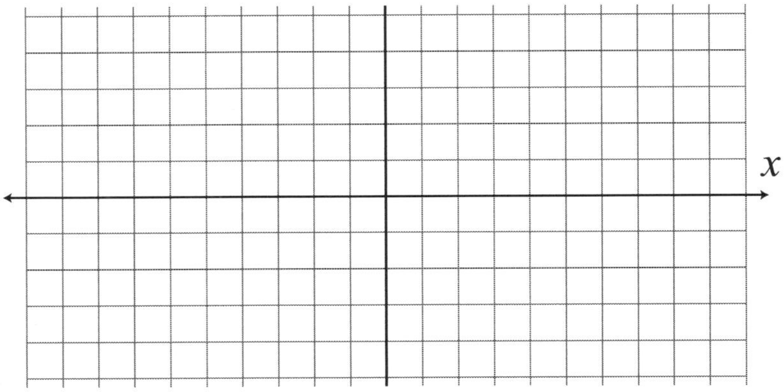

Describe the Error

Problem 2

Solve the following equation for y:

$$3x - y = 4$$

Worked Solution *(Find the error!)*

$$3x - y = 4$$
$$-y = 4 - 3x$$
$$y = -4 - 3x$$

Correct Process

Describe the Error

Problem 3

Graph the line $y = 2$

Worked Solution *(Find the error!)*

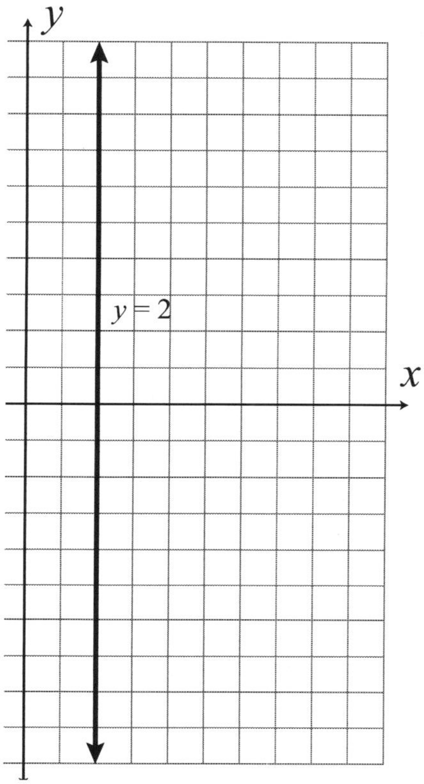

Correct Process

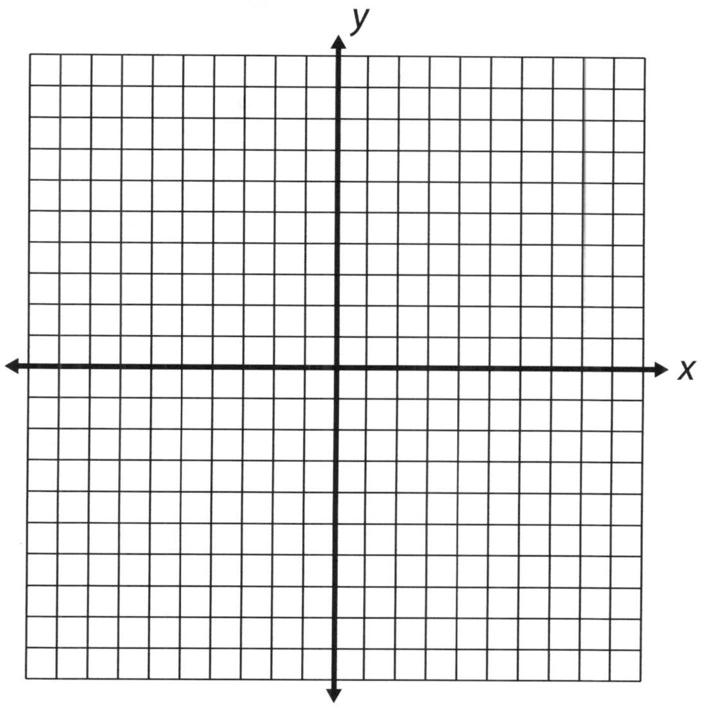

Describe the Error

Problem 4

Is the ordered pair (–2, 4) on the graph of the equation

$y = 3x + 10$?

Worked Solution *(Find the error!)*

$$y = 3x + 10$$
$$-2 = 3 \bullet 4 + 10$$
$$-2 = 12 + 10$$
$$-2 = 22$$

No, the ordered pair (–2, 4) is not on the graph of the equation $y = 3x + 10$.

Correct Process

Describe the Error

Problem 5

Graph the line $x = 3$

Worked Solution *(Find the error!)*

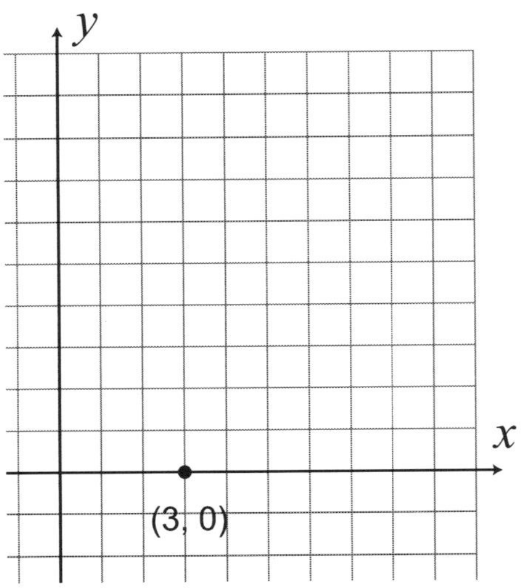

Correct Process

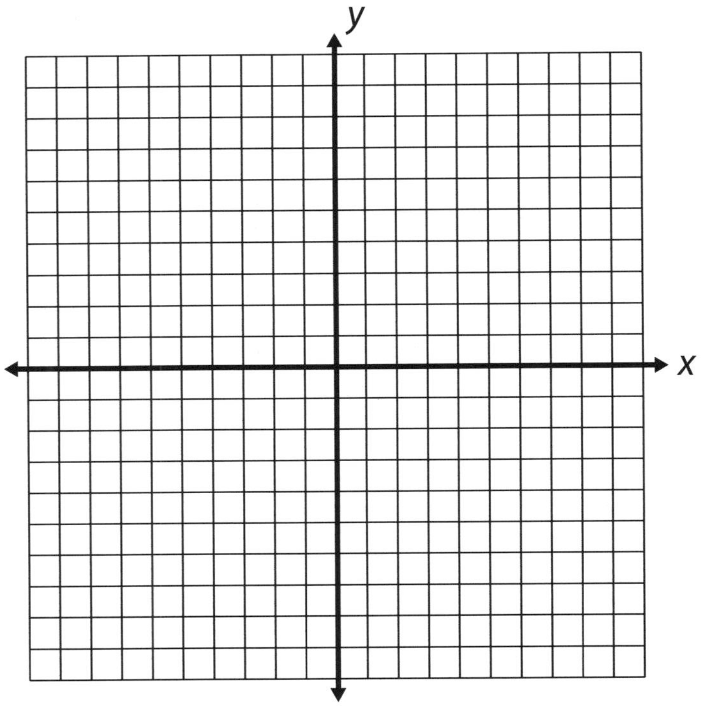

Describe the Error

What Did You Learn About Learning Math? 3.3

1. How do you remember a mathematical process like solving a linear equation for y?

2. How did visualization help you to learn to graph a point?

3. How do you validate that a point on the coordinate plane is a solution to an equation?

Section 3.4 Graphing Equations in Two Variables Using Slope

Why? 3.4

Equations in two variables often occur in situations where there is a relationship between two quantities or variables. For example, suppose you are going to the store to buy 20 pounds of chicken and beef for a party with a fixed amount of money. If you know the price per pound of chicken is \$3 and the price per pound of beef is \$6, then you know that you can buy 2 pounds of chicken for the same price you can buy a pound of beef for. In a situation like this where one quantity varies at a constant rate with respect to the second quantity, you can represent the relationship between the quantities with a line.

What Do You Already Know? 3.4

What are ordered pairs?

How do you generate ordered pairs from a T-table?

How do you decide which variable is on the horizontal axis and which variable is on the vertical axis when graphing ordered pairs?

How do you compute a constant rate of change from the T-table of a linear equation?

How do you compute slope from a graph of a linear equation?

Goals 3.4

When you complete this activity, you should be able to:

1. Understand the relationship between the concepts of a constant rate of change and slope
2. Graph linear equations in two variables using slope

Building Mathematical Language 3.4

New Terms: slope-intercept form $y = mx + b$ form

Key Concepts 3.4

If the rate of change between two variables is constant, the relationship between the two variables can be described by a linear equation. For example, say at the beginning of a snowstorm there is 3 inches of snow on the ground. During that snowstorm, snow is accumulating on the ground at the constant rate of $\frac{1}{2}$ inch per hour. The relationship between the number of hours of snowfall and the amount of snow on the ground can be described by the linear equation

$$\text{depth of snow in inches} = \frac{1}{2}\,\frac{\text{inches}}{\text{hours}} \bullet \text{hours of snowfall} + 3 \text{ inches}$$

or symbolically: $d = \frac{1}{2} \bullet h + 3$ where d is the depth of the snow and h is the number of hours of snowfall.

During any one hour, the depth of the snow increases $\frac{1}{2}$ inch. The graph of this relationship is given below.

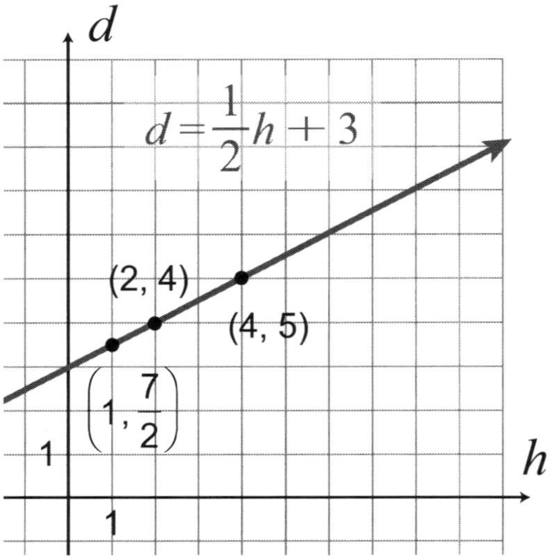

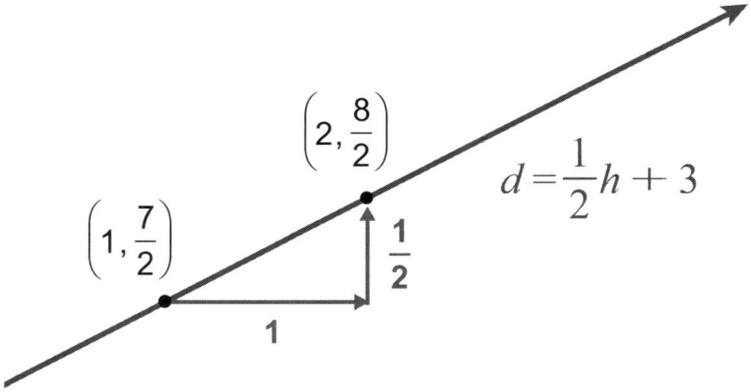

This graph looks like a line drawing of a hillside where, for example, for a 1-foot move horizontally, the hill rises $\frac{1}{2}$ foot vertically. The number in the position of $\frac{1}{2}$ in a linear equation is called the ***slope of an equation*** or, perhaps more appropriately, the slope of the line. The second coordinate of the y-intercept, (0, 3), is 3 — the initial depth of the snow when $h = 0$.

The equation $d = \frac{1}{2} \bullet h + 3$ is in ***slope-intercept form***, with the dependent variable on the left and the slope times the independent variable plus the second coordinate of the y-intercept on the right. This form is also known as the ***y* = mx + b *form*** of a linear equation, with m as the slope and b the number associated with the y-intercept. The number m, the slope of a linear equation, is the constant rate of change of such an equation, as discussed in the previous section. You can use m and b to graph a linear equation as shown in the following methodology.

Methodology 3.4

Graphing a Linear Equation Using the Constant Rate of Change (Slope)

The graph of an equation can reveal significant information about that equation.

Limitation/Caution: Different graphing windows for a graph can result in very different views of a graph, making it possible for significant information to be overlooked.

	Example 1	**Example 2**	**Your Turn**
Graph:	$2x + 3y = 6$	$s = 2p - 3$	$3x - 4y = 8$

Steps	**Discussion**
1 Write the equation in slope-intercept form	**Write the equation in *y* = *mx* + b form with *y* as the dependent variable and *x* as the independent variable.**

Ex 1

$$2x+3y=6$$
$$3y=-2x+6$$
$$y=\frac{-2}{3}x+2$$

Ex 2

$s = 2p - 3$

Already done.

Your Turn

Steps	**Discussion**
2 Determine *m* and *b*	***m* is the coefficient of the independent variable and *b* is the constant term.**

Ex 1

$m=\frac{-2}{3}$ and $b=2$

Ex 2

$m = 2$ and $b = -3$

Your Turn

Steps	Discussion
3 Plot the point on the y-axis	**Plot the point (0, *b*) — the y-intercept.**

Ex 1

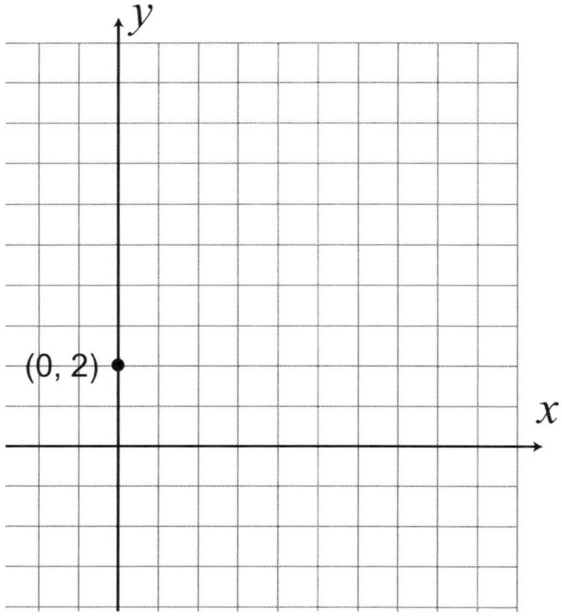

Ex 2

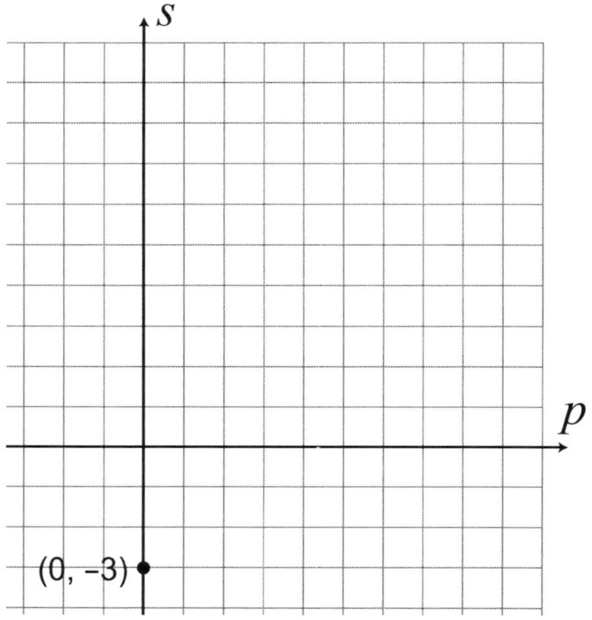

Your Turn

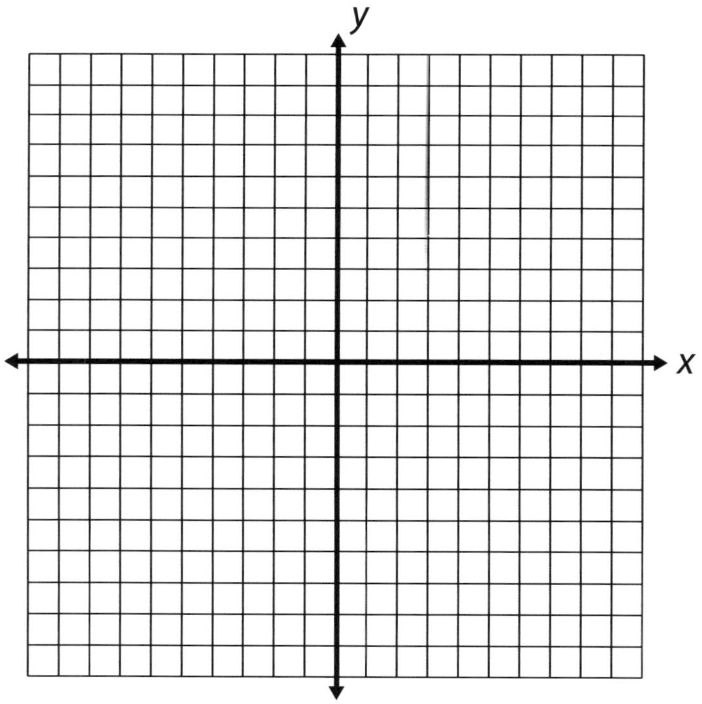

Steps	Discussion
4 Place another point on the graph	Use the slope to determine another point by writing the slope as a fraction. Starting at the first point, move horizontally by the denominator and vertically by the numerator.

Ex 1

$m = \frac{-2}{3}$

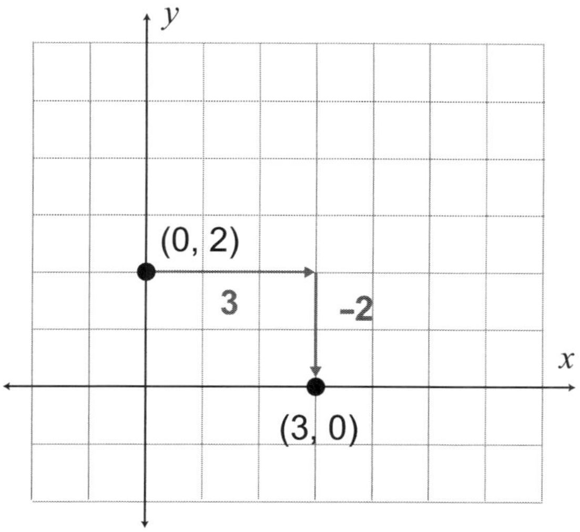

Ex 2

$m = \frac{2}{1}$

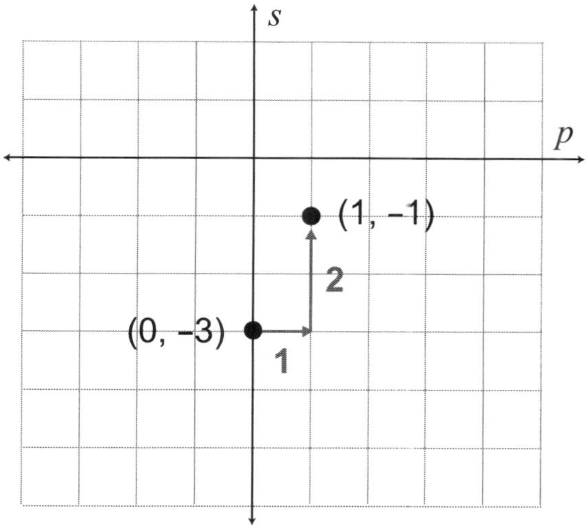

Your Turn

Use your previous graph.

5 Graph the line through these two points	The line passing through these two points will have the correct slope and the correct y-intercept.

Ex 1

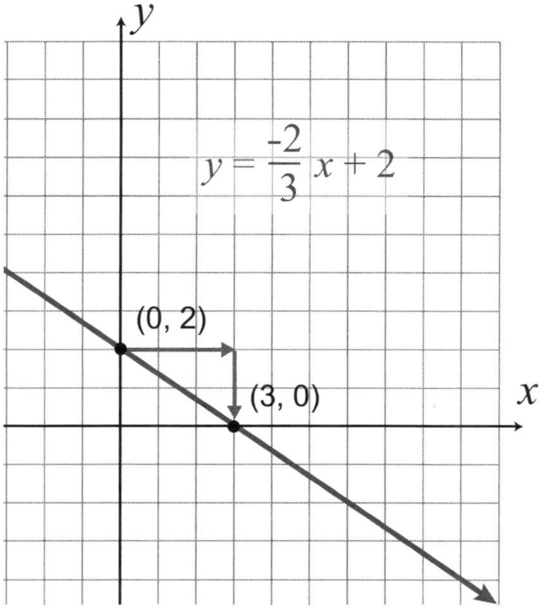

Ex 2

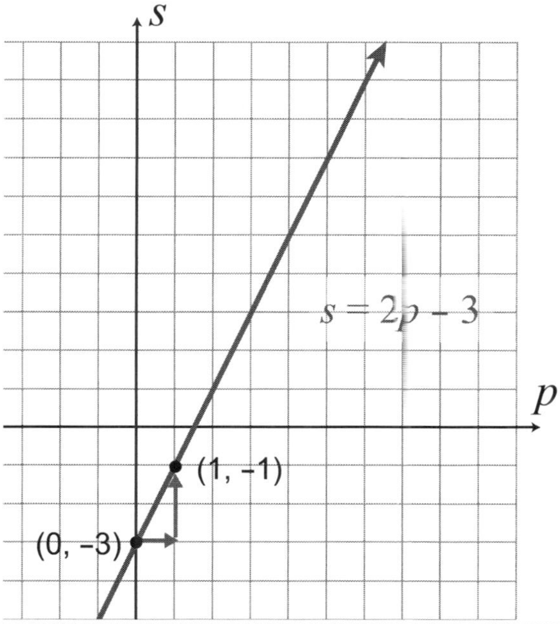

Your Turn

Use your previous graph.

Steps	Discussion
6 Validate	Estimate another point on the line and check that it satisfies the equation.

Ex 1

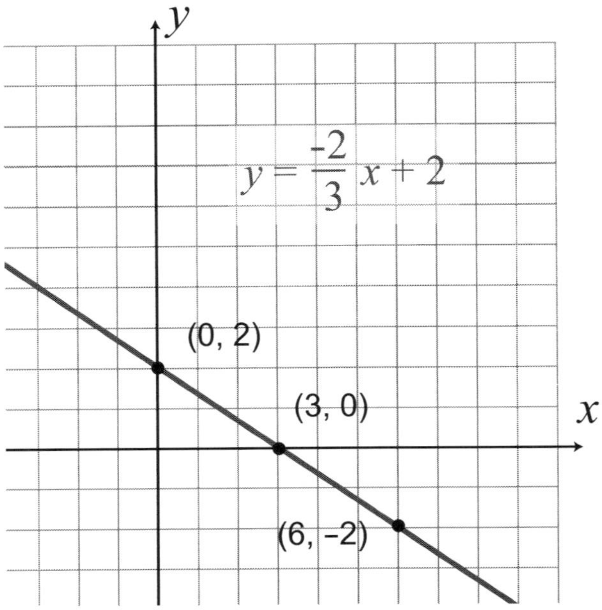

$x = 6$ and $y = -2$

$2x + 3y \stackrel{?}{=} 6$

$2(6) + 3(-2) \stackrel{?}{=} 6$

$12 - 6 \stackrel{?}{=} 6$

$6 = 6$ ✓

Ex 2

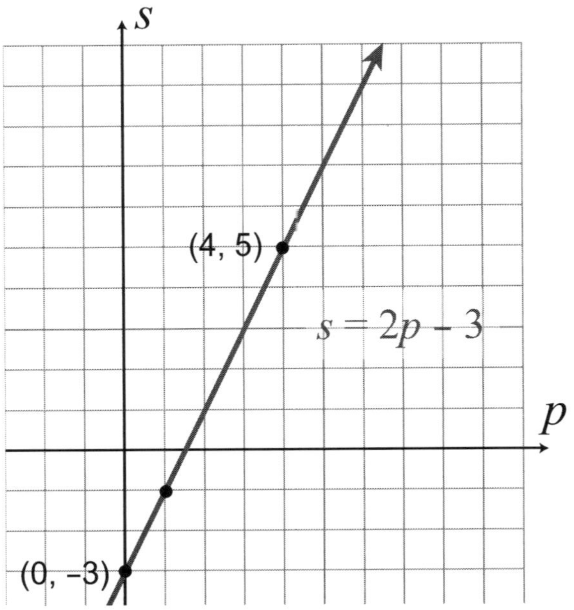

$p = 4$ and $s = 5$

$s \stackrel{?}{=} 2p - 3$

$5 \stackrel{?}{=} 2(4) - 3$

$5 \stackrel{?}{=} 8 - 3$

$5 = 5$ ✓

Your Turn

Use your previous graph.

Addressing Common Errors 3.4

The following table shows some of the most common errors that learners tend to make with the kinds of problems covered in this section.

ERROR: **Incorrectly graphing an equation by reversing *m* and *b* in the slope-intercept form of the equation**

Incorrect Process

Graph $y = 4x - 2$

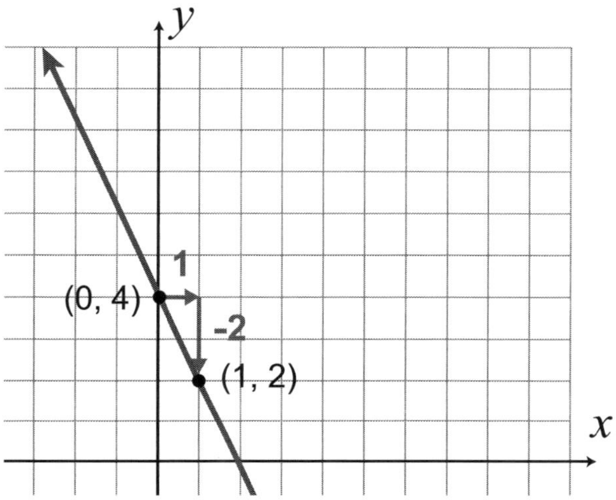

Correct Process

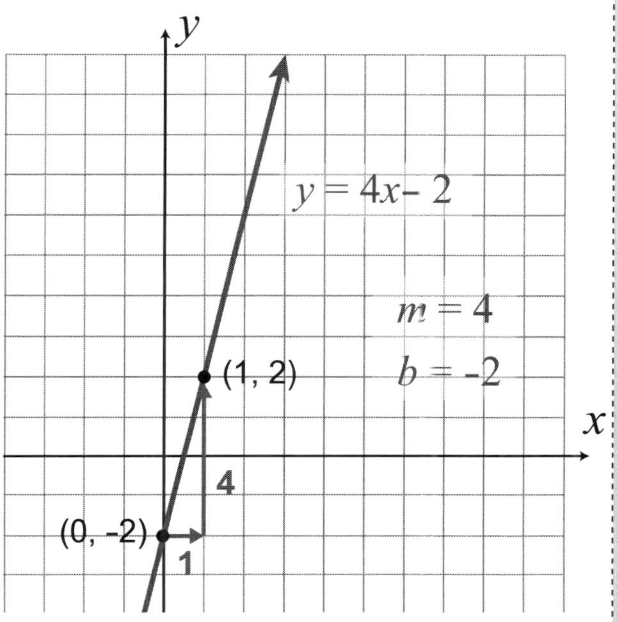

Resolution

Plot the independent variable (x) on the horizontal axis and the dependent (y) on the vertical axis.

Validation

Choose a point on your graph of the equation. Check that it satisfies the equation.

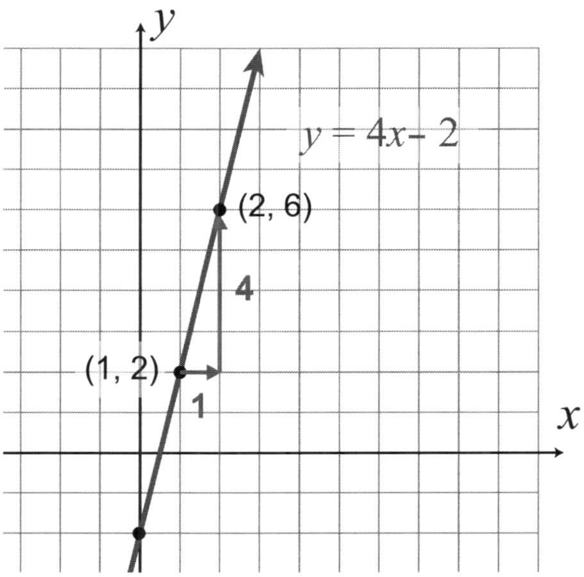

Choose (2, 6).

$y = 4x - 2$

$6 \overset{?}{=} 4(2) - 2$

$6 \overset{?}{=} 8 - 2$

$6 = 6$ ✓

ERROR: Incorrectly determining the slope from the equation

Incorrect Process

Find the slope of the linear equation $2x + y = 5$

$m = \text{slope} = 2$

Resolution

Put the equation into slope-intercept form before reading the coefficient of x as the slope.

Correct Process

$$2x + y = 5$$
$$y = -2x + 5$$
$$m = \text{slope} = -2$$

Validation

Create a table of values and select two points on the line. Use the slope formula to determine the slope of the line from those points.

$2x + y = 5$ Let $x = 2$ and $x = 4$

x	y
2	1
4	–3

$x = 2$, $2(2) + y = 5$, $y = 5 - 4 = 1$, $(2,1)$

$x = 4$, $2(4) + y = 5$, $y = 5 - 8 = -3$, $(4,-3)$

$$m = \frac{-3-1}{4-(2)} = \frac{-4}{2} = -2 \checkmark$$

ERROR: Incorrectly determining the slope-intercept form of your equation

Incorrect Process

Determine the slope of the linear equation $2x + 3y = 6$

$$2x + 3y = 6$$
$$2x = -3y + 6$$
$$x = \frac{-3}{2}y + \frac{6}{2}$$
$$x = \frac{-3}{2}y + 3$$

The slope is $\frac{-3}{2}$.

Resolution

Solve for the dependent variable y in terms of the independent variable x to determine the slope-intercept form. The coefficient of x is the slope.

Correct Process

$$2x + 3y = 6$$
$$3y = -2x + 6$$
$$y = \frac{-2}{3}x + 2$$

The slope is $\frac{-2}{3}$.

Validation

Create a table of values and select two points on the line. Use the slope formula to determine the slope of the line from those two points.

$2x + 3y = 6$ Let $x = -3$ and $x = 6$

x	y
–3	4
6	–2

$x = -3$, $2x + 3y = 6$, $2(-3) + 3y = 6$, $3y = 6 + 6$, $3y = 12$, $y = 4$, $(-3,4)$

$x = 6$, $2x + 3y = 6$, $2(6) + 3y = 6$, $3y = 6 - 12$, $y = \frac{-6}{3} = -2$, $(6,-2)$

$$m = \frac{-2-4}{6-(-3)} = \frac{-6}{9} = \frac{-2}{3} \checkmark$$

Preparation Inventory 3.4

I can...	OR *Here's my question...*
☐ change a linear equation into the slope-intercept form.	☐
☐ determine *m* and *b* from the slope-intercept form.	☐
☐ graph a linear equation using *m* and *b*.	☐
☐ validate that I have graphed a linear equation correctly using *m* and *b*.	☐

Critical Thinking Questions 3.4

1. What is slope-intercept form?

2. How do you determine the values of m and b for a linear equation in two variables?

3. How would you determine the linear equation of the line passing through (0, –3) with slope $\frac{1}{4}$?

4. How would you determine the y-intercept of the linear equation with slope 3 that passes through the point (–4, 2)?

5. How would you find the linear equation that is satisfied by the two points (–3, 2) and (1, –6)?

6. What does the graph of a linear equation look like if its slope is 0?

7. How would you modify the Methodology for Graphing a Linear Equation Using the Constant Rate of Change if, rather than first plotting the y-intercept, you first plotted some other point that satisfied the equation?

8. How is the slope of a linear equation related to its constant rate of change?

9. How would you write $y = 3$ in slope-intercept form?

10. Why can't you write the linear equation $x = 2$ in slope-intercept form?

Demonstrate Your Understanding 3.4

1. Find the slope-intercept form of the linear equation $3y - 4x = 2$.

2. Find the linear equation of the line passing through the points (3, 5) and (–3, –2).

3. Find the linear equation with slope $-\frac{3}{2}$ passing through the point (5, 2).

4. Find a point on the line that is the graph of $2x - 3y = 11$.

5. Find the slope of the equation $ax + by = c$. How would you determine the slope of any linear equation in two variables in this standard form?

6. Graph the linear equations $y = 3$ and $x = 2$.

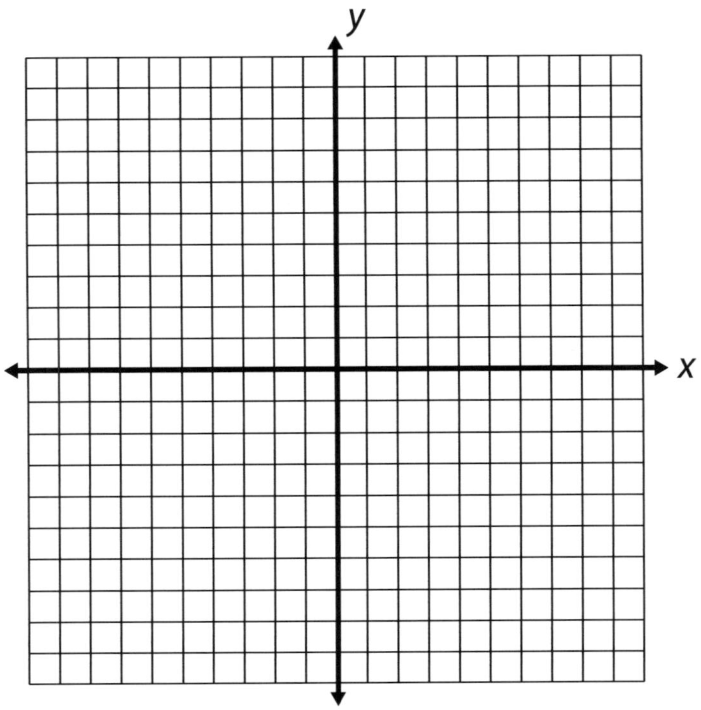

Create and Solve the Hardest Problem 3.4

Your Hardest Problem

Problem Solution

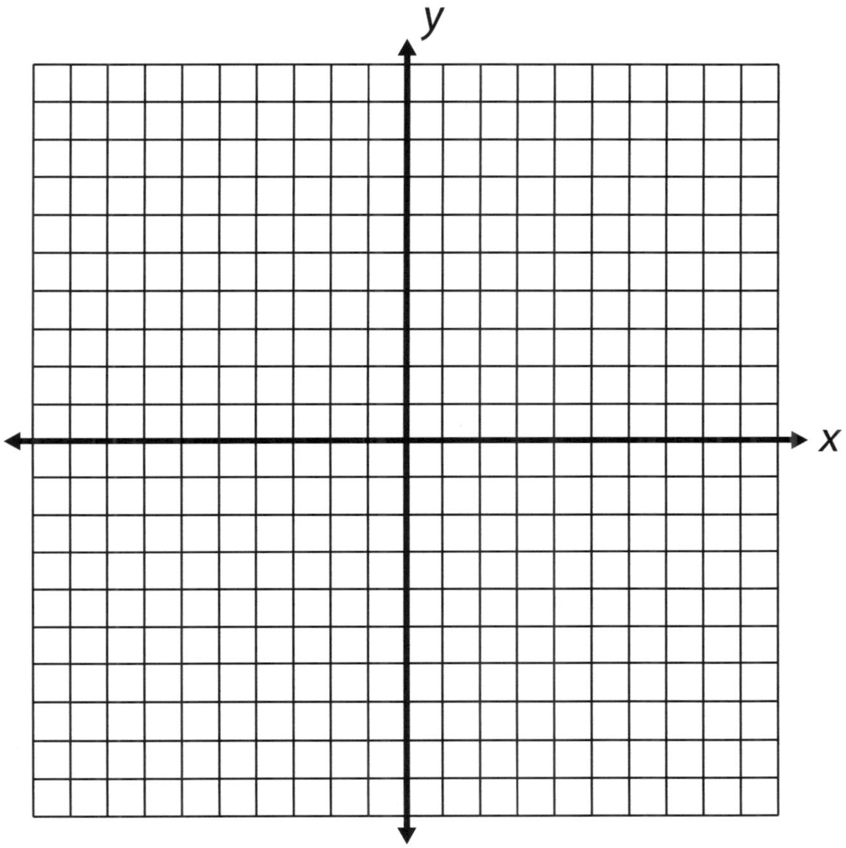

What makes a problem where you must graph an equation in two variables using slope a hard problem to solve?

Identify, Describe, & Correct the Errors 3.4

Describe the error you found in the worked solution in the appropriate space. Solve the problem correctly under "Correct Process." You are welcome to solve the problem first and then describe the error afterward.

Problem 1

Graph the equation $y = 3x - 2$

Worked Solution *(Find the error!)*

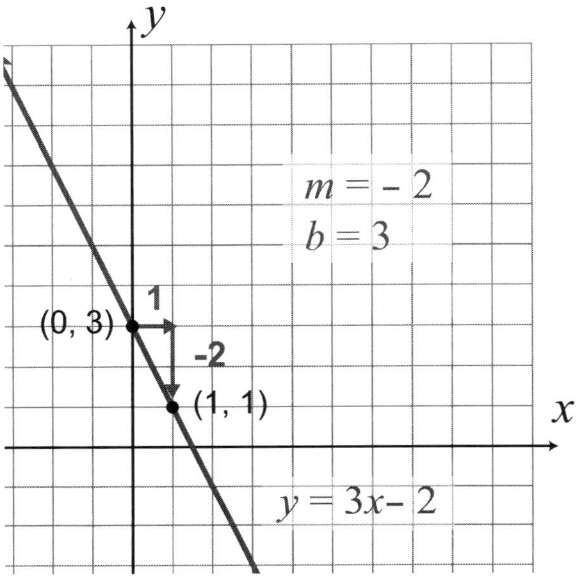

Correct Process

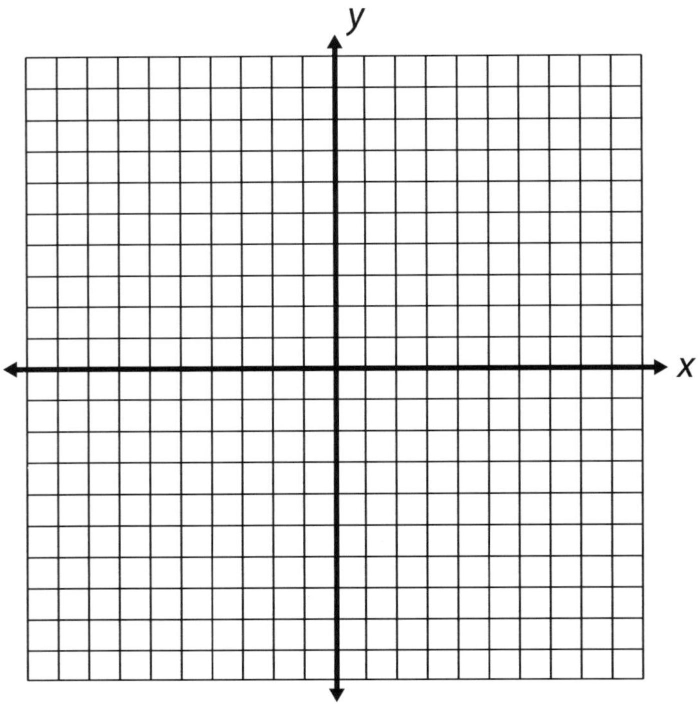

Describe the Error

Problem 2

Graph the equation $2y - 4x = 7$

Worked Solution *(Find the error!)*

$$2y - 4x = 7$$
$$-4x = 7 - 2y$$
$$x = \frac{-2y}{-4} + \frac{7}{-4}$$
$$x = \frac{1}{2}y + \frac{-7}{4}$$

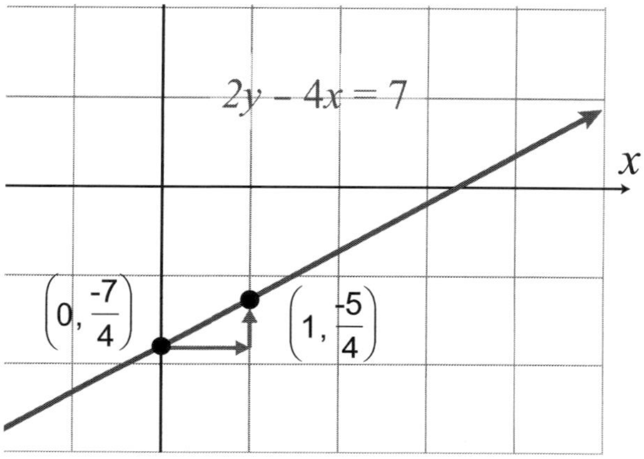

Correct Process

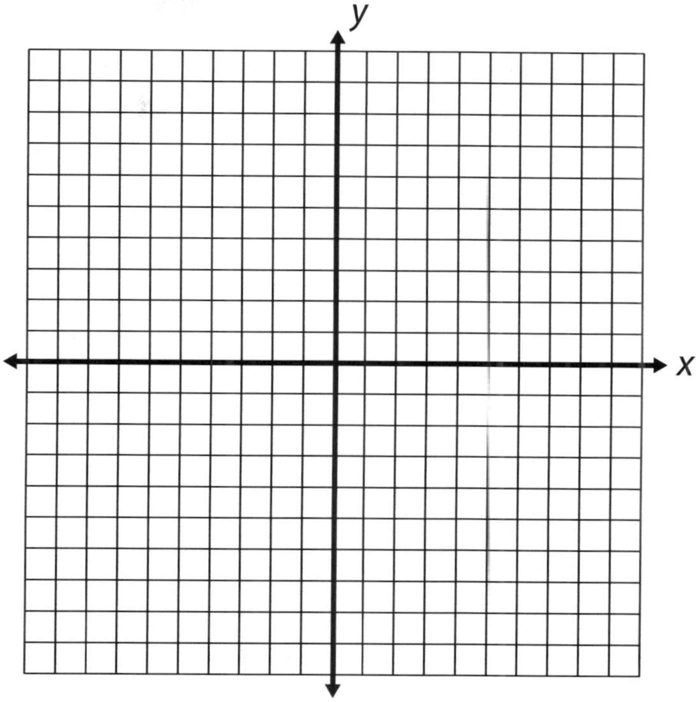

Describe the Error

Problem 3	Correct Process	Describe the Error
Find the constant rate of change (slope) of the linear equation whose graph passes through the points (2, –3) and (–4, 1)		
Worked Solution *(Find the error!)*		
$m = \frac{1-(-3)}{2-(-4)} = \frac{4}{6} = \frac{2}{3}$		

What Did You Learn About Learning Math? 3.4

1. How does visualizing the relationship between two variables help in understanding that relationship?

2. How did the comparison between the constant rate of change and slope of a hillside help in graphing linear equations?

3. What prior knowledge did you use in understanding how to graph a linear equation using constant rate of change?

Section 3.5 Graphing Other Equations in Two Variables

Why? 3.5

There are numerous real-world examples of two variables being related in a way that can be described by an equation or a graph. For example, over the years, the number of animals used in laboratory experiments, the average time of major league baseball games, the amount of radioactivity remaining after a nuclear explosion or accident, or the amount of money earned in a savings account can all be described with equations and pictured with graphs.

What Do You Already Know? 3.5

How do you create a T-table (table of values) from an equation?

Describe vertical and horizontal lines and their equations.

How do you distinguish an independent variable from a dependent one?

Goals 3.5

When you complete this activity, you should be able to:

1. Accurately graph an equation in two variables
2. Select an appropriate graphing window

Building Mathematical Language 3.5

New Terms: graphing window graphical representation

Key Concepts 3.5

When ***graphing an equation in two variables***, one variable depends on the other. The independent variable is associated with the horizontal axis and the dependent variable is associated with the vertical axis. Each value of the independent variable generates at least one value for the dependent variable.

Limitation/Caution: It is not always clear which variable is independent and which is dependent. Although straight lines are easy to graph, other types of equations may require more thought.

Example 1 Graph $y = x^2$

In this equation, ***x*** is the independent variable and ***y*** is the dependent variable.

Example 2 Graph $a = \sqrt{b}$

In this equation, ***b*** is the independent variable and ***a*** is the dependent variable.

In T-tables, the independent variable is usually in the left column, the dependent variable in the right column.

$y = x^2$

x (independent)	***y*** (dependent)
0	0
1	1
2	4
3	9
–1	1
–2	4
–3	9

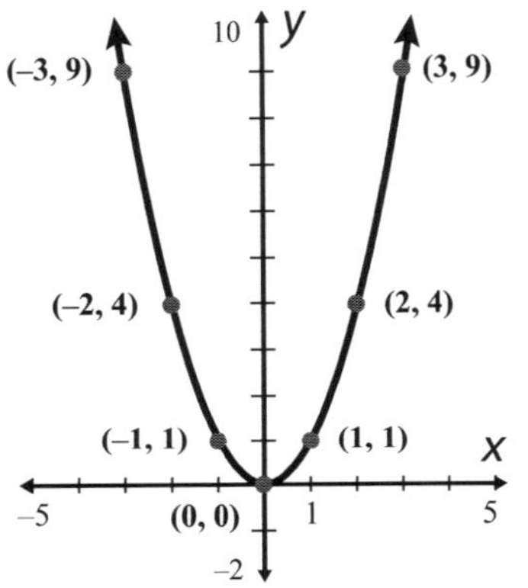

$a = \sqrt{b}$

b (independent)	***a*** (dependent)
0	0
1	1
4	2
9	3

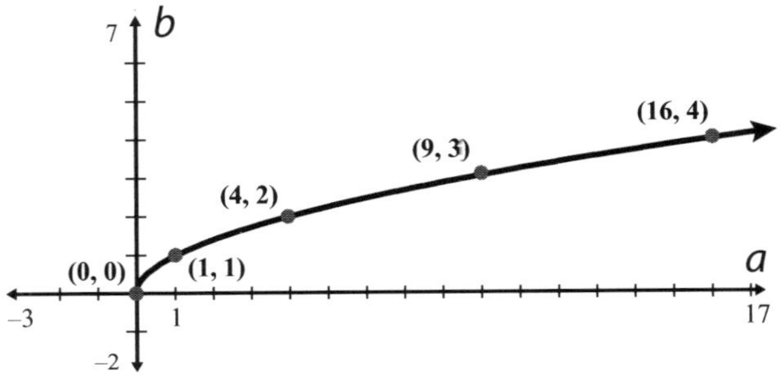

Note that *a* can only be non-negative.

A graphical representation of any equation in two variables can be created with the methodology given in this section. This is the method used by most graphing calculators and computer programs. The following types of equations can be graphed in other ways based on their characteristics.

absolute value	exponential	logarithmic	rational
conic section	linear	quadratic	transformations

The ***graphing window*** is a rectangular part of the coordinate plane where you display a part of the graph of a function and is defined by an *x* (horizontal) range and a *y* (vertical) range of values. The graphing window of the graph for Example 1 is: $-5 \le x \le 5$ and $-2 \le y \le 10$. The figure at right shows a graphing window with *x*-range interval [–9, 9] and *y*-range interval [–6, 6].

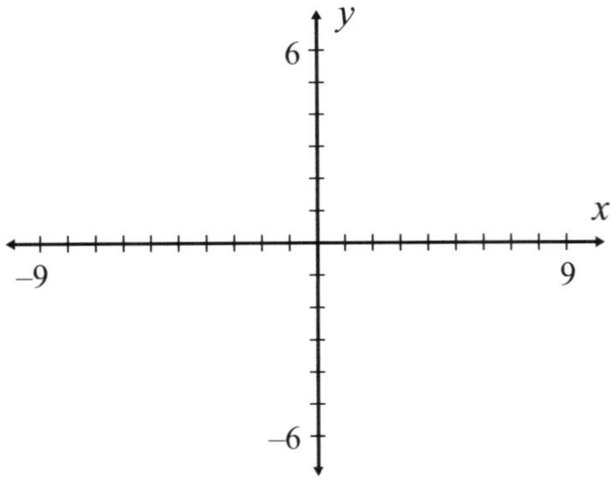

Limitation/Caution: Because not all of a function is displayed in the graphing window, not all of the *behavior* of the function may be displayed in the graphing window.

Convention

To ***approximate the graph of an equation*** means to draw the curve or line that best represents the space in between the points.

Limitation/Caution: Sometimes, the number of points you have used may not be sufficient to accurately approximate the graph of the equation.

Methodology 3.5

Graphing an Equation of Two Variables

Graphs of equations can give significant information about an equation.

Limitation/Caution: Different graphing windows for a graph can result in very different graphs that can cause significant information to be overlooked.

	Example 1	Example 2	Your Turn
Graph:	$x = 2y + 6$	$x^2 + y^2 = 4$	$4x^2 + 2y = 5$

Steps	Discussion
1 Define axes	**Define the horizontal axis variable and the vertical axis variable using the standard conventions of independent and dependent variables.**

Ex 1

x is the independent variable and y is the dependent variable

Ex 2

x is the independent variable and y is the dependent variable

Your Turn

Steps	Discussion
2 Transform equation	**Transform the equation so the dependent variable is isolated (for example, y = expression in x).**

Ex 1

$$x = 2y + 6$$
$$-2y = -x + 6$$
$$y = \frac{1}{2}x - 3$$

Ex 2

$$x^2 + y^2 = 4$$
$$y^2 = 4 - x^2$$
$$y = \pm\sqrt{4 - x^2}$$

Your Turn

Steps	Discussion
3 Define domain	**Determine all possible values that the independent variable can take on.**

Ex 1 Domain is all real numbers

Ex 2 Domain is [–2, 2]

Your Turn

Steps	Discussion
4 Build a table	**Build a table of ordered pairs.**

Ex 1

x	y
0	–3
2	–2
–2	–4

Ex 2

x	y
0	± 2
1	$\pm \sqrt{3}$
2	0

Your Turn

Steps	Discussion
5 Plot the pairs	**Plot the ordered pairs carefully.**

Ex 1

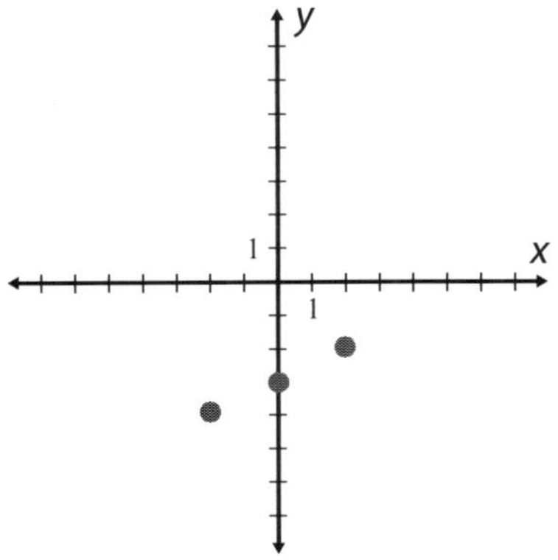

Ex 2

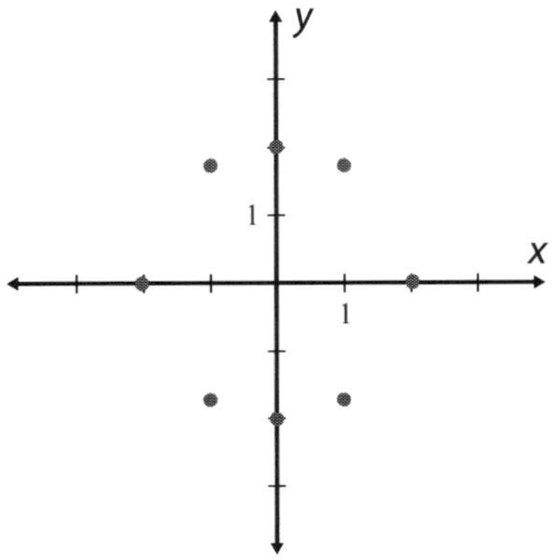

Steps	Discussion

Your Turn

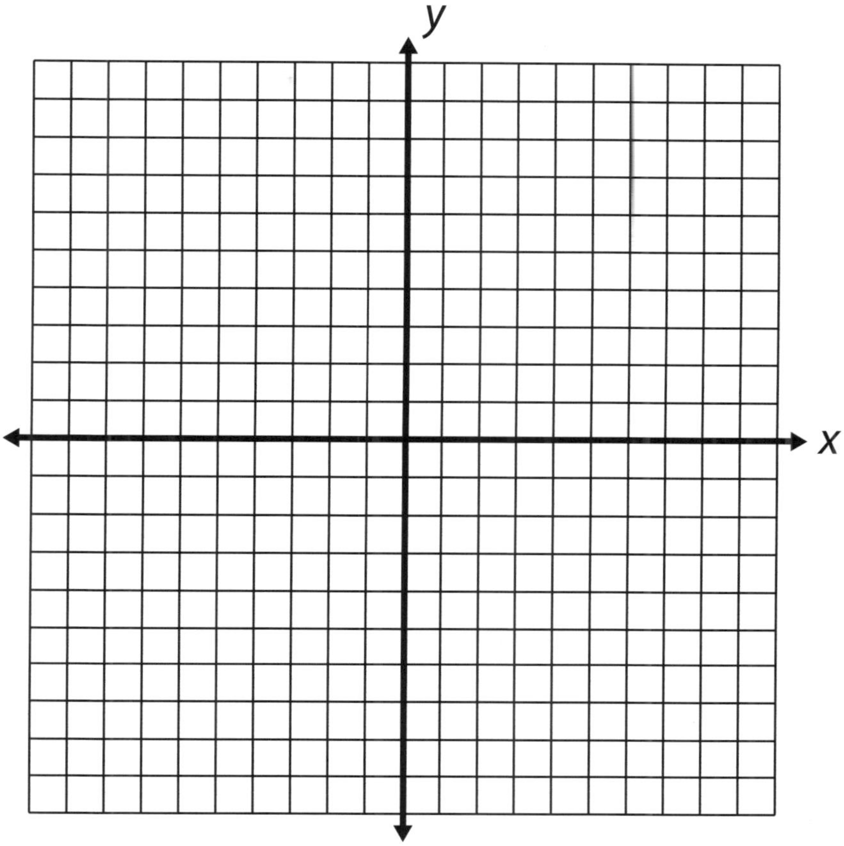

6 Draft the line or curve	Draw a curve or line that passes through the points in a way that might approximate the other infinitely many points on the graph.

Ex 1

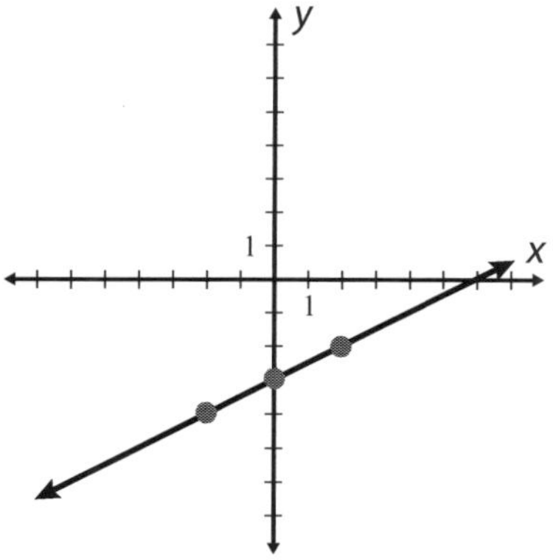

Ex 2

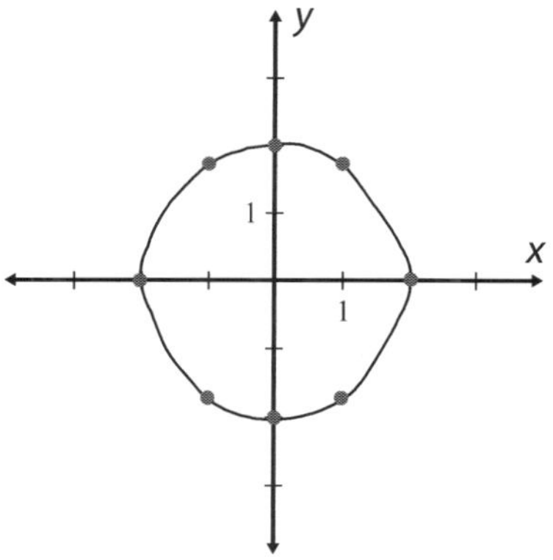

Your Turn

Use your previous graph.

Steps	Discussion
7 Evaluate accuracy	**Did you list enough ordered pairs in the table to effectively draft the line or curve?**

Ex 1

There are enough points for us to determine that this is a straight line.

Label points on the graph.

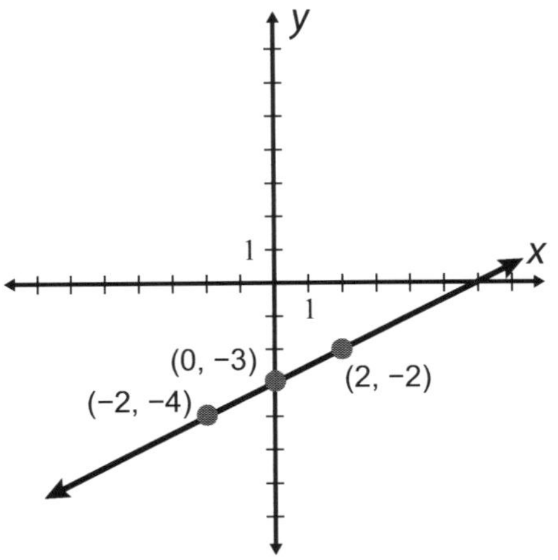

Ex 2

There are not enough points. Add more points and create a new graph.

x	y
$\frac{3}{2}$	$\pm\sqrt{\frac{7}{4}}$
$-\frac{3}{2}$	$\pm\sqrt{\frac{7}{4}}$

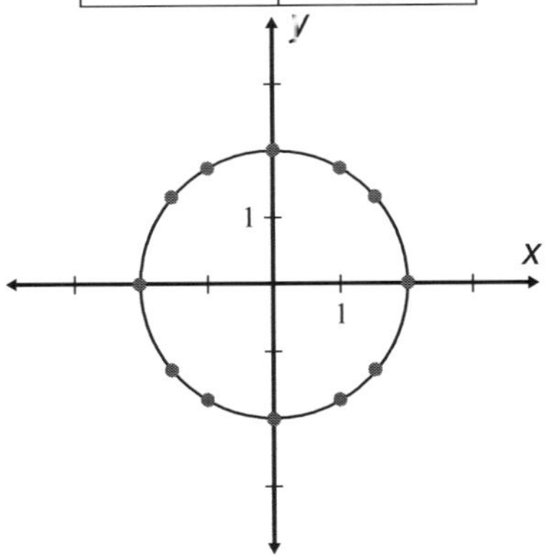

Your Turn

Use your previous graph.

8 Validate the graph	**Technique 1:** **Pick an ordered pair on the line or curve, not in the table, and determine if it satisfies the equation.** **Technique 2:** **Use a graphing calculator to determine if the graphs are very close.**

Ex 1

The point (6, 0) appears to be on the graph. Substitute to see if this point satisfies the equation:

$$x = 2y + 6$$

$$6 \stackrel{?}{=} 2(0) + 6$$

$$6 = 6 \checkmark$$

Ex 2

Pick $\left(\frac{1}{2}, 1.8\right)$

$$\left(\frac{1}{2}\right)^2 + (1.8)^2 \stackrel{?}{=} 4$$

$$\frac{3}{4} + 3.24 \stackrel{?}{\approx} 4$$

$$.75 + 3.24 \stackrel{?}{\approx} 4$$

$$3.99 \approx 4 \checkmark$$

(very close)

Your Turn

Addressing Common Errors 3.5

The following table shows some of the most common errors that learners tend to make with the kinds of problems covered in this section.

ERROR: Connecting points across a location that is not in the domain

Incorrect Process

x	y
0	undef.
1	1
2	$\frac{1}{4}$
3	$\frac{1}{9}$
$\frac{1}{2}$	4
$\frac{1}{3}$	9
–1	1
–2	$\frac{1}{4}$
–3	$\frac{1}{9}$
$-\frac{1}{2}$	4
$-\frac{1}{3}$	9

Graph $\frac{1}{x^2}$

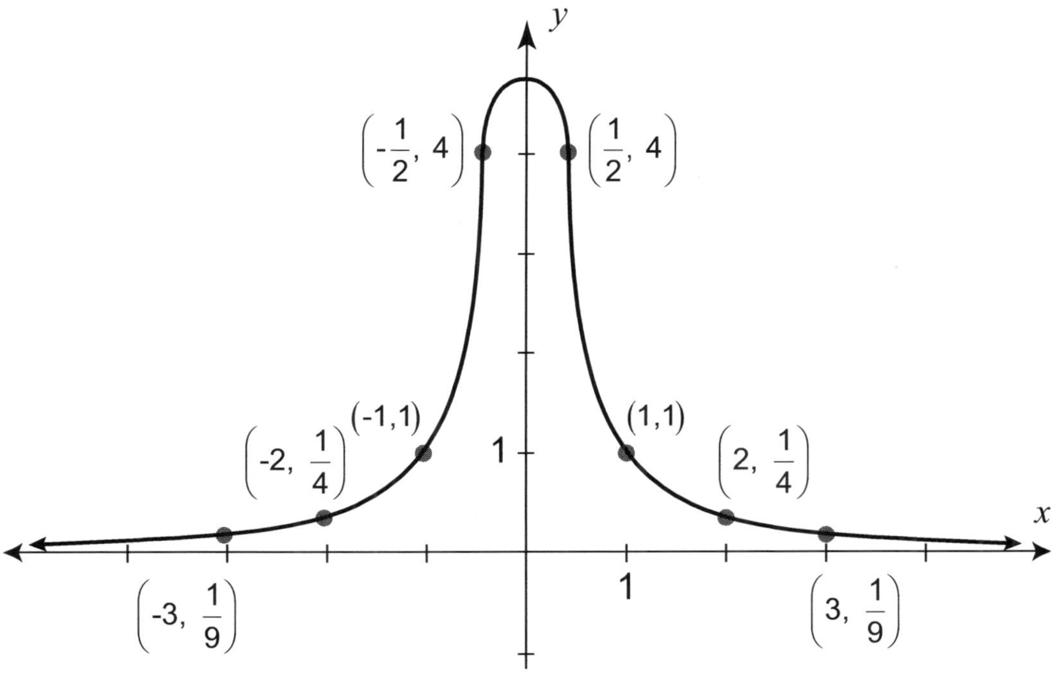

Resolution

Note that x = 0 is not in the domain, so the graph cannot cross the y-axis. This graph will have two separate "pieces."

ERROR: **Connecting points across a location that is not in the domain (continued)**

Correct Process

Graph $\frac{1}{x^2}$

x	*y*
0	undef
1	1
2	$\frac{1}{4}$
3	$\frac{1}{9}$
$\frac{1}{2}$	4
$\frac{1}{3}$	9
–1	1
–2	$\frac{1}{4}$
–3	$\frac{1}{9}$
$-\frac{1}{2}$	4
$-\frac{1}{3}$	9

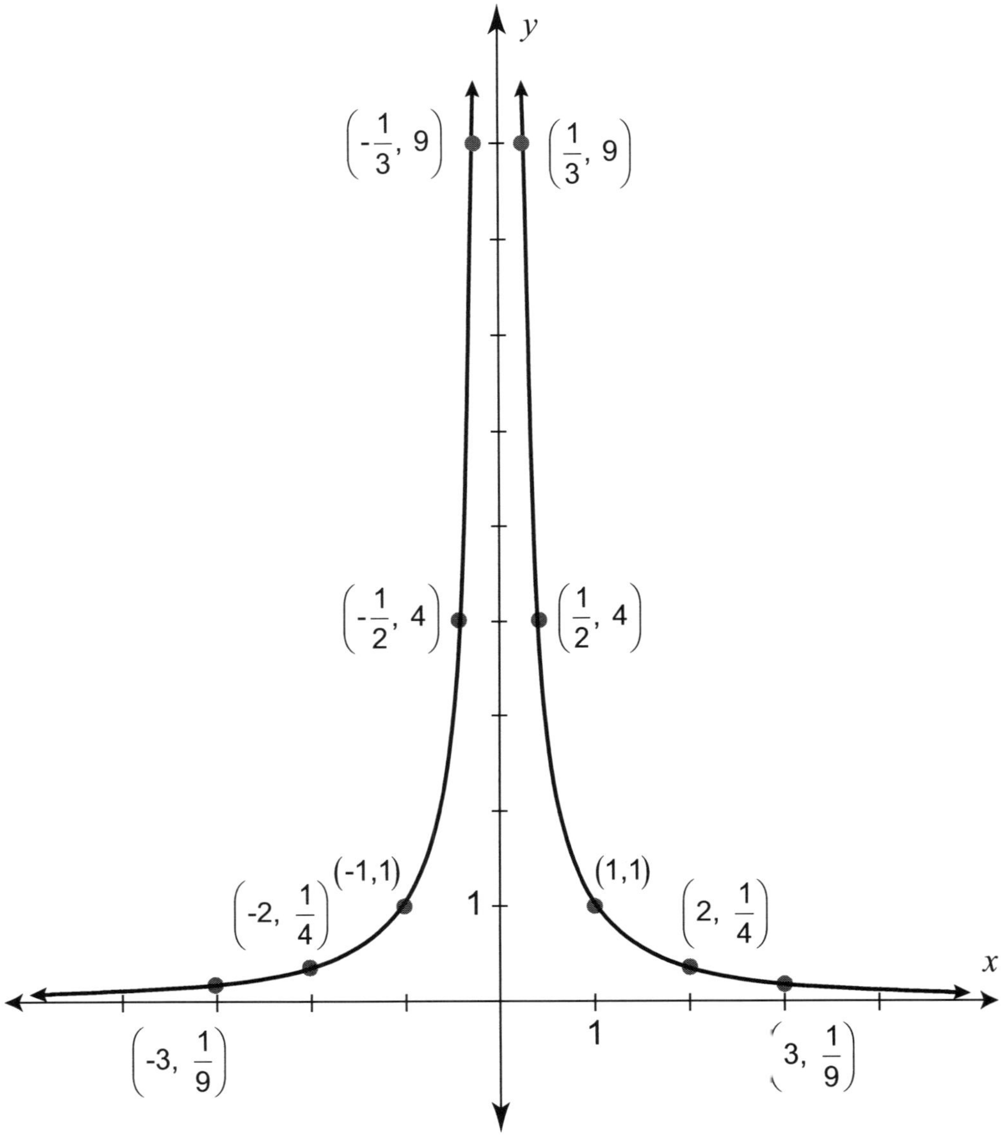

Validation

Use a graphing calculator with the standard window to verify that your graph is accurate.

ERROR: Not graphing enough points

Incorrect Process

Graph $y = \frac{1}{x}$

x	y
1	1
2	$\frac{1}{2}$
3	$\frac{1}{3}$

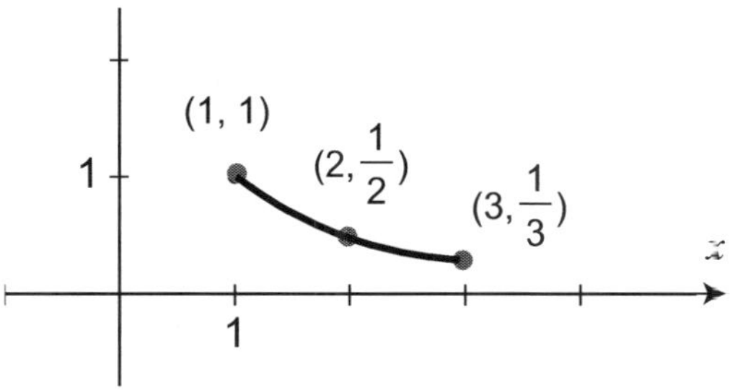

Resolution

Think about the domain (all real numbers except zero) and select various kinds of values for x (fractions, positives, negatives, large or small numbers). It also helps to identify the basic kind of graph you are expecting to see, before you create your graph.

Correct Process

Graph $y = \frac{1}{x}$

x	y
0	undef.
1	1
2	$\frac{1}{2}$
4	$\frac{1}{4}$
$\frac{1}{2}$	2
$\frac{1}{4}$	4

x	y
–1	–1
–2	$-\frac{1}{2}$
–4	$-\frac{1}{4}$
$-\frac{1}{2}$	–2
$-\frac{1}{4}$	–4

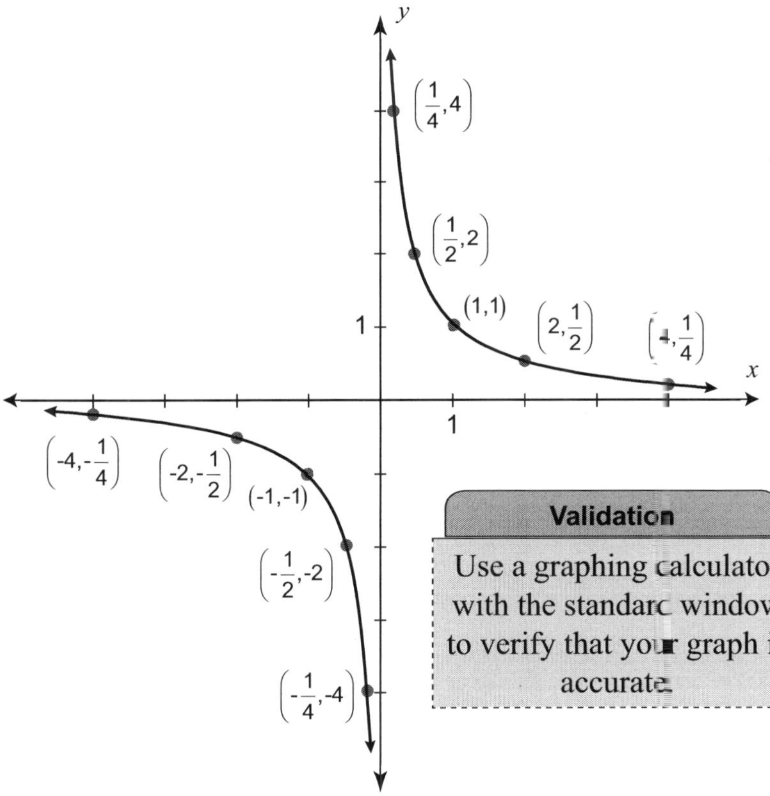

Validation

Use a graphing calculator with the standard window to verify that your graph is accurate.

Preparation Inventory 3.5

I can...	OR *Here's my question...*
☐ create a T-table from an equation.	☐
☐ determine the domain of an independent variable.	☐
☐ accurately plot points on a graph.	☐

Critical Thinking Questions 3.5

1. What are the traditional labeling conventions for the horizontal and vertical axes?

2. How do you prepare an equation for graphing?

3. How do you know how many values your T-table should contain before you begin your graph?

4. After you build a table of values, how do you determine the graphing window for your graph?

5. What are the two ways to validate that your graph is accurate?

6. What is the best strategy to visualize the curve or line through points you have graphed?

7. What is an acceptable number of points to use to produce an acceptable graph?

8. What is the relationship between a table of values, an equation, and a graph?

Demonstrate Your Understanding 3.5

Graph each of the following equations by building a table of values, determining an appropriate graphing window, and accurately plotting points.

1. $2x^2 - 3x - y = 4$

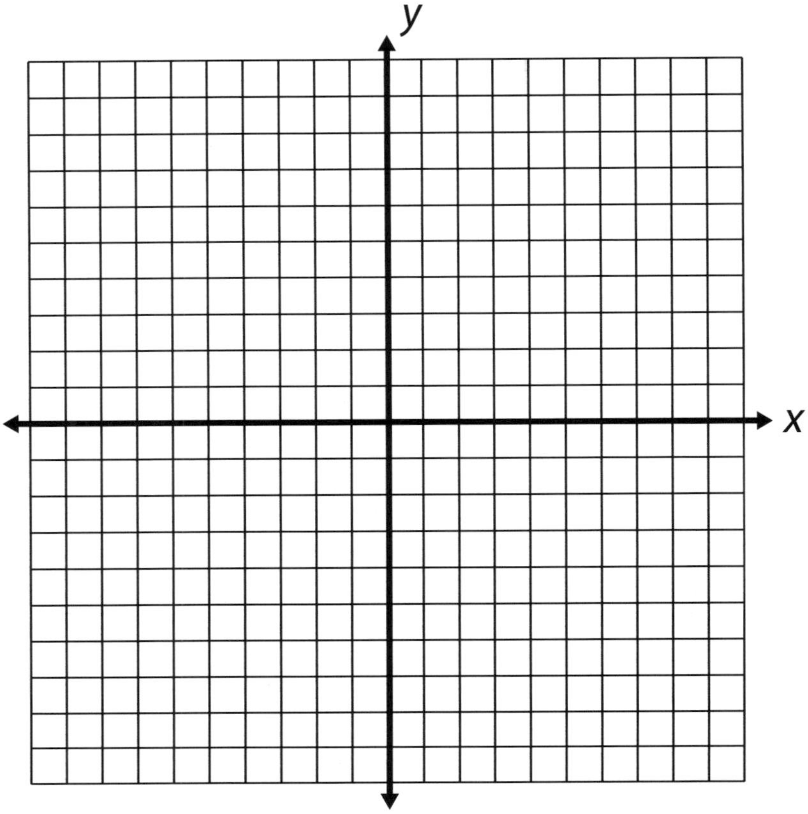

2. $x = 2$

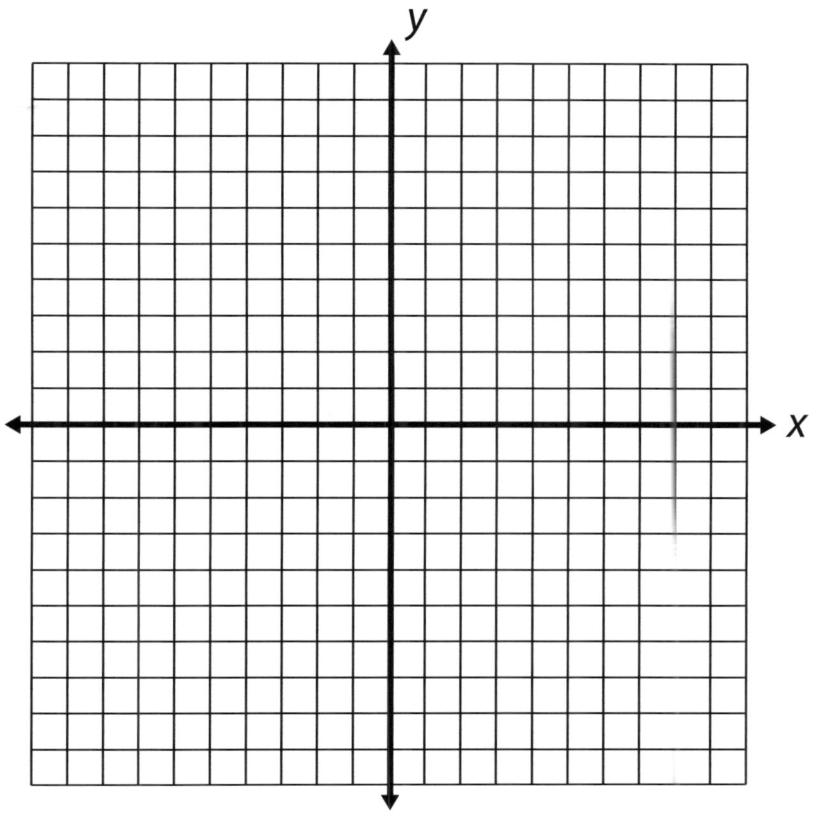

3. $a + 2b = 6$

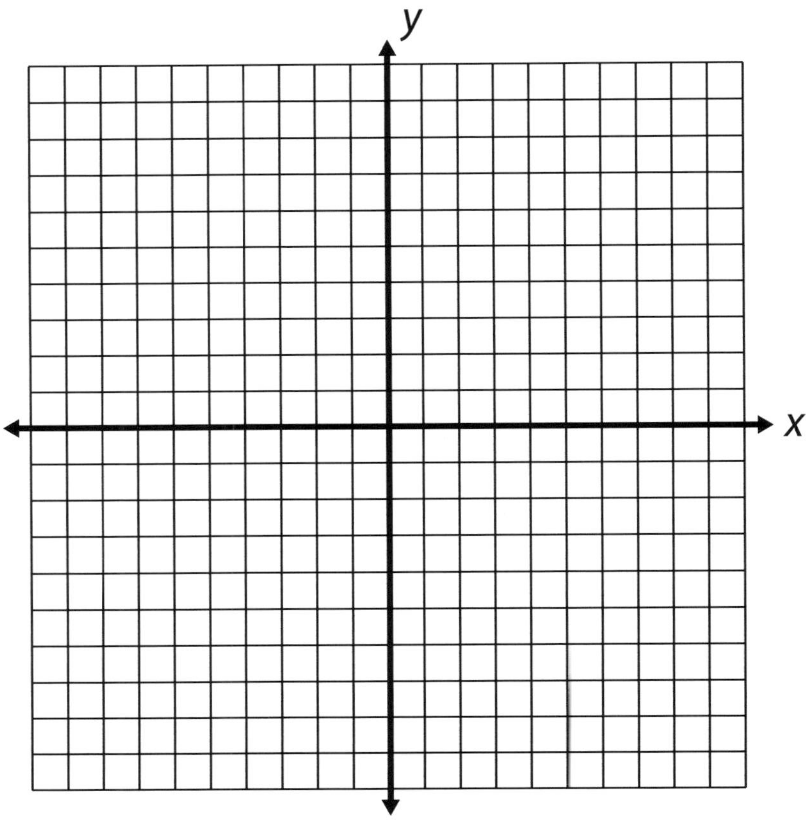

4. $y + \frac{1}{x} = 2(y - 5)$

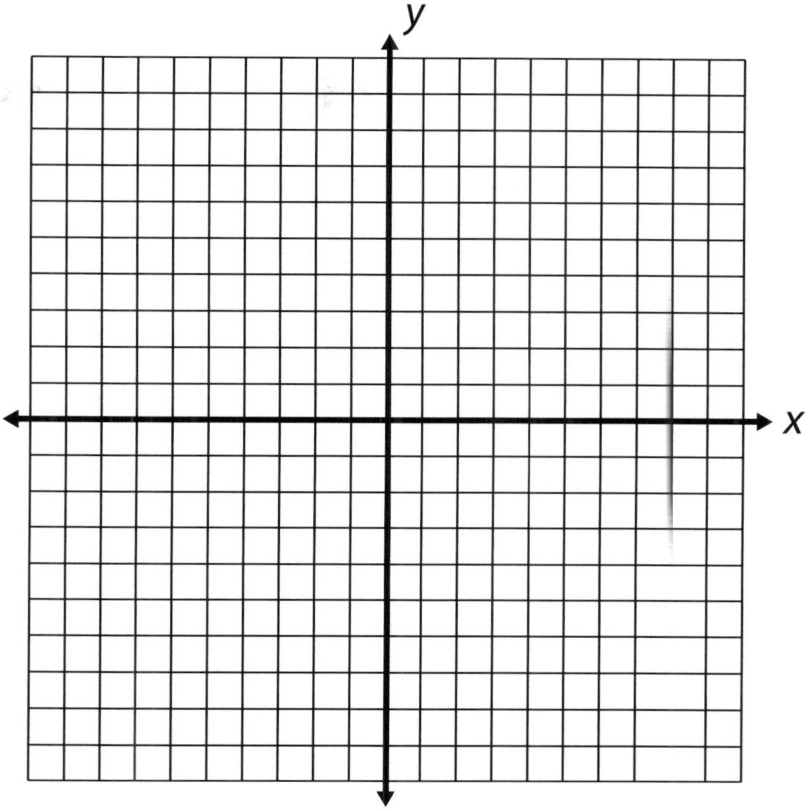

Create and Solve the Hardest Problem 3.5

Your Hardest Problem

Problem Solution

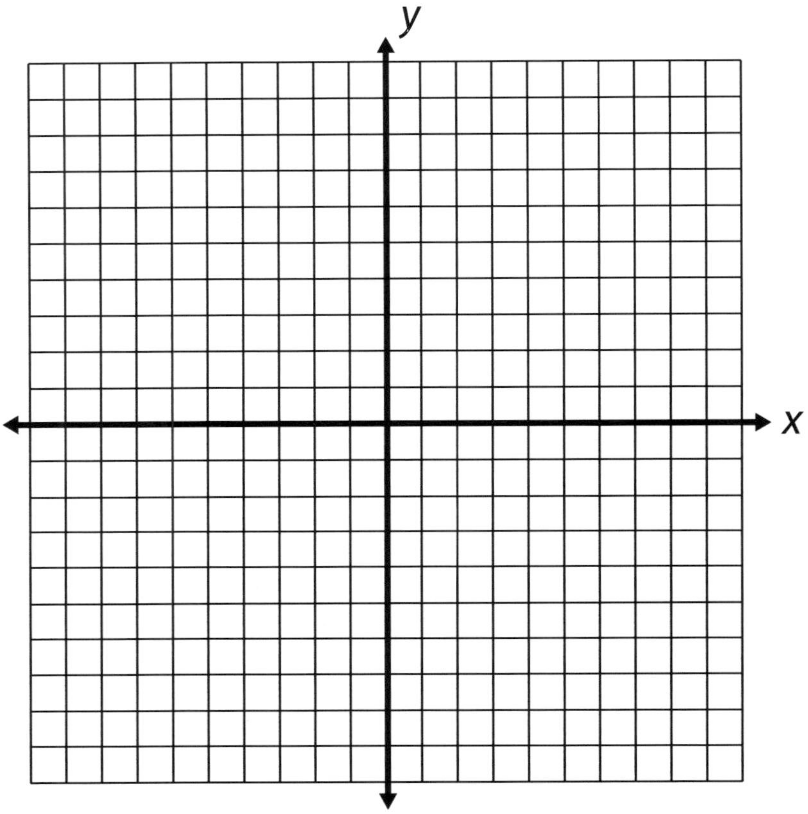

What makes a problem where you must graph equations in two variables a hard problem to solve?

Identify, Describe, & Correct the Errors 3.5

Describe the error you found in the worked solution in the appropriate space. Solve the problem correctly under "Correct Process." You are welcome to solve the problem first and then describe the error afterward.

Problem 1

Graph the equation $y = x^2 + 2x$

Worked Solution *(Find the error!)*

x	y
0	0
1	3
–1	–1
–2	0
–3	3

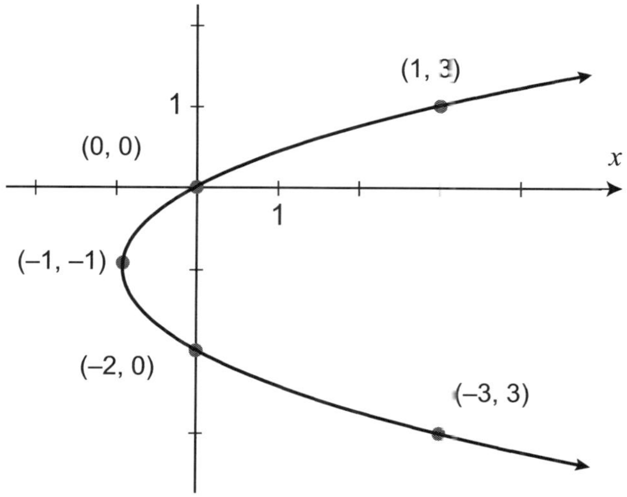

Describe the Error

Correct Process

Graph the equation $y = x^2 + 2x$

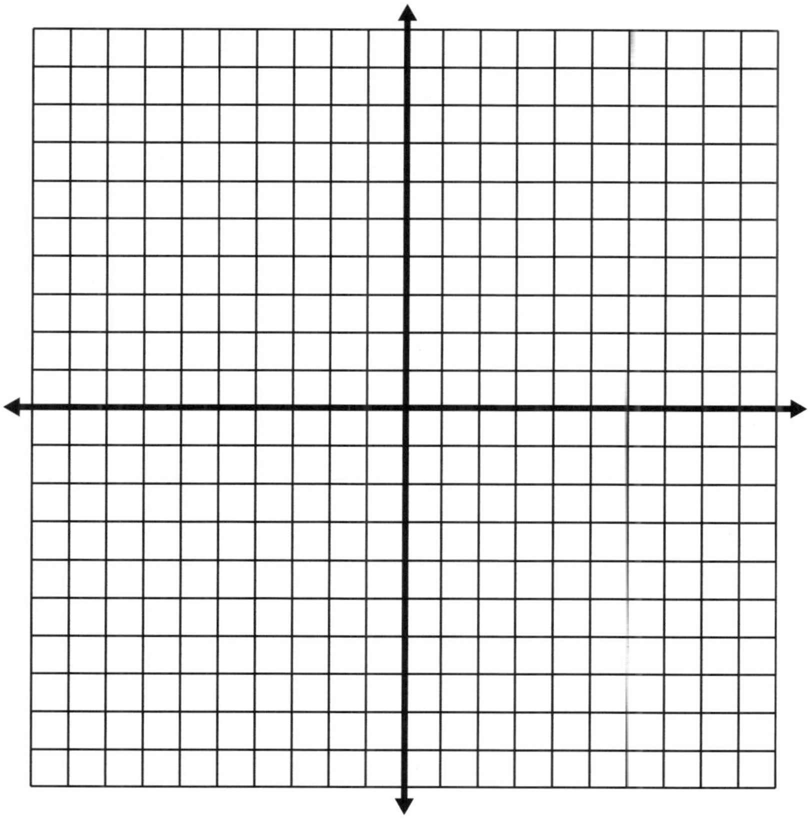

Problem 2

Graph $y = \frac{1}{x^2} + 1$

Describe the Error

Worked Solution *(Find the error!)*

x	y
3	0.9
2	1.25
1	2
0	undef.
–1	2
–2	1.25
–3	0.9

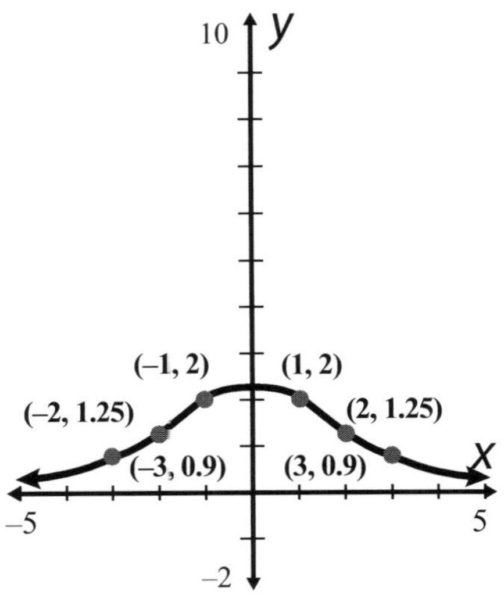

Correct Process

Graph $y = \frac{1}{x^2} + 1$

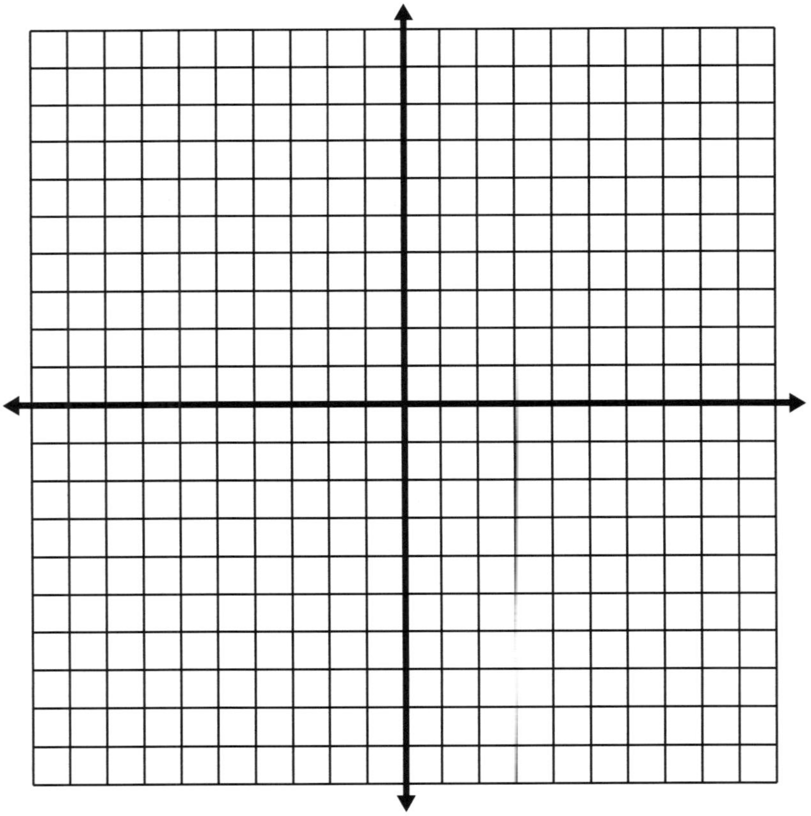

Problem 3

Graph: $2x + |x| = y$

Describe the Error

Worked Solution *(Find the error!)*

x	y
0	0
1	3
2	6
3	9

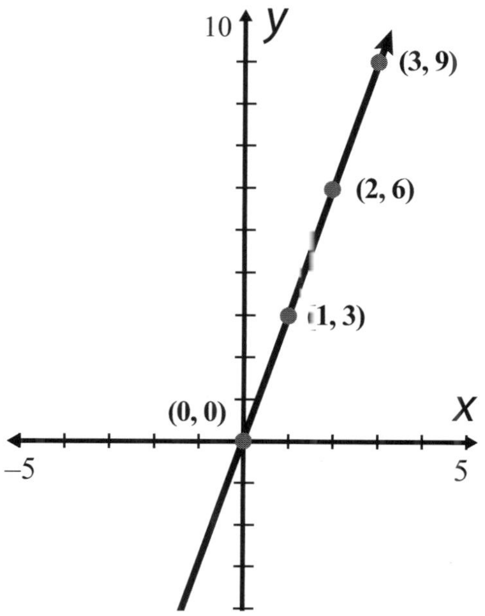

Correct Process

Graph: $2x + |x| = y$

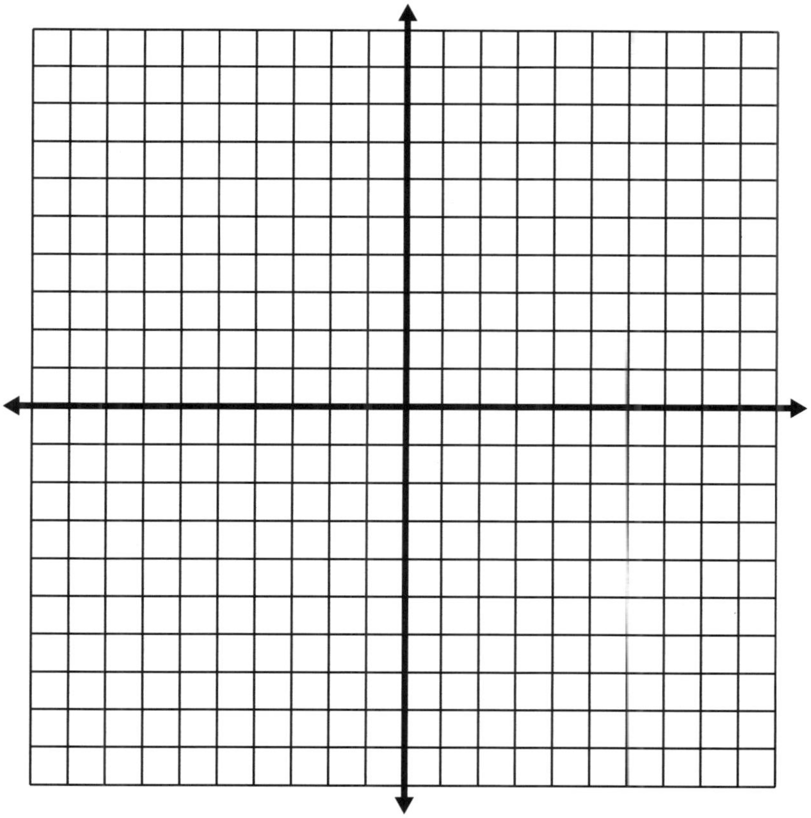

What Did You Learn About Learning Math? 3.5

1. How do you use your prior knowledge about making graphs to help you create these new graphs?

2. How can you determine in advance approximately what your graph should look like?

3. How did you find the error in Problem #1 of Identify, Describe, and Correct the Errors?

Section 3.6 Solving Inequalities in Two Variables

Why? 3.6

There are many real life situations in which it is useful to know when a certain quantity will exceed or fall below a certain value. For example, in what year could we expect global warming to exceed the point at which the polar ice caps melt? How fast does an Olympic runner need to go in order to have a time less than the world record? Such situations can be modeled with inequalities.

What Do You Already Know? 3.6

How does an inequality differ from an equation?

What does it mean for a point (x, y) to be a solution to an inequality in two variables?

Goals 3.6

When you complete this activity, you should be able to:

1. Identify the equation associated with an inequality
2. Represent the boundary line of the graphed solution to an inequality
3. Graph the solutions of a linear inequality in two variables on the coordinate plane

Building Mathematical Language 3.6

New Terms: associated equation boundary line

Key Concepts 3.6

The ***associated equation*** for a linear inequality in two variables is the original inequality with an equal sign replacing the inequality sign. The associated equation's linear graph is the ***boundary line*** of the section of the coordinate plane that is the solution set to the inequality. This line may or may not be included in the solution set, depending on whether the inequality is a strict inequality ($<$ or $>$) or not ($\leq$ or $\geq$). The strict inequality sign ($>$) means greater than; the sign $\geq$ means greater than or equal to and is not a strict inequality sign.

Methodology 3.6

Solving Linear Inequalities in Two Variables

Linear inequalities in two variables can be solved using the ideas from the Methodology for Solving Linear Equations in Two Variables. The solution set of such an inequality, from a graphical perspective, is a set of points, specifically a section of a plane.

Limitation/Caution: Care must be taken in determining if the boundary line of the solution set (half-plane) is included or not included in the solution set.

	Example 1	**Example 2**	**Your Turn**
Solve the inequality:	$x + 3y \le 4$	$y < -2x + 3$	$2x - y > 4$

Steps	**Discussion**
1 Determine the inequality's associated equation	**The associated equation of the inequality is the inequality with the inequality sign replaced by an equal sign.**

Ex 1	Ex 2	Your Turn
$x + 3y = 4$	$y = -2x + 3$	

Steps	**Discussion**
2 Graph the associated equation	**The graph of the associated equation is the boundary of the plane section that is the solution set of the inequality.**

Ex 1

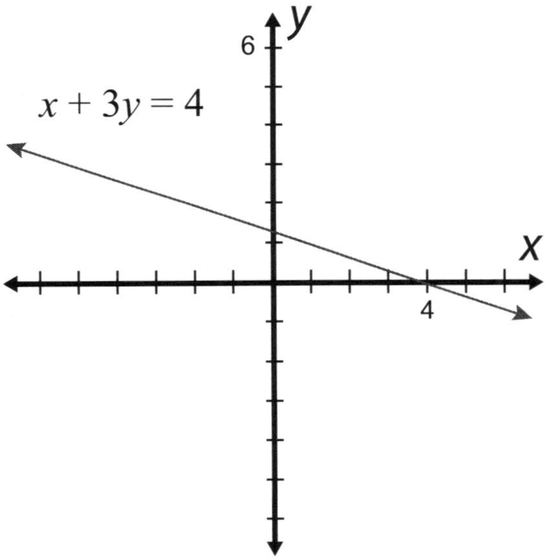

Ex 2

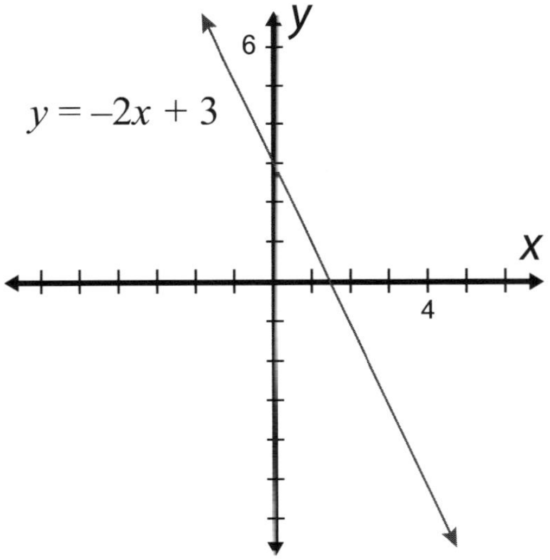

Steps	Discussion

Your Turn

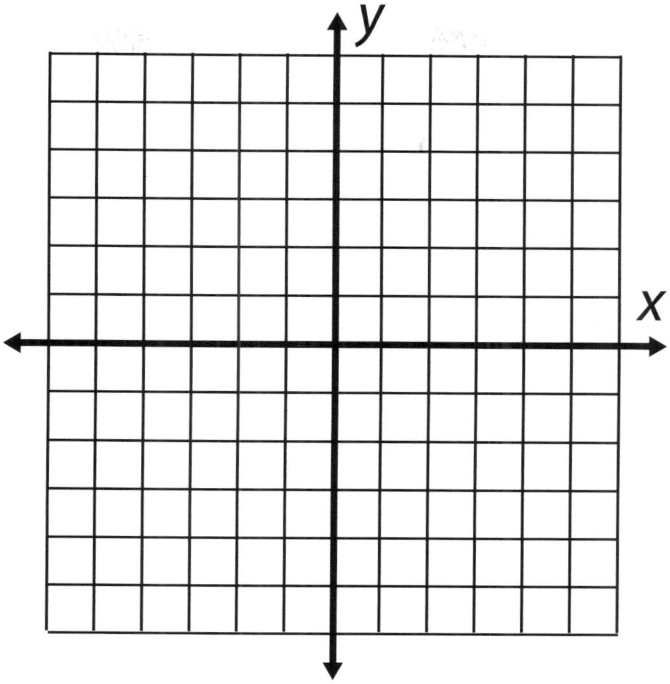

3 Select a point not on the line	Use the selected point to determine which side of the line will be the plane section that is the solution set. The point (0,0) is a good point to choose if it is not on the line.

Ex 1

Select (0, 0).

Ex 2

Select (2, 0).

Your Turn

4 Decide if the point satisfies the inequality	If the point satisfies the inequality, then all the points on that side of the boundary line are in the solution set. If not, the points on the other side of the line are in the solution set.

Ex 1

$x + 3y \leq 4$

$0 + 3(0) \leq 4$

$0 + 0 \leq 4$

$0 \leq 4$ (true)

Ex 2

$y < -2x + 3$

$1 < -2(2) + 3$

$1 < -4 + 3$

$1 < -1$ (false)

Your Turn

Steps	Discussion
5 Shade the correct plane section	**Use shading to indicate the plane section that is in the solution set.**

Ex 1

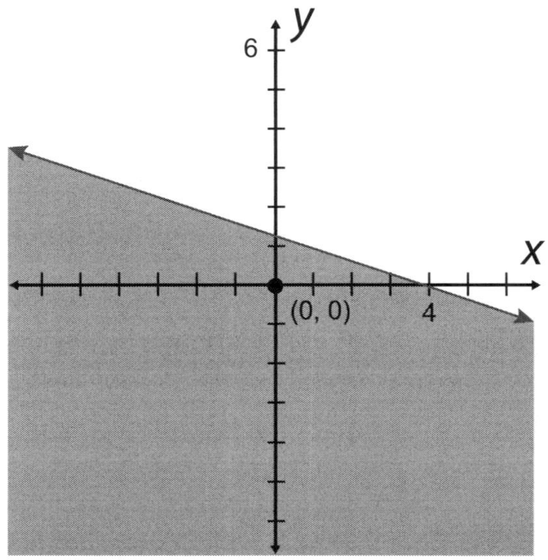

Ex 2

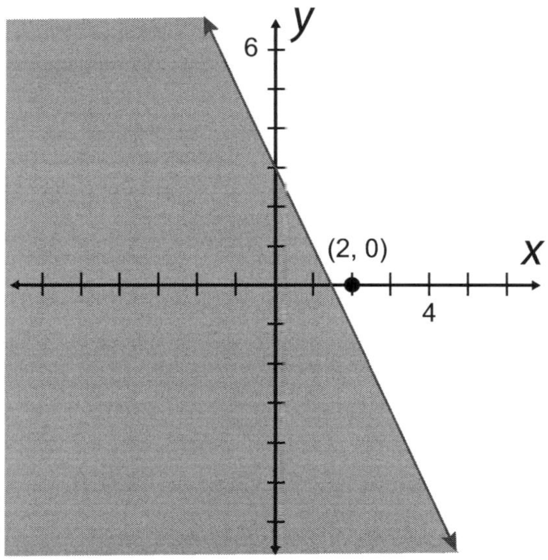

Your Turn

Use your previous graph.

6 Determine if the line is in the solution set	**Determine if the line is in the solution set. The line is in the solution if it is not a strict inequality (Ex 1) and not in the solution if it is a strict inequality (Ex 2).**

Ex 1

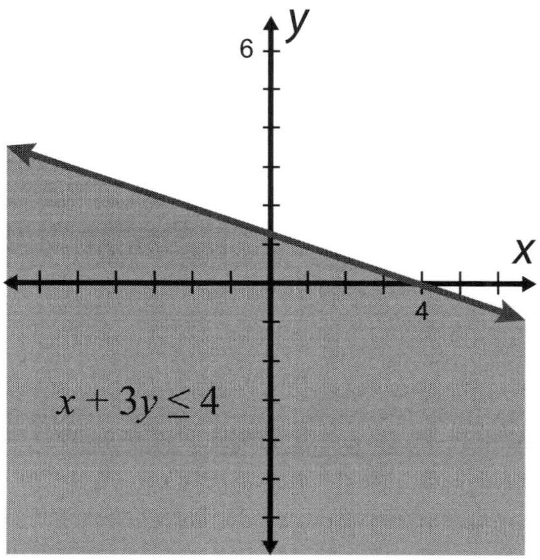

Ex 2

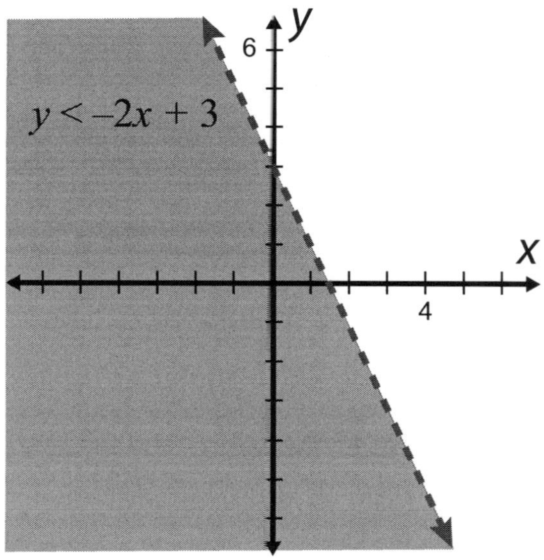

Your Turn

Use your previous graph.

Steps	Discussion
7 Validate	Select points to use for validation: one in the solution set but not on the line, one on the line, and one not on the line and not in the solution set.

Ex 1

Select $(-2,0),\ (-5,3),\ (1,2)$

$x+3y \le 4$

$(-2)+3(0) \overset{?}{\le} 4$

$-2 \le 4$ (true)

$(-5)+3(3) \overset{?}{\le} 4$

$-5+9 \overset{?}{\le} 4$

$4 \le 4$ (true, the line is in)

$(1)+3(2) \overset{?}{\le} 4$

$1+6 \overset{?}{\le} 4$

$7 \le 4$ (false) ✓

Ex 2

Select $(-2,1),\ (2,-1),\ (3,0)$

$y < -2x+3$

$1 \overset{?}{<} -2(-2)+3$

$1 \overset{?}{<} 4+3$

$1 < 7$ (true)

$-1 \overset{?}{<} -2(2)+3$

$-1 \overset{?}{<} -4+3$

$-1 < -1$ (false, the line is not in)

$0 \overset{?}{<} -2(3)+3$

$0 \overset{?}{<} -6+3$

$0 < -3$ (false) ✓

Your Turn

Model 1 3.6

Solve the inequality: $y > -2x + 1$

Step 1 Determine the associated equation $y = -2x + 1$

Step 2 Graph the associated equation

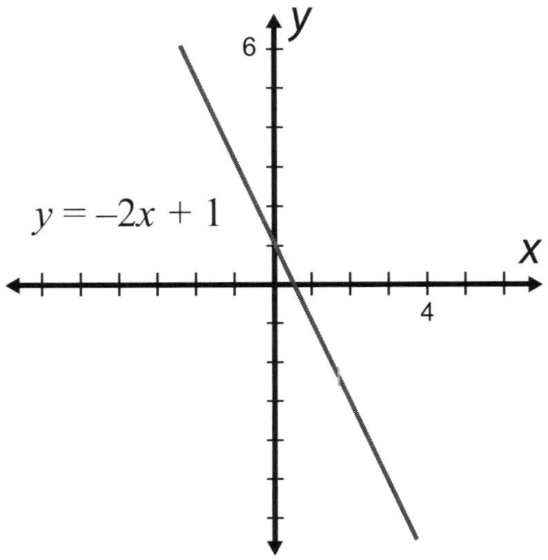

Step 3 Select a point not on the line Select (1, 1)

Step 4 Decide if the point satisfies the inequality

$y > -2x + 1$

$1 > -2(1) + 1$

$1 > -2 + 1$

$1 > -1$ (true)

Step 5 Shade the correct plane section

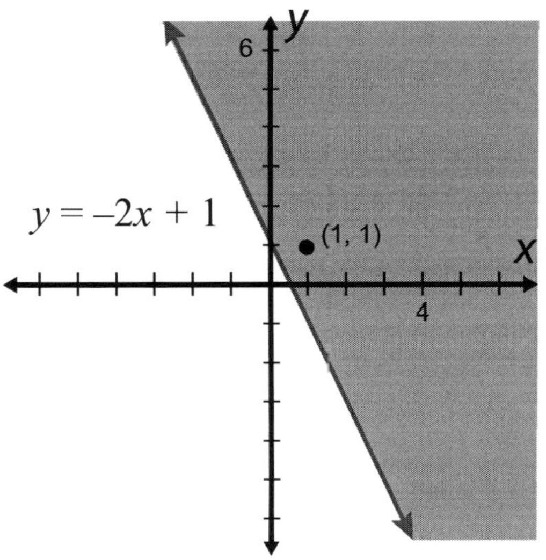

Step 6 Determine if the line is in the solution set

The inequality is a strict inequality. The line is not in the solution set and should be dashed.

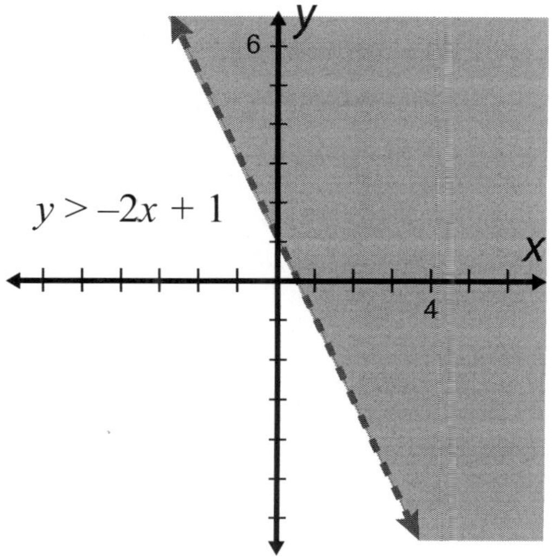

Step 7 Validate

Select: (1, 2), (–1, 3), (–2, 0)

$y > -2x + 1$

$2 \overset{?}{>} -2(1) + 1$

$2 \overset{?}{>} -2 + 1$

$2 > -1$ (true)

$3 \overset{?}{>} -2(-1) + 1$

$3 \overset{?}{>} 2 + 1$

$3 > 3$ (false, the line is not in)

$0 \overset{?}{>} -2(-2) + 1$

$0 \overset{?}{>} 4 + 1$

$0 > 5$ (false) ✓

Model 2 3.6

Solve the two variable inequality: $x > 2$

Step 1 Determine the associated equation $x = 2$

Step 2 Graph the associated equation

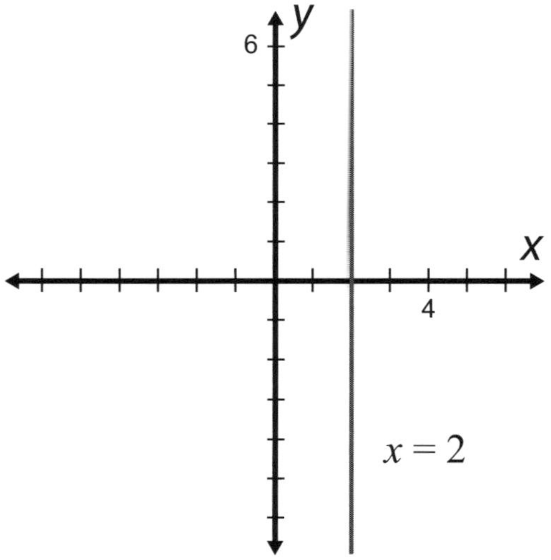

Step 3 Select a point not on the line Select (3, 0)

Step 4 Decide if the point satisfies the inequality

$x > 2$

$3 > 2$ (true)

Step 5 Shade the correct plane section

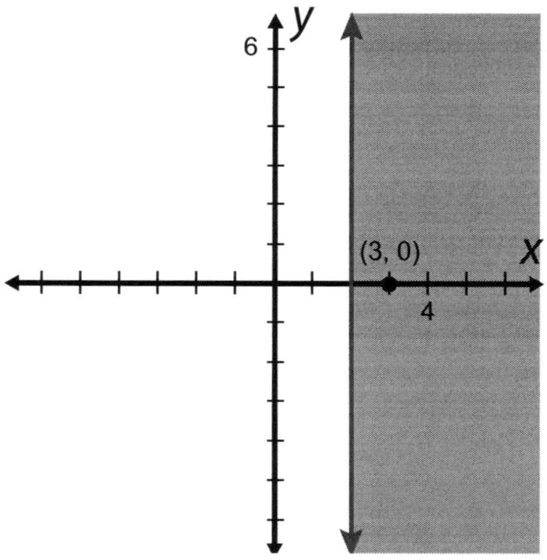

Step 6 Determine if the line is in the solution set

The inequality is a strict inequality. The line is not in the solution set and should be dashed:

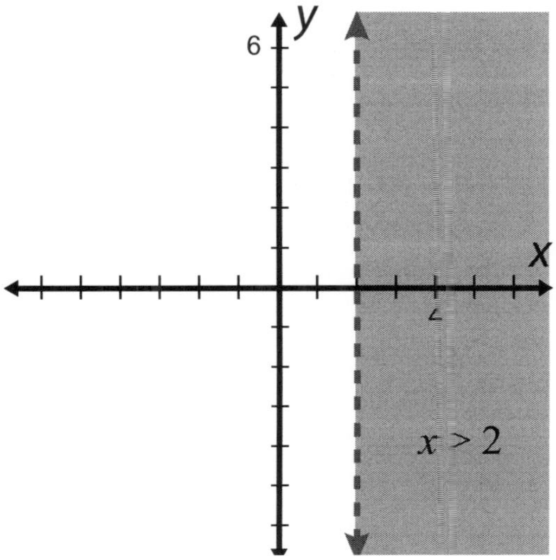

Step 7 Validate

Select: (–1, 0), (2, 0), (4, 0)

$x > 2$

$-1 > 2$ (false)

$2 > 2$ (false, the line is not in)

$4 > 1$ (true) ✓

Addressing Common Errors 3.6

The following table shows some of the most common errors that learners tend to make with the kinds of problems covered in this section.

ERROR: **Shading the wrong side of the line**

Incorrect Process

Solve the inequality: $2x - 4y < 7$

$2x - 4y < 7$

$2x - 4y = 7$

$(2, -4)$

$2(2) - 4(-4) < 7$

$20 < 7$ (false)

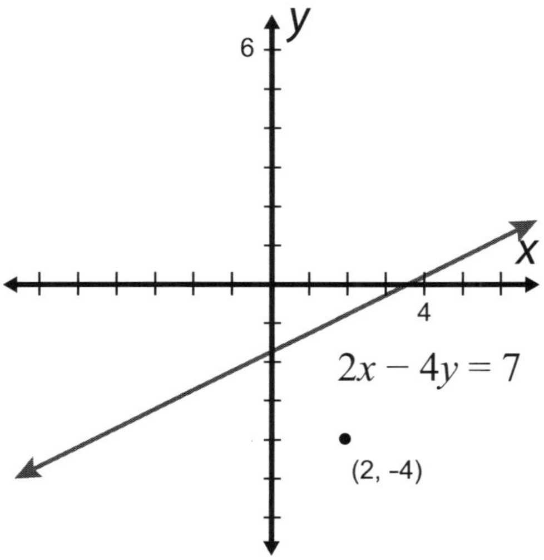

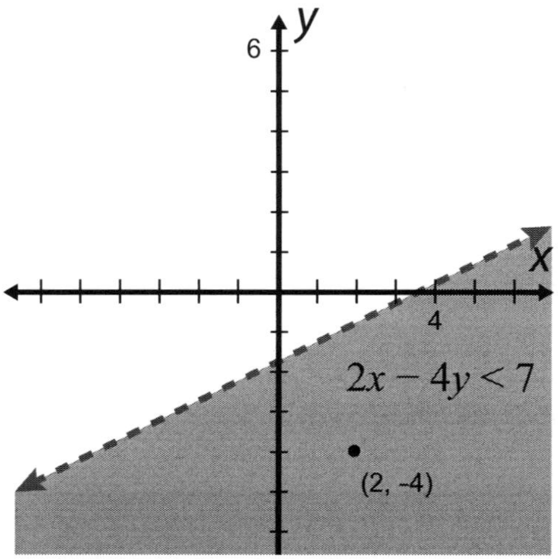

Resolution

Choose the correct side of the line, opposite the point (2, –4).

Correct Process

Solve the inequality: $2x - 4y < 7$

$2x - 4y < 7$

$2x - 4y = 7$

$(2, -4)$

$2(2) - 4(-4) < 7$

$20 < 7$ (false)

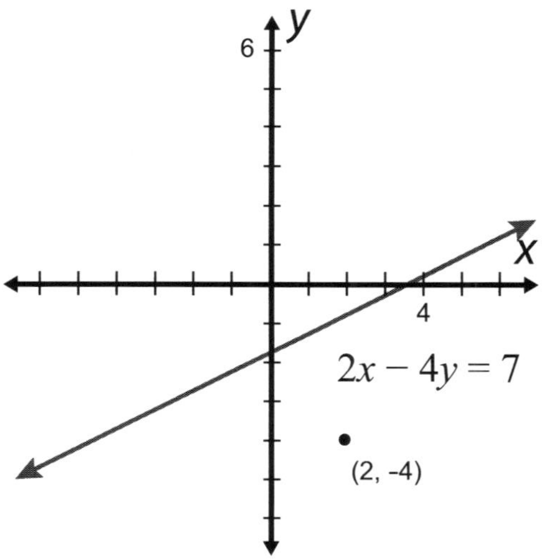

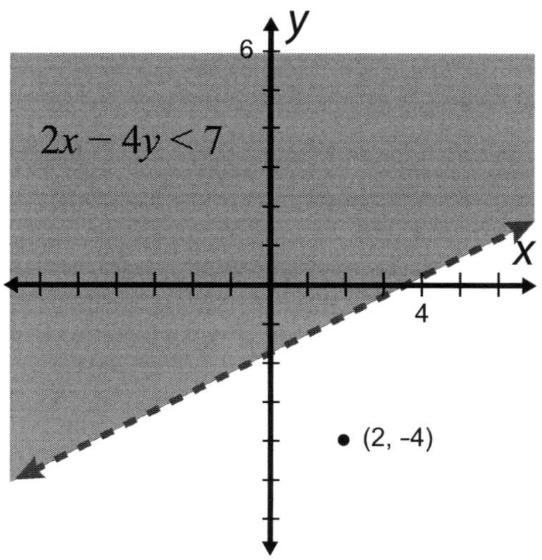

Validation

Use the point (0, 0) in the shaded area.

$2x - 4y < 7$

$2(0) - 4(0) \overset{?}{<} 7$

$0 < 7$ (true) ✓

ERROR: **Incorrectly representing the boundary of the solution set**

Incorrect Process

Solve the inequality: $2y - 3x < 4$

$2y - 3x < 4$

$2y - 3x = 4$

$(1, 1)$

$2(1) - 3(1) \overset{?}{<} 4$

$2 - 3 \overset{?}{<} 4$

$-1 < 4$ (true) ✓

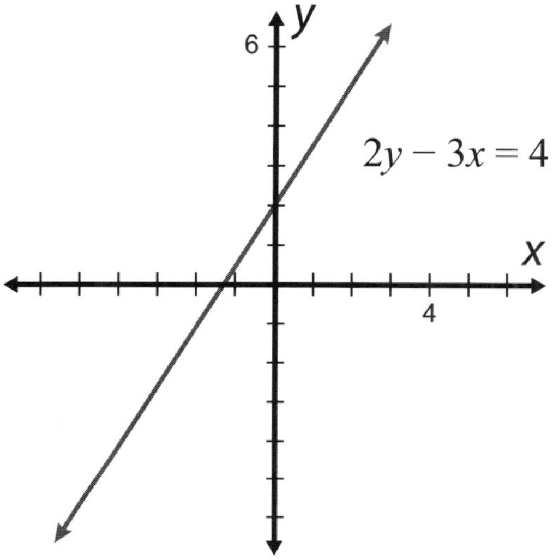

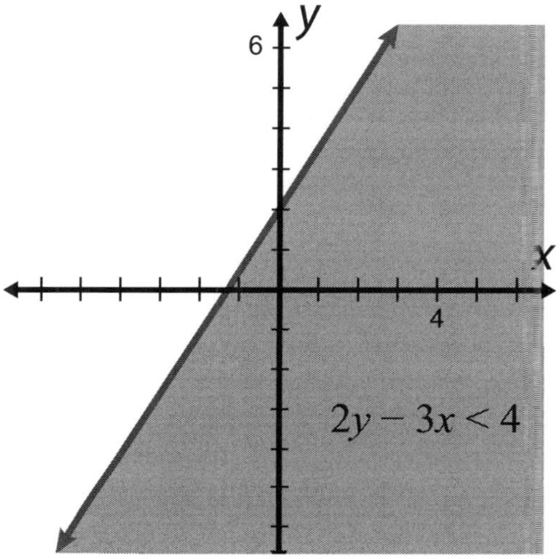

Resolution

Correctly determine that the inequality is not a strict inequality and that the boundary line should be dashed.

Correct Process

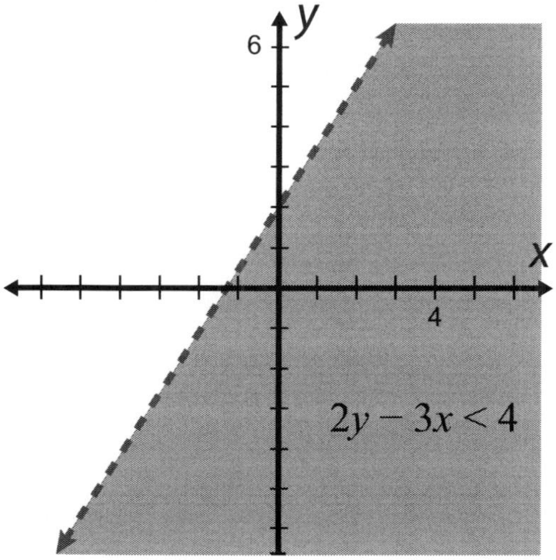

Validation

The point (0,0) satisfies the equation and the inequality is not strict, so the line should be dashed.

Preparation Inventory 3.6

I can...	OR *Here's my question...*
☐ solve a linear equation.	☐
☐ graph a linear equation.	☐
☐ decide if a point is a solution to an inequality.	☐

Critical Thinking Questions 3.6

1. What are the strict inequality signs? (Provide their symbols and names.)

2. How do you determine which section of the coordinate plane to shade as the solution set to a linear inequality in two variables?

3. How do you know whether the boundary line should be included or should be not included in the solution set to an linear inequality in two variables?

4. How do you designate that the boundary line is not included in the solution set to a linear inequality in two variables?

5. Why use a point not on the graph of the associated equation to determine which half-plane to shade?

6. How do you validate that your graph of the solution set to an inequality in two variables is correct?

7. How does graphing a linear inequality in two variables differ from graphing a linear equation in two variables?

8. How does the solution to the the inequality $x > 2$ as an inequality in one variable differ from its solution as an inequality in two variables?

9. If an inequality is written in the form $y < ax + b$, will its solution set always be the region BELOW the line $y = ax + b$? Explain.

10. If an inequality is written in the form $y > ax + b$, will its solution set always be the region ABOVE the line $y = ax + b$? Explain.

Demonstrate Your Understanding 3.6

Graph each of the following equations or inequalities.

1. Graph the solution set of

$$y \le 3x - 5$$

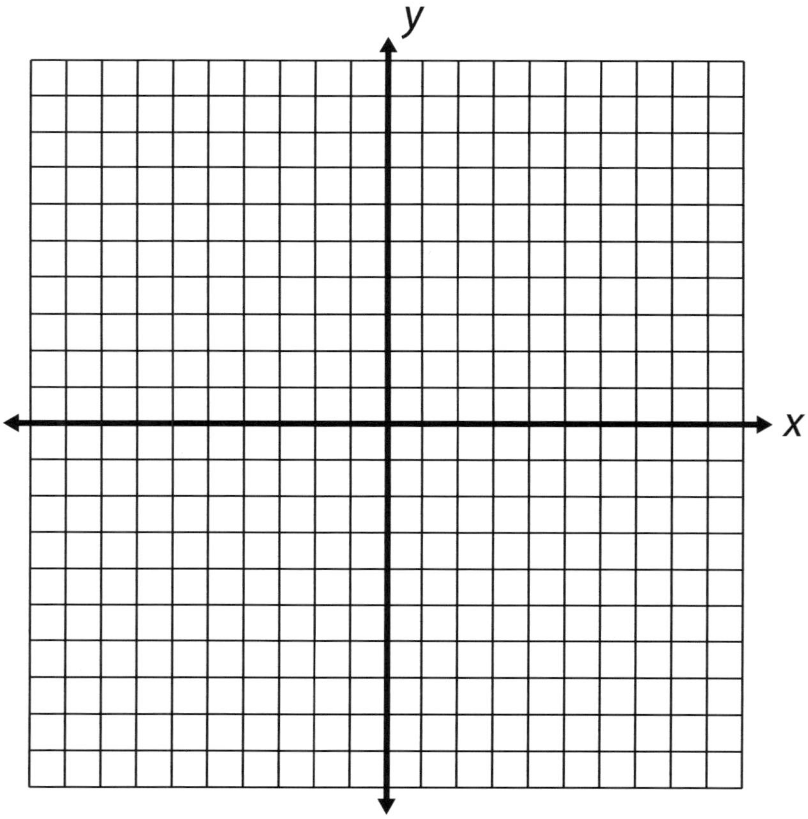

2. Graph the solution set of

$2x + 3y > 7$

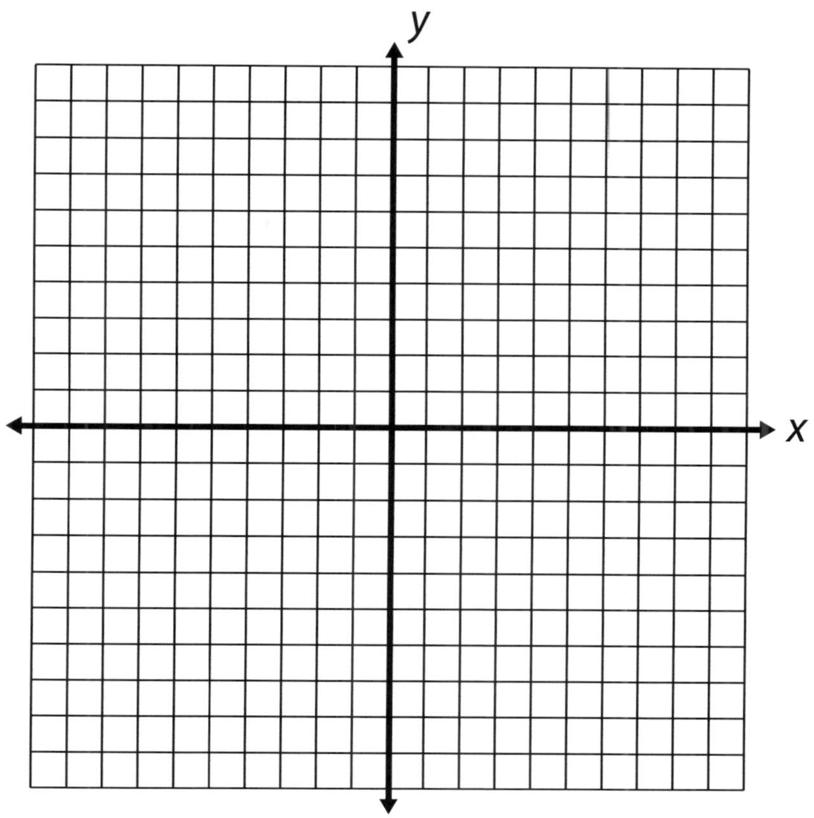

3. Graph the solution set for the two variable inequality $y \leq 3$

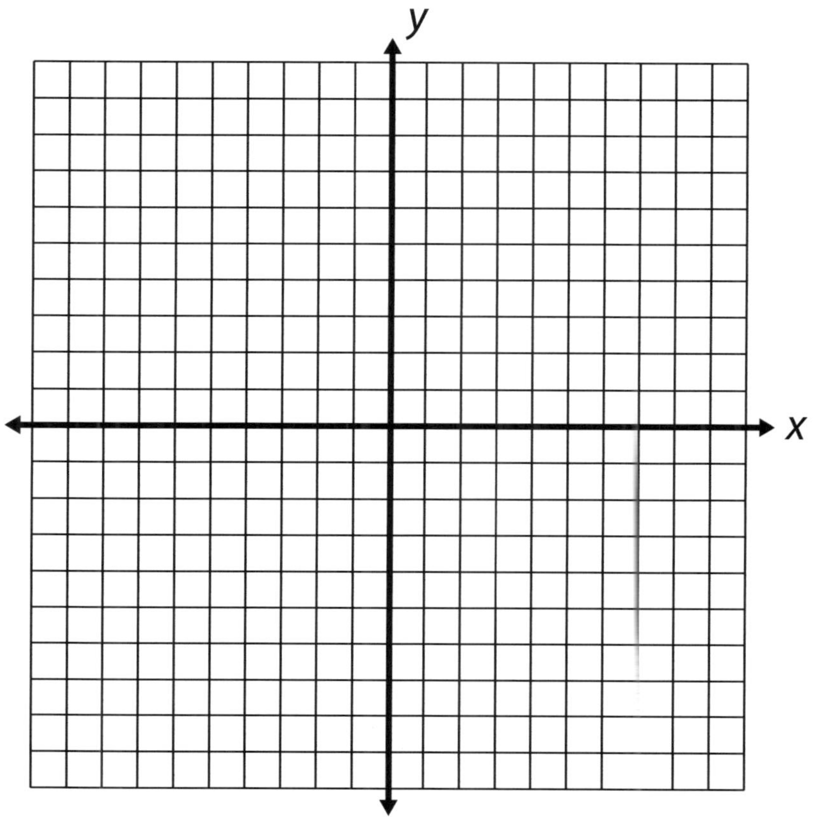

4. Graph the solution set for

   $x - 2y > 3$

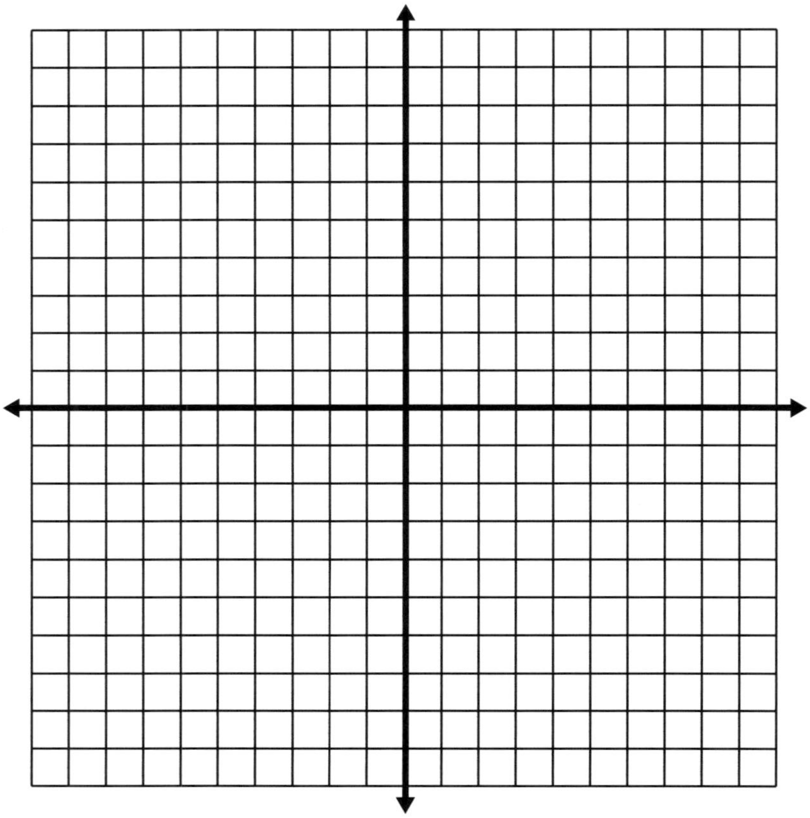

5. Graph the solution set for

   $x + 4 \leq 2y$

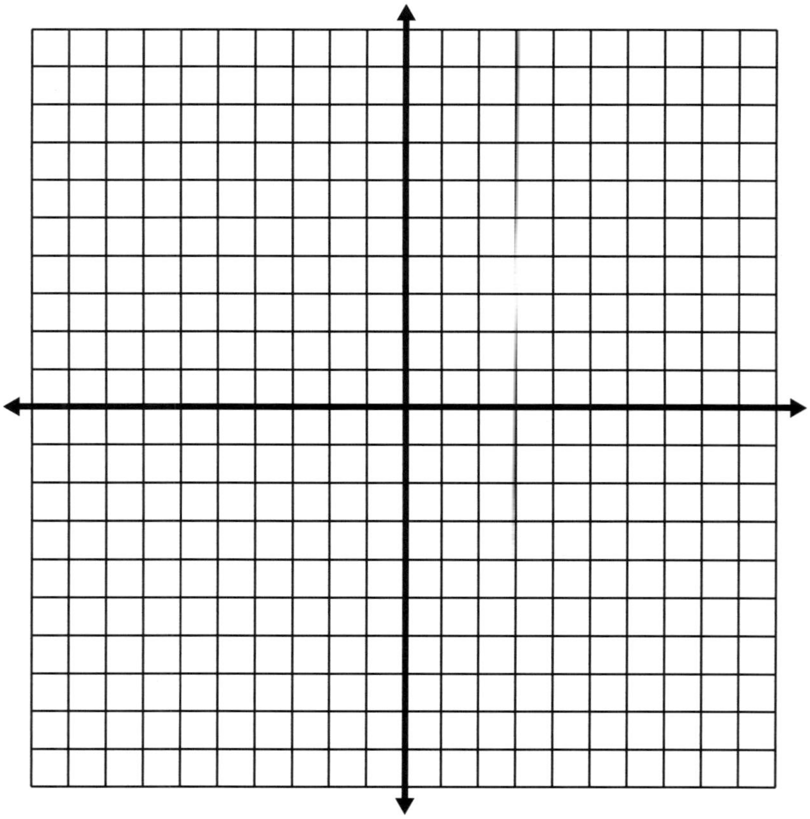

Create and Solve the Hardest Problem 3.6

Your Hardest Problem

Problem Solution

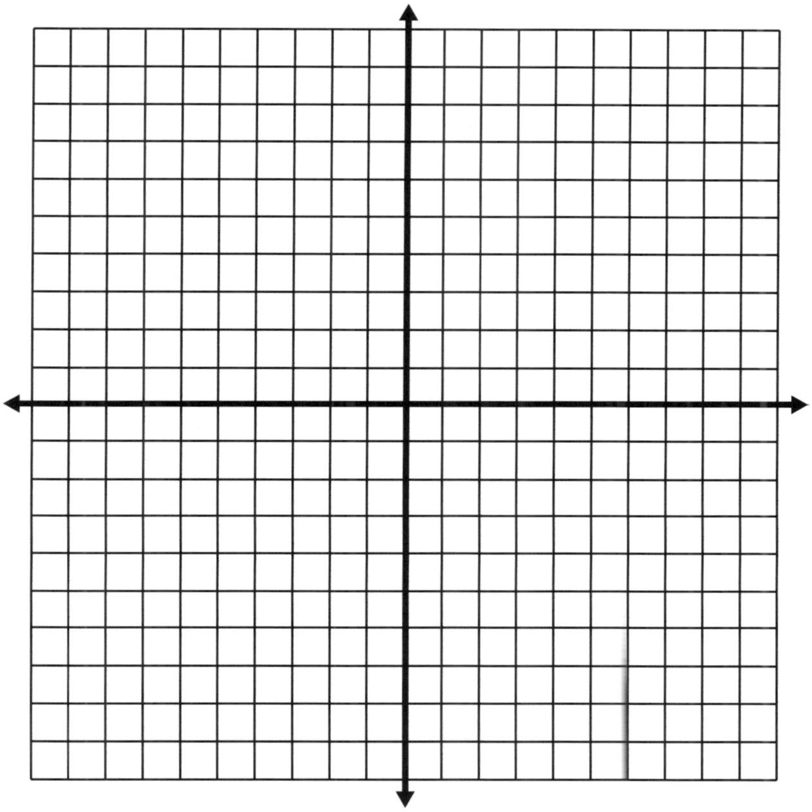

What makes a problem where you must solve an inequality in two variables a hard problem to solve?

Identify, Describe, & Correct the Errors 3.6

Describe the error you found in the worked solution in the appropriate space. Solve the problem correctly under "Correct Process." You are welcome to solve the problem first and then describe the error.

Problem 1

Graph the solution set of $x + y \geq 5$

Worked Solution *(Find the error!)*

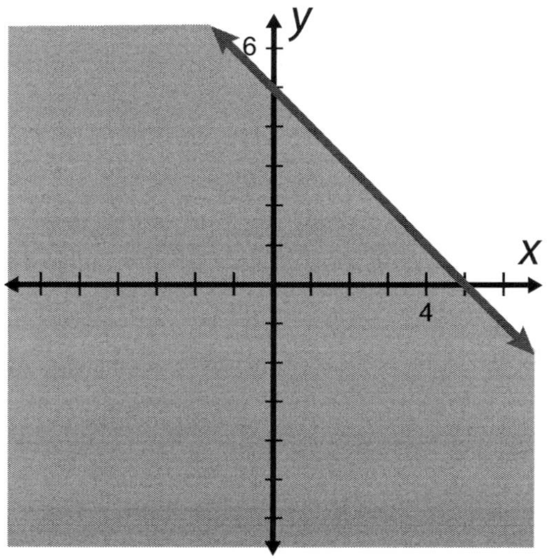

Describe the Error

Correct Process

Graph the solution set of $x + y \geq 5$

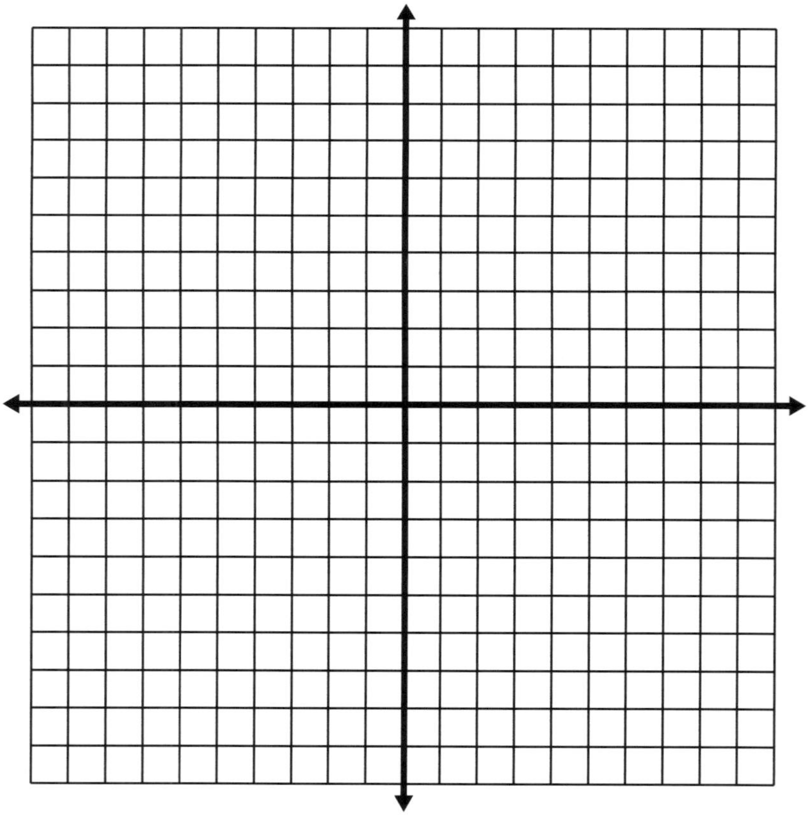

Problem 2

Graph the solution set of $y < 2x - 3$

Worked Solution *(Find the error)*

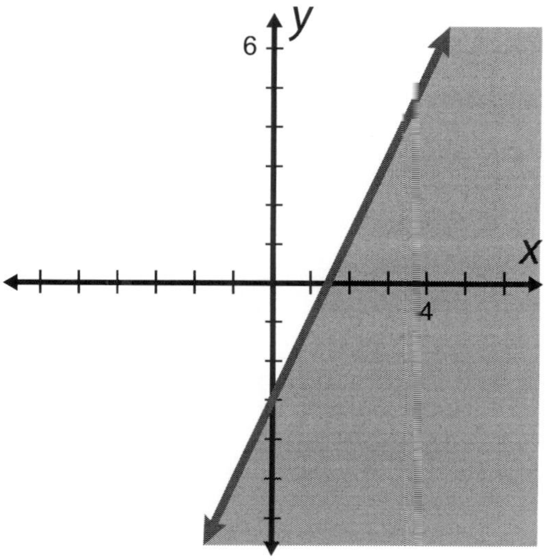

Describe the Error

Correct Process

Graph the solution set of $y < 2x - 3$

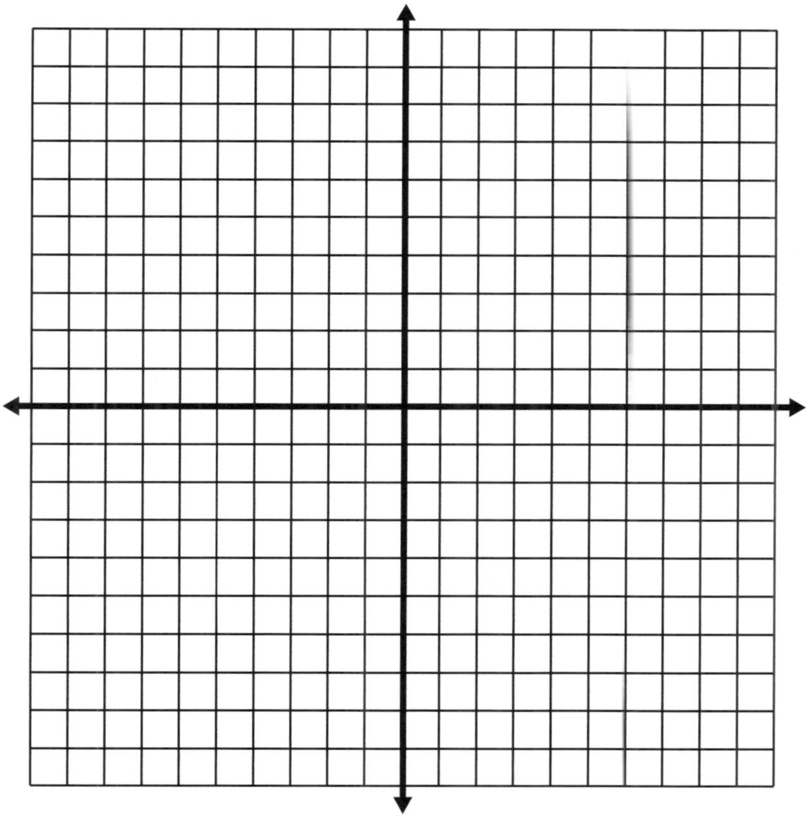

Problem 3

Graph the solution set of $x - 2y \geq 6$

Describe the Error

Worked Solution *(Find the error!)*

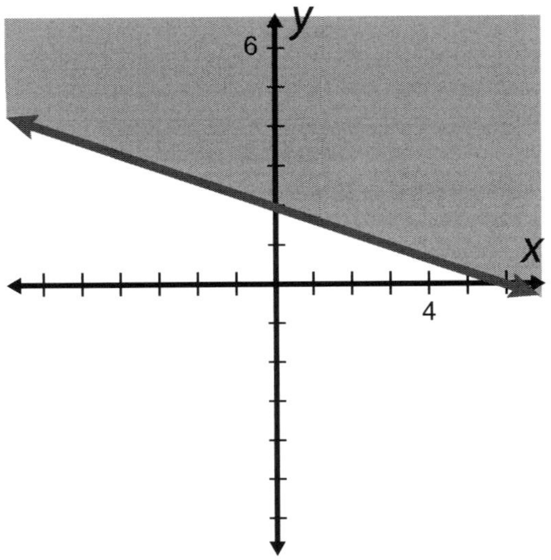

Correct Process

Graph the solution set of $x - 2y \geq 6$

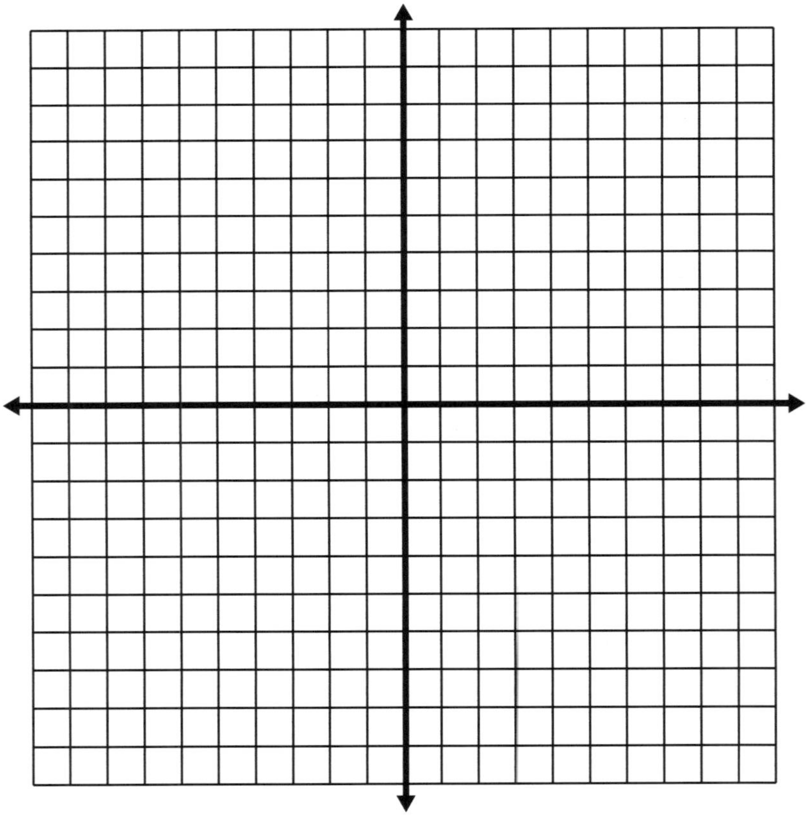

What Did You Learn About Learning Math? 3.6

1. How did you use what you know about graphing the solutions to equations to help you graph the solutions to inequalities?

2. How do you apply your understanding of validation when graphing the solution to an inequality?

3. How do you think about what the subproblems of solving an inequality in two variables are?

Concept Map — Chapter 3

Terminology from the chapter
check the terms used in your map

__absolute value inequality

__associated equation

__boundary line

__closed interval

__compound inequality

__dependent variable

__graphical representation

__graphing window

__independent variable

__interval notation

__Multiplicative Law of Inequalities

__open interval

__origin

__rate of change

__slope

__slope-intercept form

__standard form of a linear equation in two variables

__table of values (T-table)

__three-part inequality

__x-intercept

__y = mx + b form

__y-intercept

Your instructor will specify how many terms from the chapter you should use in your concept map.

CHAPTER 4

This chapter develops methods for working with polynomials and solving quadratic and other types of polynomial equations. In addition, you will learn how to use factoring and the Lowest Common Multiple (LCM) to simplify rational expressions.

Section 4.1 Dividing Polynomials

Why? 4.1

You know how to add, subtract, and multiply polynomials. In this section, you will learn how to divide polynomials to complete your knowledge of the standard polynomial operations.

What Do You Already Know? 4.1

What is the leading term in a polynomial?

In the long division problem below, can you identify each of the following?

divisor =

dividend =

quotient =

remainder =

$$\begin{array}{r} 32 \\ 23\overline{)745} \\ \underline{69} \\ 55 \\ \underline{46} \\ 9 \end{array}$$

How do you validate the result of a numerical long division problem?

Goals 4.1

When you complete this activity, you should be able to:

1. Divide a polynomial by a monomial
2. Use long division to divide a polynomial by a polynomial

Building Mathematical Language 4.1

New Terms:

division algorithm

missing term(s) in a polynomial

factoring

rational expression

Key Concepts 4.1

A ***rational expression*** is a quotient of polynomials. Such expressions can be thought of as "fractions" with polynomials in the numerator and denominator.

The ***division algorithm*** states that for polynomials a and b, there exist polynomials q and r such that $a = bq + r$ where $r = 0$ or the degree of r is less than the degree of a. The polynomial q is called the quotient and r is called the remainder.

$$\text{In long division form: } b\overline{)\,a}^{\,q+R} \quad \text{where } R = \frac{r}{b}$$

When dividing a polynomial by a polynomial, where neither polynomial is a monomial, you first set up the division in a form that uses the division algorithm. The quotient will be a constant or a polynomial with a remainder that can be presented in two ways.

Missing terms in a polynomial are the terms whose coefficients are zero and are omitted when the polynomial is written.

Example: The polynomial $5x^4 - 2x^2 - 7$ is missing the terms $0x^3$ and $0x$. This polynomial can also be written as $5x^4 + 0x^3 - 2x^2 + 0x - 7$.

Factoring can be used to reduce the degrees of the polynomials involved in the division ***if*** the polynomials can be factored (see section 4.2).

Methodology 4.1

Dividing a Polynomial By a Monomial

Sometimes you can simplify an algebraic expression by dividing a polynomial by a monomial. Such simplifications can yield information about the expression that is not obvious from the original expression.

Limitation/Caution: When dividing a polynomial by a monomial, you must be careful to divide each term of the numerator by the denominator.

	Example 1	**Example 2**	**Your Turn**
Divide:	$6x^3 - 3x^2 - 12x$ by $3x$	$2py^4 - 8qy^2 + 4ry - 5$ by $2y$	$2ab^2 - 3ab + 5a - 2$ by a

NOTE: Division is equivalent to multiplication by the reciprocal of the divisor, which means that the distributive property can be applied to this situation.

$$\text{Example: } \left(6x^4 - 4x^3 + 10x\right) \div 2x = \left(6x^4 - 4x^3 + 10x\right) \bullet \frac{1}{2x} = \frac{6x^4}{2x} - \frac{4x^3}{2x} + \frac{10x}{2x}$$

Steps	Discussion
1 Set up the division as a rational expression	**Write the dividend as the numerator and the divisor as the denominator.**

Ex 1	Ex 2	Your Turn
$\dfrac{6x^3 - 3x^2 - 12x}{3x}$	$\dfrac{2py^4 - 8qy^2 + 4ry - 5}{2y}$	

Steps	Discussion
2 Write the expression as separate fractions	The Distributive Property can be used to write a quotient as a sum. $\left[\frac{b+c}{a}=\frac{b}{a}+\frac{c}{a}\right]$

Ex 1

$$\frac{6x^3}{3x}-\frac{3x^2}{3x}-\frac{12x}{3x}$$

Ex 2

$$\frac{2py^4}{2y}-\frac{8qy^2}{2y}+\frac{4ry}{2y}-\frac{5}{2y}$$

Your Turn

Steps	Discussion
3 Simplify each term	Each term can be reduced by dividing out common factors.

Ex 1

$$2x^2-x-4$$

Ex 2

$$py^3-4qy+2r-\frac{5}{2y}$$

Your Turn

Steps	Discussion
4 Validate	The result in Step 3 multiplied by the original denominator should equal the numerator of the original rational expression.

Ex 1

$$3x\left(2x^2-x-4\right)$$
$$=6x^3-3x^2-12x \checkmark$$

Ex 2

$$2y\left(py^3-4qy+2r-\frac{5}{2y}\right)$$
$$=2py^4-8qy^2+4ry-5 \checkmark$$

Your Turn

Methodology 4.1

Dividing Polynomials

Rewriting the division of two polynomials can be important when graphing equations in two variables.

Limitation/Caution: Polynomial division cannot be used to divide one expression (dividend) by another (divisor) if the degree of the divisor is greater than the degree of the dividend.

	Example 1	**Your Turn**
Divide:	$9x^3 - 10x + 1$ by $3x - 2$	$x^3 - 2x^2 - 5x - 7$ by $x + 2$

Steps	**Discussion**
1 Set up the division	**Write the polynomial being divided as the dividend, inserting 0 for missing terms, if needed. The dividend polynomial is divided by the divisor polynomial.**

Ex 1

$$3x-2\overline{)9x^3+0x^2-10x+1}$$

Your Turn

Steps	**Discussion**
2 Find the degree of the quotient	**Take the degree of the dividend and subtract the degree of the divisor.**

Ex 1

$$3-1=2$$

The degree of the quotient will be 2.

Your Turn

Steps	**Discussion**
3 Calculate the ratio of the leading terms	**Divide the leading term of the dividend by the leading term in the divisor.**

Ex 1

$$\frac{9x^3}{3x}=3x^2$$

Your Turn

Steps	**Discussion**
4 Create the new term for the quotient	**Write this term as a term in the quotient.**

Ex 1

$$\begin{array}{r} 3x^2 \\ 3x-2\overline{)9x^3+0x^2-10x+1} \end{array}$$

Your Turn

Steps	Discussion
5 Multiply	Multiply the divisor by the new term of the quotient and place the result under the dividend.

Ex 1

$$\begin{array}{r} 3x^2 \\ 3x-2\overline{)9x^3+0x^2-10x+1} \\ \underline{9x^3-6x^2} \end{array}$$

Your Turn

Steps	Discussion
6 Subtract	Subtract the product in Step 5 from the dividend. (Change the signs and add.)

Ex 1

$$\begin{array}{r} 3x^2 \\ 3x-2\overline{)9x^3+0x^2-10x+1} \\ \underline{-9x^3+6x^2} \\ +6x^2-10x+1 \end{array}$$

Your Turn

Steps	Discussion
7 Repeat Steps 3 through 6 as needed	Repeat steps 3 through 6 until the expression after the subtraction is zero or its degree is less than the degree of the divisor.

Ex 1

$$\begin{array}{r} 3x^2+2x-2 \\ 3x-2\overline{)9x^3+0x^2-10x+1} \\ \underline{-9x^3+6x^2} \\ +6x^2-10x+1 \\ \underline{-6x^2+4x} \\ -6x+1 \\ \underline{+6x-4} \\ -3 \end{array}$$

Your Turn

Steps	Discussion
8 Present the quotient	Write the result, using R (for remainder) or with a rational expression.

Ex 1

The quotient is $3x^2+2x-2 \quad \text{R} -3$

OR

$$3x^2+2x-2+\frac{-3}{3x-2}$$

Your Turn

Steps	Discussion
9 Validate your work	If you used R, validate your work by multiplying the polynomial by the divisor and adding the remainder. If you used a rational expression, multiply the quotient with the rational expression by the divisor.

Ex 1

$(3x^2+2x-2)(3x-2)+(-3)=$

$9x^3+6x^2-6x-6x^2-4x+4-3=$

$9x^3-10x+1$ ✓

OR

$\left(3x^2+2x-2+\dfrac{-3}{3x-2}\right)(3x-2)=$

$9x^3+6x^2-6x-6x^2-4x+4-3=$

$9x^3-10x+1$ ✓

Your Turn

Addressing Common Errors 4.1

The following table shows some of the most common errors that learners tend to make with the kinds of problems covered in this section.

ERROR: Cancelling terms

Incorrect Process

Simplify: $\dfrac{3x^2-5x+2}{2x}$

$\dfrac{3\cancel{x^2}^{x}-5x+\cancel{2}^{1}}{\cancel{2}^{1}\cancel{x}^{1}}$

$3x-5x+1=-2x+1$

Resolution

Terms were cancelled rather than factors. Divide each term in the numerator by the single factor in the denominator.

Correct Process

Simplify: $\dfrac{3x^2-5x+2}{2x}$

$\dfrac{3x^2}{2x}-\dfrac{5x}{2x}+\dfrac{2}{2x}$

$\dfrac{3\cancel{x^2}^{x}}{2\cancel{x}}-\dfrac{5\cancel{x}^{1}}{2\cancel{x}^{1}}+\dfrac{\cancel{2}^{1}}{\cancel{2}^{1}x}=\dfrac{3x}{2}-\dfrac{5}{2}+\dfrac{1}{x}$

Validation

$2x\left(\dfrac{3x}{2}-\dfrac{5}{2}+\dfrac{1}{x}\right)$

$2x\bullet\dfrac{3x}{2}-2x\bullet\dfrac{5}{2}+2x\bullet\dfrac{1}{x}$

$\cancel{2}^{1}x\bullet\dfrac{3x}{\cancel{2}^{1}}-\cancel{2}^{1}x\bullet\dfrac{5}{\cancel{2}^{1}}+2\cancel{x}^{1}\bullet\dfrac{1}{\cancel{x}^{1}}$

$3x^2-5x+2$ ✓

ERROR: **Cancelling terms**

Incorrect Process

Simplify: $\dfrac{2a^3+3a^2-6a-9}{2a+3}$

$$\frac{a^2(2a+3)-6a-9}{2a+3}$$

$$\frac{a^2\cancel{(2a+3)}-6a-9}{\cancel{2a+3}}$$

$$a^2-6a-9$$

Resolution

Do not cancel binomials that are not factors of the numerator. Use the Dividing Polynomials Methodology.

Correct Process

Simplify: $\dfrac{2a^3+3a^2-6a-9}{2a+3}$

$$\begin{array}{r} a^2+0a-3 \\ 2a+3\,\overline{)\,2a^3+3a^2-6a-9} \\ \underline{2a^3+3a^2} \\ 0a^2-6a \\ \underline{0a^2-0a} \\ -6a-9 \\ \underline{6a+9} \\ 0 \end{array}$$

Validation

$$(2a+3)(a^2-3)$$

$$2a^3-6a+3a^2-9$$

$$2a^3+3a^2-6a-9 \checkmark$$

ERROR: Failing to make space for missing terms

Incorrect Process

Divide: $\dfrac{3x^3+4x^2+4}{x-2}$

$$
\begin{array}{r}
3x^2+10x+24+\dfrac{48}{x-2} \\
x-2\,\overline{)\,3x^3+4x^2+4} \\
\underline{3x^3-6x^2} \\
+10x^2+4 \\
\underline{+10x^2-20x} \\
+24x \\
\underline{+24x-48} \\
+48
\end{array}
$$

Correct Process

Divide: $\dfrac{3x^3+4x^2+4}{x-2}$

$$
\begin{array}{r}
3x^2+10x+20+\dfrac{44}{x-2} \\
x-2\,\overline{)\,3x^3+4x^2+0x+4} \\
\underline{3x^3-6x^2} \\
+10x^2+0x \\
\underline{+10x^2-20x} \\
+20x+4 \\
\underline{+20x-40} \\
+44
\end{array}
$$

Resolution

Place zero terms to represent missing terms.

Validation

$(x-2)\left(3x^2+10x+20+\dfrac{44}{x-2}\right)$

$3x^3+10x^2\cancel{+20x}-6x^2\cancel{-20x}-40+(x-2)\bullet\dfrac{44}{x-2}$

$3x^3+4x^2+0x-40+44$

$3x^3+4x^2+4$ ✓

Preparation Inventory 4.1

I can...	OR *Here's my question...*
☐ understand the division algorithm for polynomials.	☐
☐ divide and validate division of a polynomial by a monomial.	☐
☐ use long division to divide and validate division of a polynomial by a polynomial inequality.	☐

Critical Thinking Questions 4.1

1. What are the names of the four parts of a division problem?

2. How do you know when you are done dividing?

3. What are some key issues with dividing polynomials?

4. In what two ways can you present the remainder of a polynomial division problem?

5. What is the degree of the quotient in relationship to the degrees of the divisor and dividend?

6. How are the coefficients of the quotient determined?

7. How do you validate that your division is correct?

Demonstrate Your Understanding 4.1

1. Divide $9y^5 - 3y^2 - 6y + 12$ by $3y$.

2. Divide $x^3 - 4x^2 + 2x - 5$ by $x - 4$.

3. Divide $x^3 - 5x^2 - 29x + 108$ by $x - 7$.

4. Divide $12x^3 + 23x^2 - 164x + 98$ by $3x - 7$.

Create and Solve the Hardest Problem 4.1

Your Hardest Problem

Problem Solution

What makes a problem where you must divide polynomials a hard problem to solve?

Identify, Describe, & Correct the Errors 4.1

Describe the error you found in the worked solution in the appropriate space. Solve the problem correctly under "Correct Process." You are welcome to solve the problem first and then describe the error.

Problem 1

Divide $2x^2 - 4x - 5$ by $x - 4$

Worked Solution *(Find the error!)*

$$\begin{array}{r} 2x-12 \quad R43 \\ x-4\,\overline{)\,2x^2-4x-5} \\ \underline{2x^2-8x} \\ -12x-5 \\ \underline{-12x+48} \\ +43 \end{array}$$

Correct Process

Describe the Error

Problem 2

Divide: $\dfrac{x^3-5x+2}{x-2}$

Worked Solution *(Find the error!)*

$$\begin{array}{r} x^2-3+\dfrac{8}{x-2} \\ x-2\,\overline{)\,x^3-5x+2} \\ \underline{-x^3+2x^2} \\ -3x+2 \\ \underline{3x+6} \\ +8 \end{array}$$

Correct Process

Describe the Error

Problem 3

Divide: $\dfrac{x^4+2x^3-7x^2-5x+2}{x^3+2x^2-6x-2}$

Describe the Error

Worked Solution *(Find the error!)*

$$
\begin{array}{r}
x^{-1} \qquad\qquad Rx+3-2x^{-1} \\
x^4+2x^3-7x^2-5x+2 \,\big)\, \overline{x^3+2x^2-6x-2} \\
\underline{x^3+2x^2-7x-5+2x^{-1}} \\
x+3-2x^{-1}
\end{array}
$$

Correct Process

Problem 4

Divide: $2x^2 - 4x - 10$ by x

Worked Solution *(Find the error!)*

$$\frac{2x^2-4x-10}{x}$$

$$\frac{2\cancel{x^2}^{\,x}-4x-10}{\cancel{x}^{\,1}}$$

$$2x-4x-10$$

$$-2x-10$$

Correct Process

Describe the Error

What Did You Learn About Learning Math? 4.1

1. How did you use prior knowledge when learning about polynomial division?

2. What difficulties did you encounter in transferring your knowledge about the long division of numbers to the long division of polynomials?

3. How could you generalize what you have learned about long division to divide a 5th degree polynomial by a 3rd degree polynomial with a missing 2nd degree term?

Section 4.2 Factoring Trinomials

Why? 4.2

Factoring polynomials can assist you in solving real-world problems found in fields such as construction, architecture, and legal and financial planning. Being able to factor polynomials will also help you to simplify rational expressions and graph equations.

What Do You Already Know? 4.2

What are the differences between a monomial, binomial, trinomial, and polynomial?

What is a factor?

What does GCF mean?

How do you multiply two binomials?

Goals 4.2

When you complete this activity, you should be able to:

1. Factor out the GCF of the terms in any polynomial
2. Factor some trinomials

Building Mathematical Language 4.2

New Terms: completely factored factor (verb)

Key Concepts 4.2

The ***product of two binomials*** can be seen as a trinomial with coefficients related to the coefficients of each of the binomials.

To ***factor (verb)*** a sum of terms means to rewrite the sum as an equivalent product of factors (noun).

A polynomial is ***completely factored*** when none of the factors can be factored further. Conventionally, when we say "factor" we mean "factor completely."

Table of Possibly Factorable Polynomials

Type	Factorable in the Real Number System	Example	Factors
Sum of Two Squares *(2 terms)*	Never	$a^2 + b^2$	Prime
Difference of Two Squares *(2 terms)*	Always	$a^2 - b^2$	$(a + b)(a - b)$
Sum of Two Cubes *(2 terms)*	Always	$a^3 + b^3$	$(a + b)(a^2 - ab + b^2)$
Difference of Two Cubes *(2 terms)*	Always	$a^3 - b^3$	$(a - b)(a^2 + ab + b^2)$
Quadratic Trinomial *(3 terms)*	Sometimes	$a^2 - 5a - 6$	$(a - 6)(a + 1)$
Quadrinomial *(4 terms)*	Sometimes	$a^2 + 3a - 2a - 6$	$a^2 + 3a - 2a - 6$ $= (a^2 + 3a) - (2a + 6)$ $= a(a + 3) - 2(a + 3)$ $= (a + 3)(a - 2)$

Note: The sum $a^2 + b^2$ is factorable in the Complex Number System. Only quadratic trinomials will be discussed in this section. The remaining polynomials will be covered in Section 4.3.

Methodology 1 — 4.2

Factoring Simple Trinomials: $ax^2 + bx + c$

This methodology works well for trinomials $ax^2 + bx + c$ where the coefficients a and c are prime or have few factors.

	Example 1	Example 2	Your Turn
Factor:	$x^2 + 7x + 10$	$15x^3 + 39x^2 - 18x$	$2x^2 + 5x - 3$

Steps	Discussion
1 Look for common factors	Factor out any common monomial factors.

Ex 1	There are no common factors.	Ex 2	$3x\,(5x^2 + 13x - 6)$	Your Turn	

Steps	Discussion
2 List possibilities	Identify a and c in the quadratic factor and list the factors of each.

Ex 1	$a = 1$: 1 (only factor) $c = 10$: 1 & 10 or 2 & 5	Ex 2	$a = 5$: 5 & 1 $c = -6$: 6 & 1 or 3 & 2 (one +, one –)

Steps	Discussion
Your Turn	
3 Set up binomials	Put two factors of a as the coefficients of x.
Ex 1: $(x\ )(x\ )$ · Ex 2: $(5x\ )(x\ )$ · Your Turn: $(x\ )(x\ )$	
4 Complete the binomials	Put two factors of c as the constant terms.
Ex 1: $(x+2)(x+5)$ · Ex 2: $(5x+3)(x-2)$ · Your Turn:	
5 Validate	Multiply the binomials to confirm that the product is the quadratic factor.
Ex 1: $x^2+2x+5x+10$ $x^2+7x+10$ ✓	Ex 2: $5x^2-10x+3x-6$ $5x^2-7x-6$ Not right
Your Turn	
6 Repeat steps 4 and 5 until validation occurs	Reverse signs or factors or try another pair of factors for a and/or c.
Ex 1: Already validated.	Ex 2: $(5x-2)(x+3)$ $5x^2+15x-2x-6$ $5x^2+13x-6$ ✓
Your Turn	

Steps	Discussion
7 Present the answer	Include factors from steps 1 and 6.

Ex 1	$(x + 2)(x + 5)$	Ex 2	$3x(5x - 2)(x + 3)$
Your Turn			

Methodology 2 — 4.2

Factoring Trinomials $ax^2 + bx + c$

You can create a table to factor a trinomial.

Limitation/Caution: This method fails if you do not first factor out the GCF. Most trinomials cannot be factored.

	Example 1	**Example 2**	**Your Turn**
Factor:	$6x^2 - x - 40$	$-40x^2 + 134x - 112$	$12x^2 + 17x - 40$

Steps	Discussion
1 Look for common factors	Factor out any common monomial factors; factor out –1 if a < 0.

Ex 1	There are no common factors.	Ex 2	$-2(20x^2 - 67x + 56)$	Your Turn	

Steps	Discussion
2 Set up the table to factor the trinomial factor	Start by putting ax^2 and the constant of the trinomial factor in the corners of the box.

Ex 1

$6x^2$	
	-40

Ex 2

$20x^2$	
	56

Your Turn

Steps	Discussion
3 Find the two missing terms	**Find two numbers whose product = ac and whose sum = b. (This is the key step that makes everything work.)**

Ex 1

Find two numbers m, n so that

mn = –240 and

m + n = –1

m and n are –16 and 15.

(See the calculator tip which follows this methodology.)

Ex 2

Note that ac = 1120 and b = –67.

Find m,n so that

mn = 1120 and

m + n = –67

m and n are –32 and –35.

Your Turn

Steps	Discussion
4 Complete the table.	**Enter mx and nx in the remaining places in the box**

Ex 1

$6x^2$	$-16x$
$15x$	-40

Ex 2

$20x^2$	$-32x$
$-35x$	56

Your Turn

Steps	Discussion
5 Find GCFs	**Find the GCF of each column and each row. In the second column and second row only, *if the x term is negative*, use a negative GCF.**

Ex 1

$3x$	-8	
$6x^2$	$-16x$	$2x$
$15x$	-40	5

Ex 2

$5x$	-8	
$20x^2$	$-32x$	$4x$
$-35x$	56	-7

Your Turn

Steps	Discussion
6 Read off the factors	One factor is above the box and the other is on the right.

Ex 1	$(3x - 8)(2x + 5)$	Ex 2	$(5x - 8)(4x - 7)$
Your Turn			

7 Present the answer	Include factors from steps 1 and 6.

Ex 1	$(3x - 8)(2x + 5)$	Ex 2	$-2(5x - 8)(4x - 7)$
Your Turn			

8 Validate	Multiply the factors.

Ex 1	$6x^2 + 15x - 16x - 40$ $6x^2 - x - 40$ ✓	Ex 2	$-2\ (20x^2 - 35x - 32x + 56)$ $-40x^2 + 134x - 112$ ✓
Your Turn			

Tool 4.2

If ac is a large number, it may be difficult, or at least time consuming, to find values of m and n such that mn = ac and m + n = b. A calculator with a TABLE feature can be used to make this an easy task.

Because we want two numbers whose product is ac, using the calculator variables x and y (instead of m and n), we want $xy = \text{ac}$. So $y = \dfrac{\text{ac}}{x}$.

Enter this equation in the calculator (including the negative sign if ac is negative). Then scroll up or down the table to look for the two numbers whose sum is b.

Justification of methodology:

Suppose that $ax^2 + bx + c$ factors into $(px + q)(rx + s)$, where GCF (p, q) = 1 and GCF (r, s) = 1 (i.e., all common monomial factors have been removed). If the relationships between a, b, and c, and p, q, r, and s, can be identified, these relationships can then be used to find p, q, r, and s, given a, b, and c.

$ax^2 + bx + c = (px + q)(rx + s)$

$ax^2 + bx + c = prx^2 + psx + qrx + qs.$

$ax^2 + bx + c = prx^2 + (ps + qr)x + qs.$

Setting corresponding coefficients equal, $a = pr$, $b = ps + qr$, $c = qs$.

Let $m = ps$ and $n = qr$. Then $m + n = b$ and $mn = ac$ [both mn and ac are equivalent to pqrs].

If numbers m and n can be found that satisfy these two conditions, then p, q, r, and s can be found in the following manner.

ax^2	nx
mx	c

prx^2	qrx
psx	qs

Note that px and q are the GCFs of the columns, while rx and s are the GCFs of the rows.

px	q	
prx^2	qrx	rx
psx	qs	s

The factors can easily be read off and written $(px + q)(rx + s)$

Addressing Common Errors 4.2

The following table shows some of the most common errors that learners tend to make with the kinds of problems covered in this section.

ERROR: Forgetting to factor out the GCF before factoring anything else

Incorrect Process

Factor $2x^2 + 10x + 12$

$(2x + 4)(x + 3)$

Resolution

Factor out all common factors first.

Correct Process

Factor $2x^2 + 10x + 12$

$= 2(x^2 + 5x + 6)$

$= 2(x + 2)(x + 3)$

Validation

$2(x + 2)(x + 3)$

$= 2(x^2 + 5x + 6)$

$= 2x^2 + 10x + 12$ ✓

ERROR: Finding m and n incorrectly (only satisfying one of the two conditions)

Incorrect Process

Factor $10x^2 + 13x - 30$

ac = –300

Choose m and n = 30, –10

$10x$	-30	
$30x^2$	$-90x$	$3x$
$-100x$	300	-10

Factors are $(10x - 30)(3x - 10)$

Resolution

Select m, n so that BOTH conditions are met: $mn = ac$ AND $m + n = b$.

Correct Process

Factor $10x^2 + 13x - 30$

ac = –300

Choose m and n = 25, –12

$5x$	-6	
$10x^2$	$-12x$	$2x$
$25x$	-30	5

Factors are $(5x - 6)(2x + 5)$

Validation

$(5x - 6)(2x + 5)$

$= 10x^2 + 25x - 12x - 30$

$= 10x^2 + 13x - 30$ ✓

ERROR: Omitting validation (and hence not recognizing incorrect answers)

Incorrect Process

Factor $x^2 + 5x - 6$

$(x - 3)(x - 2)$

Resolution

Multiply factors to validate. (The product must match the original trinomial.)

Correct Process

Factor $x^2 + 5x - 6$

$(x + 6)(x - 1)$

Validation

$(x + 6)(x - 1) = x^2 + 5x - 6$ ✓

Preparation Inventory 4.2

I can...	OR *Here's my question...*
☐ factor out the GCF from a polynomial.	☐
☐ factor a simple trinomial.	☐
☐ factor any (factorable) trinomial.	☐
☐ use a calculator table to identify m and n.	☐

Critical Thinking Questions 4.2

1. What is the standard form for trinomials?

2. What are m and n?

3. Why do we choose m and n such that, when multiplied, their product is equal to a • c?

4. Why do we choose m and n such that m + n = b (the middle term of the quadratic expression)?

5. How do you ensure that the signs in the factored form are correct?

6. How do you validate that the factors are correct?

7. When writing mx and nx in their two places in the table, does it matter which term is written in which place? Give an example to justify your answer.

Demonstrate Your Understanding 4.2

Factor the following.

1. $2x^2 - 4x - 6$
2. $3xy^2 - 15xy + 18x$

3. $-5a^2 + 10a - 13$

4. $-3x^2 + 6x - 3$

5. $24x^2 + 37x - 72$

6. $-18x^3 + 21x^2 + 60x$

Create and Solve the Hardest Problem 4.2

Your Hardest Problem

Problem Solution

What makes a problem where you must factor trinomials a hard problem to solve?

Identify, Describe, & Correct the Errors 4.2

Describe the error you found in the worked solution in the appropriate space. Solve the problem correctly under "Correct Process." You are welcome to solve the problem first and then describe the error afterward.

Problem 1

Factor: $x^2 - 5x - 6$

Worked Solution *(Find the error!)*

$x^2 - 5x - 6$

$(x - 3)(x - 2)$

Correct Process

Describe the Error

Problem 2

Factor: $-3b^2 - 6b + 24$

Worked Solution *(Find the error!)*

$$-3b^2 - 6b + 24$$
$$-3(b^2 + 2b - 8)$$
$$(b-2)(b+4)$$

Correct Process

Describe the Error

Problem 3

Factor: $ax^2 - 3ax + 10a$

Worked Solution *(Find the error!)*

$$ax^2 - 3ax + 10a$$
$$a(x^2 - 3x + 10)$$
$$a(x-5)(x+2)$$

Correct Process

Describe the Error

Problem 4

Factor: $4x^2 + 7x - 36$

Worked Solution *(Find the error!)*

$ac = -144$, $b = 7$
m and n are 16 and -9

x	4	
$4x^2$	$16x$	$4x$
$-9x$	-36	9

$$(x+4)(4x+9)$$

Correct Process

Describe the Error

Problem 5

Factor: $2x^2 - 10x + 12$

Worked Solution *(Find the error!)*

ac = 24, b = –10
m and n are –6 and –4.

$2x$	-6	
$2x^2$	$-6x$	$2x$
$-4x$	12	-4

$(2x - 6)(2x - 4)$

Correct Process

Describe the Error

What Did You Learn About Learning Math? 4.2

1. How did you use what you know about multiplying binomials to help you factor simple polynomials?

2. Why is recognizing patterns important in order to be successful at factoring polynomials?

3. Why is validation important in this activity?

Section 4.3 Factoring Other Polynomials

Why? 4.3

Factoring polynomials can assist you in solving real-world problems involving trajectories of falling objects, in simplifying rational expressions, and in graphing equations. For example, to determine how long it will take a cliff diver to enter the water when diving from a 64-foot cliff requires that the equation $-16t^2 + 64 = 0$ be solved by factoring the expression, $-16t^2 + 64$.

What Do You Already Know? 4.3

How do you factor out a common monomial from a polynomial?

How do you recognize that a number is a perfect square or a perfect cube?

Goals 4.3

When you complete this activity, you should be able to:

1. Factor special kinds of binomials
2. Factor quadrinomials of certain forms
3. Factor special polynomials of higher degrees

Building Mathematical Language 4.3

New Terms: factor by grouping quadratic form quadrinomial

Key Concepts 4.3

A ***quadrinomial*** is a four-term polynomial. Some quadrinomials can be ***factored by grouping***, which means that the first two terms are grouped together, the last two terms are grouped together, the GCF of each group is factored out, and then a common binomial factor is factored out of the two groups.

A ***polynomial in quadratic form*** is a polynomial of degree higher than two, for which a substitution can turn it into a quadratic (degree two) polynomial. For example, substituting $y = x^2$ into the polynomial $x^4 - 9$ turns it into $y^2 - 9$, which can be recognized as the difference of squares, and then factored.

In this section, you will learn to factor special binomials and some specially formed trinomials. The following table is similar to the table in Section 4.2.

Patterns to Factor Special Binomials

Type	Factorable in the Real Number System	Example	Factors
Sum of Two Squares *(2 terms)*	Never	$a^2 + b^2$	Prime
Difference of Two Squares *(2 terms)*	Always	$a^2 - b^2$	$(a + b)(a - b)$
Sum of Two Cubes *(2 terms)*	Always	$a^3 + b^3$	$(a + b)(a^2 - ab + b^2)$
Difference of Two Cubes *(2 terms)*	Always	$a^3 - b^3$	$(a - b)(a^2 + ab + b^2)$

Limitation/Cautions: Only selected polynomials will factor with integer, rational, or even real coefficients.

Methodology 1 — 4.3

Factoring Special Binomials

	Example 1	Example 2	Your Turn
Factor:	$36x^2 - 49$	$-16x^3 - 54$	$64x^3 - 125$

Steps	Discussion
1 Look for common factors	Factor out any common monomial factors; factor out –1 if a < 0.

Ex 1	There are no common factors.	Ex 2	$-2(8x^3 + 27)$	Your Turn	

2 Identify type of binomial	Determine if both terms of the binomial are perfect squares or cubes.

Ex 1	Both terms are perfect squares.	Ex 2	Both terms are perfect cubes.	Your Turn	

3 Write the factors of the binomial	Follow the pattern in the Table for Factoring Special Binomials.

Ex 1	$(6x + 7)(6x - 7)$	Ex 2	$(2x + 3)(4x^2 - 6x + 9)$
Your Turn			

Steps	Discussion
4 Present the answer	Include factors from Steps 1 and 3.

Ex 1	$(6x + 7)(6x - 7)$	Ex 2	$-2(2x + 3)(4x^2 - 6x + 9)$
Your Turn			

Steps	Discussion
5 Validate	Multiply the factors.

Ex 1	$(6x + 7)(6x - 7)$ $= 36x^2 - 49$ ✓	Ex 2	$-2(2x + 3)(4x^2 - 6x + 9)$ $= -2(8x^3 + 27)$ $= -16x^3 - 54$ ✓
Your Turn			

Tool 4.3

Using a Calculator to Identify Perfect Squares and Cubes

If only a single number is under consideration, taking the square root or cube root of the number on a calculator will determine that the number is a perfect square or cube if the result is an integer.

Entering the equation $y = x^2$ or $y = x^3$ and then creating a TABLE on the calculator will produce a list of perfect squares or cubes.

Methodology 2 — 4.3

Factoring Quadrinomials

Note: This method is often referred to as "**factoring by grouping**."

Limitation/Caution: Only very special quadrinomials of certain forms will factor by grouping.

	Example 1	**Example 2**	**Your Turn**
Factor:	$3y^2 + 2ay - 9ay - 6a^2$	$-40x^2 - 70x - 64x + 112$	$12x^2 + 8xy - 15xy - 10y^2$

Steps	Discussion
1 Look for common factors	**Factor out any common monomial factors; factor out –1 if a < 0.**
Ex 1	There are no common factors.
Ex 2	$-2(20x^2 - 35x - 32x + 56)$
Your Turn	
2 Try to factor each binomial	**Consider the first two terms of the quadrinomial as a binomial and factor out the GCF. Repeat for the last two terms.**
Ex 1	$y(3y + 2a) - 3a(3y + 2a)$
Ex 2	$5x(4x-7) + 8(-4x + 7)$
Your Turn	
3 Check signs	**The two binomials should match. If not, change the sign of one GCF.**
Ex 1	Binomials match $(3y + 2a)$
Ex 2	$5x(4x - 7) - 8(4x - 7)$
Your Turn	
4 Factor	**Factor out the common (binomial) factor.**
Ex 1	$(3y + 2a)(y - 3a)$
Ex 2	$(4x - 7)(5x - 8)$
Your Turn	

Steps	Discussion
5 Present the answer	Include factors from Steps 1 and 4.

Ex 1	$(3y + 2a)(y - 3a)$	Ex 2	$-2(4x - 7)(5x - 8)$
Your Turn			

6 Validate	Multiply the factors.

Ex 1	$(3y + 2a)(y - 3a)$ $= 3y^2 + 2ay - 9ay - 6a^2$ ✓	Ex 2	$-2(4x - 7)(5x - 8)$ $= -2(20x^2 - 35x - 32x + 56)$ $= -40x^2 - 70x - 64x + 112$ ✓
Your Turn			

Methodology 3 — 4.3

Factoring Polynomials of Higher Degrees

Sometimes polynomials of higher degree can be factored if they are in binomial or trinomial form. Following the patterns and methodologies for binomials and trinomials can result in factors for these polynomials.

Limitation/Caution: Very few higher degree polynomials can be factored with these methods.

	Example 1	Example 2	Your Turn
Factor:	$24a^4 x - 14a^2x - 20x$	$x^6 - y^6$	$81x^4 - 256$ [Hint: $(9x^2)^2 - 16^2$]

Steps	Discussion
1 Look for common factors	**Factor out any common monomial factors; factor out –1 if a < 0.**

Ex 1: $2x(12a^4 - 7a^2 - 10)$

Ex 2: There are no common factors.

Your Turn:

Steps	Discussion
2 Identify type of polynomial	**Can a substitution make a trinomial in quadratic form or a binomial that factors?**

Ex 1: Trinomial in quadratic form

$12a^4 - 7a^2 - 10 =$

$12(a^2)^2 - 7(a^2) - 10$

Ex 2: Difference of cubes

$x^6 = (x^2)^3$

$y^6 = (y^2)^3$

Your Turn:

Steps	Discussion
3 Choose a substitution	**Choose a substitution that will better show the form of the polynomial.**

Ex 1: Let $p = a^2$

Ex 2: Let $p = x^2$, $q = y^2$

Your Turn:

Steps	Discussion
4 Make the substitution	**Substitute the choice you made in step 3 into the polynomial.**

Ex 1: $12p^2 - 7p - 10$

Ex 2: $p^3 - q^3$

Steps	Discussion
Your Turn	
5 Factor the polynomial	**Use methods learned previously.**
Ex 1: $(3p+2)(4p-5)$	Ex 2: $(p-q)(p^2+pq+q^2)$
Your Turn	
6 Reverse the substitution	**Substitute back into the original variables.**
Ex 1: $(3a^2+2)(4a^2-5)$	Ex 2: $(x^2-y^2)(x^4+x^2y^2+y^4)$
Your Turn	
7 Check for further factors	**Inspect each factor to see if it factors further.**
Ex 1: $(3a^2+2)(4a^2-5)$ Done; no further factors.	Ex 2: x^2-y^2 is the difference of squares and can be factored further.
Your Turn	
8 Present the answer	**Include factors from steps 1 and 7.**
Ex 1: $2x(3a^2+2)(4a^2-5)$	Ex 2: $(x+y)(x-y)(x^4+x^2y^2+y^4)$
Your Turn	

Steps	Discussion
9 Validate	Multiply the factors.

Ex 1

$$2x(12a^4 - 7a^2 - 10)$$
$$= 24a^4x - 14a^2x - 20 \checkmark$$

Ex 2

$$(x^2 - y^2)(x^4 + x^2y^2 + y^4)$$
$$= x^6 + x^4y^2 - x^4y^2 + x^2y^4 - x^2y^4 - y^6$$
$$= x^6 - y^6 \checkmark$$

Your Turn

Addressing Common Errors 4.3

The following table shows some of the most common errors that learners tend to make with the kinds of problems covered in this section.

ERROR: Forgetting to factor out the GCF to start

Incorrect Process

Factor $8x^3 - 64$

$(2x - 4)(4x^2 + 8x + 16)$

Resolution

Always factor out the GCF before starting any other factoring process. Factor always means to "factor completely."

Correct Process

Factor $8x^3 - 64$

$8(x^3 - 8)$

$8(x - 2)(x^2 + 2x + 4)$

Validation

$$8(x - 2)(x^2 + 2x + 4)$$
$$= 8(x^3 + 2x^2 + 4x - 2x^2 - 4x - 8)$$
$$= 8(x^3 - 8) = 8x^3 - 64 \checkmark$$

ERROR: Factoring the sum of squares

Incorrect Process

Factor $4x^2 + 9$

$(2x + 3)(2x + 3)$

Resolution

Note that the sum of squares is prime in the real number system. It will not factor with real coefficients.

Correct Process

Factor $4x^2 + 9$

$4x^2 + 9$ is prime in the real number system

Validation

This binomial is the sum of two squares as shown in the table at the beginning of the section.

ERROR: **Omitting validation (and hence not recognizing incorrect solutions)**

Incorrect Process	Resolution
Factor $a^4 + 5a^2 - 6$ $(a^2 + 3)(a^2 + 2)$	*Validate your solution by multiplying the factors together. If the product is not the original expression, your factors are incorrect.*

Correct Process	Validation
Factor $a^4 + 5a^2 - 6$ $(a^2 + 6)(a^2 - 1)$ $(a^2 + 6)(a + 1)(a - 1)$	$(a^2 + 6)(a + 1)(a - 1)$ $= (a^2 + 6)(a^2 - 1)$ $= a^4 - a^2 + 6a^2 - 6$ $= a^4 + 5a^2 - 6$ ✓

Preparation Inventory 4.3

I can...	OR *Here's my question...*
☐ factor out the GCF from a polynomial.	☐
☐ identify perfect square and cube numbers.	☐
☐ follow the patterns to factor special binomials.	☐
☐ factor a quadrinomial by grouping.	☐
☐ factor selected polynomials of higher degree.	☐

Critical Thinking Questions 4.3

1. What should you always do first when factoring any polynomial?

2. Why doesn't the sum of squares factor in the real numbers system?

3. What is the difference between the patterns for the sum of two cubes and the difference of two cubes?

4. When you write your solution and multiply the factors together, how do you know if you have "validated" your solution?

5. How would you explain to a classmate how to "match" a given binomial with one of the types described in the table of "**Patterns to Factor Special Binomials**"?

6. What method do you know for factoring quadrinomials?

7. If you consider $x^6 - y^6$ as the difference of squares, rather than the difference of cubes as in Example 2 of the Methodology for Factoring Polynomials of Higher Degrees, will the resulting factors be the same after the binomial is completely factored? If not, are they equivalent?

Demonstrate Your Understanding 4.3

Factor the following.

1. $3x^2 - 3ax + 5x - 5a$

2. $x^4 - 3x^2 - 4$

3. $3x^3 - 81$

4. $p^4 - q^4$

Create and Solve the Hardest Problem 4.3

Your Hardest Problem
Problem Solution

What makes a problem where you must factor other polynomials a hard problem to solve?

Identify, Describe, & Correct the Errors 4.3

Describe the error you found in the worked solution in the appropriate space. Solve the problem correctly under "Correct Process." You are welcome to solve the problem first and then describe the error afterward.

Problem 1

Factor: $5y^3 - 5z^3$

Worked Solution *(Find the error!)*

$$5y^3 - 5z^3$$
$$5(y^3 - z^3)$$
$$5(y - z)^3$$

Correct Process

Describe the Error

Problem 2

Factor: $y^4 - 4y^2 - 5$

Worked Solution *(Find the error!)*

$$y^4 - 4y^2 - 5$$
$$y^2\left(y^2 - 4\right) - 5$$
$$y^2(y+2)(y-2) - 5$$

Correct Process

Describe the Error

Problem 3

Factor $4x^2 + 81$

Worked Solution *(Find the error!)*

$$4x^2 + 81$$

$$(2x + 9)(2x + 9)$$

Correct Process

Describe the Error

Problem 4

Factor $4x^3 - 2x^2 - 6x + 3$

Worked Solution *(Find the error!)*

$$4x^3 - 2x^2 - 6x + 3$$

$$2x^2(2x - 1) + 3(-2x - 1)$$

$$(2x - 1)(2x^2 + 3)$$

Correct Process

Describe the Error

Problem 5

Factor $4x^2 - 36$

Worked Solution *(Find the error!)*

$$4x^2 - 36$$

$$(2x - 6)(2x + 6)$$

Correct Process

Describe the Error

What Did You Learn About Learning Math? 4.3

1. How do you recognize that a binomial is a difference of squares or cubes?

2. Why is recognizing patterns especially important in order to be successful at factoring polynomials of higher degree?

3. What aspect of validation is important when factoring polynomials?

Section 4.4 Simplifying Rational Expressions

Why? 4.4

Throughout the rest of this course and future courses, rational expressions will come into play and your ability to manipulate or rewrite such expressions into a more useful form will allow you to perform more complex problem solving. Four primary manipulations occur: **simplifying**, **combining** (adding/subtracting), **multiplying**, and **dividing**.

What Do You Already Know? 4.4

How would you create a fraction that is equivalent to 2/3 and has a numerator of 28?

How does equivalence for expressions differ from equivalence for equations?

What is meant by *Least Common Denominator (LCD)* and *Least Common Multiple (LCM)*? Give an example of each.

Goals 4.4

When you complete this activity, you should be able to:

1. Identify a rational expression
2. Determine common multiples of denominators of rational expressions
3. Simplify a rational expression

Building Mathematical Language 4.4

New Terms: complex fraction rational expression

Key Concepts 4.4

A ***rational expression*** is the ratio of two polynomials $p(x)$ and $q(x)$, $\frac{p(x)}{q(x)}$, with $q(x) \neq 0$.

Such expressions can be rewritten by dividing out common factors of the numerator and denominator or by using long division.

Division of one number or expression by another can be represented in several ways: $a \div b = a \cdot \frac{1}{b} = \frac{a}{b}$

A ***complex fraction*** is a fraction with rational expressions in the numerator and/or denominator.

Methodology 1 — 4.4

Simplifying Rational Expressions by Factoring

	Example 1	Example 2	Your Turn
Simplify:	$\dfrac{x^2-1}{x-1}$	$\dfrac{2a^2+7a-15}{a^2-25}$	$\dfrac{3x^2-9x-12}{x-4}$

Steps	Discussion
1 Factor the numerator and denominator	Use the methodologies for factoring polynomials.

Ex 1

$$\frac{x^2-1}{x-1}=\frac{(x+1)(x-1)}{x-1}$$

Numerator: $(x + 1)(x - 1)$

Denominator: $(x - 1)$

Ex 2

$$\frac{2a^2+7a-15}{a^2-25}$$

$$\frac{(2a-3)(a+5)}{(a+5)(a-5)}$$

Numerator: $(2a - 3)(a + 5)$

Denominator: $(a + 5)\ (a - 5)$

Your Turn

Steps	Discussion
2 Pull out the common factors between the numerator and denominator	Write the common factors of the numerator and denominator as a separate factor (= to 1) in the expression.

Ex 1

$$\frac{x^2-1}{x-1}=(x+1)\frac{(x-1)}{(x-1)}=x+1$$

Simplified expression: $x + 1$

Ex 2

$$\frac{2a^2+7a-15}{a^2-25}$$

$$\frac{(2a-3)}{(a-5)}\bullet\frac{(a+5)}{(a+5)}$$

$$\frac{(2a-3)}{(a-5)}$$

Simplified expression:

$$\frac{2a-3}{a-5}$$

Your Turn

Steps	Discussion
3 Determine the domain of the simplified expression	Inventory the factors pulled out to determine the domain for which the simplified expression is equivalent to the original expression.

Ex 1

$\frac{(x-1)}{(x-1)}=1$ as long as $x\neq 1$.

The expression $\frac{x^2-1}{x-1}$ is equivalent to $x+1$ for all values of x except $x=1$.

Ex 2

$\frac{(a+5)}{(a+5)}=1$ as long as $a\neq 5$.

The expression $\frac{2a^2+7a-15}{a^2-25}$ is equivalent to $\frac{2a-3}{a-5}$ for all values of a except $a=5$.

Your Turn

Steps	Discussion
4 Validate	Choose a convenient number to evaluate the original expression and the simplified expression.

Ex 1

Let $x=2$

$\frac{x^2-1}{x-1}$ and $x+1$

$\frac{(2)^2-1}{(2)-1}$ $(2)+1$

$\frac{4-1}{1}$ 3 ✓

3 ✓

Ex 2

Let $a=3$

$\frac{2a^2+7a-15}{a^2-25}$ and $\frac{2a-3}{a-5}$

$\frac{2(3)^2+7(3)-15}{(3)^2-25}$ $\frac{2(3)-3}{(3)-5}$

$\frac{18+21-15}{9-25}$ $\frac{6-3}{-2}$

$\frac{24}{-16}$ $-\frac{3}{2}$ ✓

$-\frac{3}{2}$ ✓

Your Turn

Methodology 2 — 4.4

Combining Rational Expressions

	Example 1	Example 2	Your Turn
Factor:	$\frac{3}{x}-\frac{x-1}{x+1}+\frac{4}{5x}$	$\frac{3}{x+1}-\frac{4x}{x^2-1}+\frac{5}{x-1}$	$\frac{5}{x^2-5x-6}+\frac{3x}{x-6}-\frac{2x}{x+1}$

Steps	Discussion
1 Inventory the denominators	Produce a list of non-redundant denominators.

Ex 1	Ex 2	Your Turn
$x, 5x, x+1$	$x+1, x^2-1, x-1$	

Steps	Discussion
2 Factor the denominators	Completely factor the denominators and always represent the denominators in factored form.

Ex 1	Ex 2	Your Turn
x done $5x$ done $x+1$ done	$x+1$ done $x-1$ done $x^2-1=(x-1)(x+1)$	

Steps	Discussion
3 Determine non-redundant factors	In order to calculate the LCM, you must determine which factors are non-redundant.

Ex 1	Ex 2
x is already taken care of by $5x$ Thus the only factors are $5x$ and $x+1$	x^2-1 is already taken care of by $x-1$ and $x+1$ Thus the only factors are $x-1$ and $x+1$

Your Turn

Steps	Discussion
4 Present the LCM	Leave the LCM in factored form.

Ex 1	Ex 2	Your Turn
$5x(x+1)$	$(x+1)(x-1)$	

Steps	Discussion
5 Multiply to create the common denominator	For each term, determine what fractional form of 1 will produce the desired common denominator.

Ex 1

$$\frac{3}{x}-\frac{x-1}{x+1}+\frac{4}{5x}$$

First term	Second term	Third term
$\frac{5}{5}\left(\frac{x+1}{x+1}\right)$	$\frac{5x}{5x}$	$\left(\frac{x+1}{x+1}\right)$

$$\frac{5}{5}\bullet\left(\frac{3}{x}\right)\bullet\frac{(x+1)}{(x+1)}-\frac{5x}{5x}\bullet\frac{(x-1)}{(x+1)}+\frac{4}{5x}\bullet\frac{(x+1)}{(x+1)}$$

Ex 2

$$\frac{3}{x+1}-\frac{4x}{x^2-1}+\frac{5}{x-1}$$

First term	Second term	Third term
$\frac{(x-1)}{(x-1)}$	unchanged	$\frac{(x+1)}{(x+1)}$

$$\frac{3}{x+1}\bullet\frac{(x-1)}{(x-1)}-\frac{4x}{x^2-1}+\frac{5}{x-1}\bullet\frac{(x+1)}{(x+1)}$$

Your Turn

Steps	Discussion
6 Add the fractions	Now that all your terms share a common denominator, you can add them together.

Ex 1

$$\frac{15(x+1)-5x(x-1)+4(x+1)}{5x(x+1)}$$

Ex 2

$$\frac{3(x-1)-4x+5(x+1)}{(x+1)(x-1)}$$

Your Turn

Steps	Discussion
7 Simplify the numerator	Simplify the numerator into standard form and factor if possible.

Ex 1

$$\frac{15x+15-5x^2+5x+4x+4}{5x(x+1)}$$

$$\frac{-5x^2+24x+19}{5x(x+1)}$$

Ex 2

$$\frac{3x-3-4x+5x+5}{(x+1)(x-1)}$$

$$\frac{4x+2}{(x+1)(x-1)}$$

$$\frac{2(2x+1)}{(x+1)(x-1)}$$

Your Turn

8 Simplify the result	Divide out factors that are common to the numerator and denominator.

Ex 1

No common factors.

Ex 2

No common factors.

Your Turn

Steps	Discussion
9 **Validate**	**Choose a convenient number to substitute into both the original expression and the simplified expression.**

Ex 1

Let $x = 2$ in the original expression:

$$\frac{3}{x} - \frac{x-1}{x+1} + \frac{4}{5x}$$

$$\frac{3}{2} - \frac{2-1}{2+1} + \frac{4}{5 \cdot 2}$$

$$\frac{3}{2} - \frac{1}{3} + \frac{4}{10}$$

$$= \frac{3}{2} \cdot \frac{15}{15} - \frac{1}{3} \cdot \frac{10}{10} + \frac{4}{10} \cdot \frac{3}{3}$$

$$= \frac{45-10+12}{30} = \frac{47}{30}$$

Let $x = 2$ in the simplified expression:

$$\frac{-5x^2 + 24x + 19}{5x(x+1)}$$

$$\frac{-5(2)^2 + 24 \cdot 2 + 19}{5(2)(2+1)}$$

$$= \frac{-20+48+19}{10(3)} = \frac{47}{30} \quad \checkmark$$

Ex 2

Let $x = 2$ in the original expression:

$$\frac{3}{x+1} - \frac{4x}{x^2-1} + \frac{5}{x-1}$$

$$\frac{3}{2+1} - \frac{4(2)}{2^2-1} + \frac{5}{2-1}$$

$$\frac{3}{3} - \frac{8}{3} + \frac{5}{1}$$

$$1 - \frac{8}{3} + 5 = 6 - \frac{8}{3} = \frac{18}{3} - \frac{8}{3}$$

$$= \frac{10}{3}$$

Let $x = 2$ in the simplified expression:

$$\frac{2(2x+1)}{(x+1)(x-1)}$$

$$\frac{2(2 \cdot 2 + 1)}{(2+1)(2-1)} = \frac{2(4+1)}{3(1)} = \frac{2(5)}{3}$$

$$= \frac{10}{3} \quad \checkmark$$

Your Turn

Model 4.4

Simplify: $\dfrac{2}{x+4}+\dfrac{x+2}{x-2}-\dfrac{x}{x^2+2x-8}$

Step 1 Inventory the denominators $(x+4),(x-2),(x^2+2x-8)$

Step 2 Factor denominators $(x+4),(x-2),(x+4)(x-2)$

Step 3 Determine non-redundant factors $(x+4),(x-2)$

Step 4 Present the LCM $(x+4)(x-2)$

Step 5 Multiply to create common denominator

$$\frac{2}{x+4}+\frac{x+2}{x-2}-\frac{x}{x^2+2x-8}$$

First term multiplier	Second term multiplier	Third term
$\dfrac{x-2}{x-2}$	$\dfrac{x+4}{x+4}$	unchanged

$$\frac{2}{x+4}\bullet\left(\frac{x-2}{x-2}\right)+\frac{x+2}{x-2}\bullet\left(\frac{x+4}{x+4}\right)-\frac{x}{x^2+2x-8}$$

Step 6 Add the fractions

$$\frac{2(x-2)+(x+2)(x+4)-x}{(x+4)(x-2)}$$

Step 7 Simplify the numerator

$$\frac{2x-4+x^2+6x+8-x}{(x+4)(x-2)}$$

$$\frac{x^2+7x+4}{(x+4)(x-2)}$$

Step 8 Simplify the result No common factors in the numerator and denominator.

Step 9 Validate

Let $x=-2$

Evaluate in original equation:

$$\frac{2}{x+4}+\frac{x+2}{x-2}-\frac{x}{x^2+2x-8}$$

$$\frac{2}{-2+4}+\frac{-2+2}{-2-2}-\frac{-2}{(-2)^2+2(-2)-8}$$

$$\frac{2}{2}+\frac{0}{-4}-\frac{-2}{4-4-8}$$

$$1+0-\frac{-2}{-8}=1-\frac{1}{4}=\frac{3}{4}$$

Let $x=-2$

Evaluate in simplified equation:

$$\frac{x^2+7x+4}{(x+4)(x-2)}$$

$$\frac{(-2)^2+7(-2)+4}{(-2+4)(-2-2)}$$

$$\frac{4-14+4}{2(-4)}=\frac{-6}{-8}=\frac{3}{4}\ \checkmark$$

Methodology 3 4.4

Multiplying Rational Expressions

	Example 1	Example 2	Your Turn
Multiply:	$\frac{3a-b}{2b^2}$ by $\frac{2ab^2-3b^3}{6a-2b}$	$\frac{y}{6x^2y+3x^3y^2}$ by $\frac{xy+2}{4y^2}$	$\frac{x+y}{2y}$ by $\frac{4y^2}{2x+2y}$

Steps	Discussion
1 Factor	Factor the numerators and the denominators.

Ex 1

$$\frac{3a-b}{2b^2} \bullet \frac{2ab^2-3b^3}{6a-2b}$$

$$\frac{3a-b}{2b^2} \bullet \frac{b^2(2a-3b)}{2(3a-b)}$$

Ex 2

$$\frac{y}{6x^2y+3x^3y^2} \bullet \frac{xy+2}{4y^2}$$

$$\frac{y}{3x^2y(2+xy)} \bullet \frac{xy+2}{4y^2}$$

Your Turn

Steps	Discussion
2 Multiply	Multiply the numerators and multiply the denominators, keeping everything factored.

Ex 1

$$\frac{(3a-b)b^2 \bullet (2a-3b)}{2b^2 \bullet 2(3a-b)}$$

Ex 2

$$\frac{y(xy+2)}{3x^2y(2+xy)(4y^2)}$$

Your Turn

Steps	Discussion
3 Simplify	**Simplify the result by dividing out common factors in the numerator and denominator using the Methodology for Simplifying Rational Expressions by Factoring.**

Ex 1

$$\frac{2a-3b}{4}$$

Ex 2

$$\frac{\cancel{y}\,\cancel{(xy+2)}}{3x^2\,\cancel{y}\,\cancel{(2+xy)}\,(4y^2)}$$

$$\frac{1}{12x^2y^2}$$

Your Turn

Steps	Discussion
5 Validate	**Substitute numerical values into the original expression and the simplified expression.**

Ex 1

Let $a=3$ and $b=2$

$$\frac{3a-b}{2b^2}\bullet\frac{2ab^2-3b^3}{6a-2b} \quad \text{and} \quad \frac{2a-3b}{4}$$

$$\frac{3(3)-(2)}{2(2)^2}\bullet\frac{2(3)(2)^2-3(2)^3}{6(3)-2(2)} \qquad \frac{2(3)-3(2)}{4}$$

$$\frac{(9-2)(24-24)}{8(18-4)}=\frac{7(0)}{8(14)}=0\ \checkmark \qquad \frac{6-6}{4}$$

$$0\ \checkmark$$

Ex 2

Let $x=2$ and $y=3$

$$\frac{y}{6x^2y+3x^3y^2}\bullet\frac{xy+2}{4y^2} \quad \text{and} \quad \frac{1}{12x^2y^2}$$

$$\frac{(3)}{6(2)^2(3)+3(2)^3(3)^2}\bullet\frac{(2)(3)+2}{4(3)^2} \qquad \frac{1}{12(2)^2(3)^2}$$

$$\frac{3}{3(24+72)}\bullet\frac{6+2}{36} \qquad \frac{1}{432}\ \checkmark$$

$$\frac{24}{36(3)(96)}=\frac{1}{3(3)(48)}=\frac{1}{432}\ \checkmark$$

Steps	Discussion

Your Turn

Methodology 4 — 4.4

Dividing Polynomials

	Example 1	Example 2	Your Turn
Simplify:	$\frac{2x-3}{x+1} \div x$	$\frac{x^2-3x-4}{x-2} \div \frac{x^2-2x-8}{x-1}$	$\frac{3t-s^2}{ts} \div \frac{6t-2s^2}{4s^2}$

Steps	Discussion
1 Rewrite	Rewrite the problem as a single fraction.

Ex 1

$$\frac{\frac{2x-3}{x+1}}{x}$$

Ex 2

$$\frac{\frac{x^2-3x-4}{x-2}}{\frac{x^2-2x-8}{x-1}}$$

Your Turn

Steps	Discussion
2 Factor	**Factor the numerators and the denominators.**

Ex 1

Not needed.

Ex 2

$$\frac{\frac{x^2-3x-4}{x-2}}{\frac{x^2-2x-8}{x-1}} = \frac{\frac{(x+1)(x-4)}{x-2}}{\frac{(x-4)(x+2)}{x-1}}$$

Your Turn

Steps	Discussion
3 Multiply	**Multiply the numerator and the denominator by the reciprocal of the denominator to simplify the original denominator to 1.**

Ex 1

$$\frac{\frac{2x-3}{x+1} \bullet \frac{1}{x}}{x \bullet \frac{1}{x}}$$

Ex 2

$$\frac{\frac{(x+1)(x-4)}{x-2} \bullet \frac{x-1}{(x-4)(x+2)}}{\frac{(x-4)(x+2)}{x-1} \bullet \frac{x-1}{(x-4)(x+2)}}$$

$$\frac{(x+1)(x-4)\left(x-1\right)}{\left(x-2\right)(x-4)(x+2)}$$

Your Turn

Steps	Discussion
4 Simplify	**Simplify the resulting expression using the Methodology for Simplifying Rational Expressions by Factoring**

Ex 1

$$\frac{\frac{2x-3}{x+1} \bullet \frac{1}{x}}{1} = \frac{2x-3}{x(x+1)}$$

Ex 2

$$\frac{(x+1)(x-1)}{(x-2)(x+2)}$$

Your Turn

Steps	Discussion
5 Validate	**Substitute numerical values into the original expression and the simplified expression.**

Ex 1

Let $x = -2$

$$\frac{2x-3}{x+1} \div x$$

$$\frac{2(-2)-3}{(-2)+1} \div (-2)$$

$$\frac{-4-3}{-1} \div (-2)$$

$$\frac{-7}{-1} \bullet \frac{1}{-2} = \frac{-7}{2} \checkmark$$

and

$$\frac{2x-3}{x(x+1)}$$

$$\frac{2(-2)-3}{(-2)((-2)+1)}$$

$$\frac{-4-3}{-2(-1)} = \frac{-7}{2} \checkmark$$

Ex 2

Let $x = 3$

$$\frac{x^2-3x-4}{x-2} \div \frac{x^2-2x-8}{x-1}$$

$$\frac{3^2-(3 \bullet 3)-4}{3-2} \div \frac{3^2-(2 \bullet 3)-8}{3-1}$$

$$\frac{9-(9)-4}{1} \div \frac{9-(6)-8}{2}$$

$$\frac{-4}{1} \div \frac{-5}{2} = \frac{-4}{1} \bullet \frac{2}{-5} = \frac{-8}{-5} = \frac{8}{5} \checkmark$$

and

$$\frac{(x+1)(x-1)}{(x-2)(x+2)}$$

$$\frac{(3+1)(3-1)}{(3-2)(3+2)}$$

$$\frac{(4)(2)}{(1)(5)}$$

$$\frac{8}{5} \checkmark$$

Your Turn

Methodology 5a — 4.4

Simplifying Complex Fractions

Complex fractions are fractions with fractions in the numerator or denominator. Depending on the complexity of a fraction, there are a number of ways it can be simplified..

Limitation/Caution: Many errors can occur when interpreting the meaning of such fractions because it can be difficult to determine what belongs in the numerator and what belongs in the denominator because of multiple division lines (vincula).

	Example 1	Example 2	Your Turn
Factor:	$\dfrac{\left(\dfrac{x+1}{x}\right)}{\left(\dfrac{3x^2}{x-2}\right)}$	$\dfrac{x-3}{\left(\dfrac{2}{x-1}-\dfrac{x}{x+2}\right)}$	$\dfrac{\left(2+\dfrac{1}{(y+1)}\right)}{\left(\dfrac{1}{y}+\dfrac{2}{3}\right)}$

Steps	Discussion
1 Combine	Make numerator and denominator into single fractions.

Ex 1

Already single fractions.

Ex 2

$$\frac{x-3}{\left(\dfrac{2}{x-1}-\dfrac{x}{x+2}\right)}$$

$$\frac{x-3}{\left(\dfrac{2(x+2)-x(x-1)}{(x-1)(x+2)}\right)}$$

Your Turn

Steps	Discussion
2 Multiply	Multiply numerator and denominator by the fraction that will make the denominator equal to 1 (the reciprocal of the denominator).

Ex 1

$$\frac{\left(\dfrac{x+1}{x}\right)}{\left(\dfrac{3x^2}{x-2}\right)} \bullet \frac{\left(\dfrac{x-2}{3x^2}\right)}{\left(\dfrac{x-2}{3x^2}\right)}$$

$$\frac{\left(\dfrac{(x+1)(x-2)}{3x^3}\right)}{\left(\dfrac{3x^2(x-2)}{3x^2(x-2)}\right)}$$

Ex 2

$$\frac{x-3}{\left(\dfrac{2(x+2)-x(x-1)}{(x-1)(x+2)}\right)}$$

$$\frac{x-3}{\left(\dfrac{2(x+2)-x(x-1)}{(x-1)(x+2)}\right)} \bullet \frac{\left(\dfrac{(x-1)(x+2)}{2(x+2)-x(x-1)}\right)}{\left(\dfrac{(x-1)(x+2)}{2(x+2)-x(x-1)}\right)}$$

$$\frac{\dfrac{(x-3)(x-1)(x+2)}{2(x+2)-x(x-1)}}{\left(\dfrac{[2(x+2)-x(x-1)][(x-1)(x+2)]}{[2(x+2)-x(x-1)][(x-1)(x+2)]}\right)}$$

Steps	Discussion

Your Turn

3 Simplify	Simplify the resulting expression.

Ex 1

$$\frac{\left(\frac{(x+1)(x-2)}{3x^3}\right)}{1}$$

$$\frac{(x+1)(x-2)}{3x^3}$$

Ex 2

$$\frac{\frac{(x-3)(x-1)(x+2)}{2(x+2)-x(x-1)}}{1}$$

$$\frac{(x-3)(x-1)(x+2)}{2(x+2)-x(x-1)}$$

$$\frac{(x-3)(x-1)(x+2)}{2x+4-x^2+x}$$

$$\frac{(x-3)(x-1)(x+2)}{-x^2+3x+4}$$

$$\frac{(x-3)(x-1)(x+2)}{-(x-4)(x+1)}$$

$$-\frac{(x-3)(x-1)(x+2)}{(x-4)(x+1)}$$

Your Turn

Steps	Discussion
4 Validate	**Substitute numerical values of the variables into the original expression and the simplified expression.**

Ex 1

Let $x = 3$

$$\frac{\left(\frac{x+1}{x}\right)}{\left(\frac{3x^2}{x-2}\right)} = \frac{\left(\frac{3+1}{3}\right)}{\left(\frac{3(3)^2}{3-2}\right)} = \frac{\left(\frac{4}{3}\right)}{\left(\frac{27}{1}\right)} = \left(\frac{4}{3}\right) \bullet \left(\frac{1}{27}\right) = \frac{4}{81} \checkmark$$

and

$$\frac{x^2 - x - 2}{3x^3} = \frac{3^2 - 3 - 2}{3(3)^3} = \frac{9-5}{3(27)} = \frac{4}{81} \checkmark$$

Ex 2

Let $x = 2$

$$-\frac{(x-3)(x-1)(x+2)}{(x-4)(x+1)} = -\frac{(2-3)(2-1)(2+2)}{(2-4)(2+1)} = -\frac{(-1)(1)(4)}{(-2)(3)} = -\left(\frac{-4}{-6}\right) = -\frac{4}{6} = -\frac{2}{3} \checkmark$$

and

$$\frac{x-3}{\left(\frac{2}{x-1} - \frac{x}{x+2}\right)} = \frac{2-3}{\left(\frac{2}{2-1} - \frac{2}{2+2}\right)} = \frac{-1}{\left(\frac{2}{1} - \frac{2}{4}\right)} = \frac{-1}{\left(\frac{8}{4} - \frac{2}{4}\right)} = \frac{-1}{\left(\frac{6}{4}\right)} = -\frac{4}{6} = -\frac{2}{3} \checkmark$$

Your Turn

Methodology 5b — 4.4

Simplifying Complex Fractions (Alternate Methodology)

	Example 1	Example 2	Your Turn
Simplify:	$\dfrac{\left(\dfrac{x+1}{x}\right)}{\left(\dfrac{3x^2}{x-2}\right)}$	$\dfrac{x-3}{\left(\dfrac{2}{x-1}-\dfrac{x}{x+2}\right)}$	$\dfrac{\left(2+\dfrac{1}{(y+1)}\right)}{\left(\dfrac{1}{y}+\dfrac{2}{3}\right)}$

Steps	Discussion
1 Identify denominators	**Identify all the denominators.**

Ex 1	The denominators are x and $x-2$	**Ex 2**	The denominators are $(x-1)$ and $(x+2)$	**Your Turn**	

Steps	Discussion
2 Form denominator product	**Form the product of the denominators.**

Ex 1	$x(x-2)$	**Ex 2**	$(x-1)(x+2)$	**Your Turn**	

Steps	Discussion
3 Multiply	**Multiply the numerator and denominator of the main fraction by the product found in Step 2.**

Ex 1

$$\frac{\dfrac{x+1}{x}}{\dfrac{3x^2}{x-2}} \bullet \frac{x(x-2)}{x(x-2)}$$

$$\frac{\dfrac{x+1}{x}\bullet x(x-2)}{\dfrac{3x^2}{x-2}\bullet x(x-2)} = \frac{(x+1)(x-2)}{3x^2(x)}$$

Ex 2

$$\frac{(x-3)(x-1)(x+2)}{\left(\dfrac{2}{x-1}-\dfrac{x}{x+2}\right)(x-1)(x+2)}$$

$$\frac{(x-3)(x-1)(x+2)}{2(x+2)-x(x-1)}$$

Your Turn

Steps	Discussion
4 Simplify	Simplify the resulting expression.

Ex 1

$$\frac{x^2 - x - 2}{3x^3}$$

Ex 2

$$\frac{(x-3)(x-1)(x+2)}{2x+4-x^2+x}$$

$$\frac{(x-3)(x-1)(x+2)}{-x^2+3x+4}$$

$$\frac{(x-3)(x-1)(x+2)}{-(x^2-3x-4)}$$

$$-\frac{(x-3)(x-1)(x+2)}{(x-4)(x+1)}$$

Your Turn

Steps	Discussion
5 Validate	Substitute numerical values of the variables into the original expression and the simplified expression.

Ex 1

See the validation in the previous methodology.

Ex 2

See the validation in the previous methodology.

Your Turn

Addressing Common Errors 4.4

The following table shows some of the most common errors that learners tend to make with the kinds of problems covered in this section.

ERROR: Incorrect cancelling

Incorrect Process

$$\frac{x^2+5x+6}{x+3}$$

$$\frac{\cancel{x^2}^{x}+5x+\cancel{6}^{2}}{\cancel{x}+\cancel{3}}=6x+2$$

Resolution

Only divide out factors.

Correct Process

$$\frac{x^2+5x+6}{x+3}$$

$$\frac{\cancel{(x+3)}(x+2)}{\cancel{x+3}}$$

$$x+2$$

Validation

Let $x=4$

$$\frac{x^2+5x+6}{x+3}$$

$$\frac{4^2+5(4)+6}{4+3}=\frac{42}{7}=6\ \checkmark$$

and

$$x+2$$

$$4+2$$

$$6\ \checkmark$$

ERROR: Multiplying the denominators

Incorrect Process

$$\frac{7b}{3a}+\frac{2b}{3a}$$

$$\frac{9b}{(3a)^2}=\frac{\cancel{9}b}{\cancel{9}a^2}=\frac{b}{a^2}$$

Resolution

Add rational expressions with a common denominator by adding the numerators only.

Correct Process

$$\frac{7b}{3a}+\frac{2b}{3a}$$

$$\frac{\cancel{9}^{3}b}{\cancel{3}a}$$

$$\frac{3b}{a}$$

Validation

Let $a=4$ and $b=5$

$$\frac{7b}{3a}+\frac{2b}{3a}$$

$$\frac{7(5)}{3(4)}+\frac{2(5)}{3(4)}$$

$$\frac{35}{12}+\frac{10}{12}=\frac{45}{12}=\frac{15}{4}\ \checkmark$$

and

$$\frac{3b}{a}$$

$$\frac{3(5)}{(4)}$$

$$\frac{15}{4}\ \checkmark$$

ERROR: **Adding numerators and denominators**

Incorrect Process

$$\frac{x+2}{x+3}+\frac{1}{x^2-9}$$

$$\frac{x+3}{x^2+x-6}$$

$$\frac{\cancel{x+3}}{\cancel{(x+3)}(x-2)}$$

$$\frac{1}{x-2}$$

Resolution

Write each rational expression with a common denominator.

Correct Process

$$\frac{x+2}{x+3}+\frac{1}{x^2-9}$$

$$\frac{x+2}{x+3}\bullet\frac{x-3}{x-3}+\frac{1}{(x+3)(x-3)}$$

$$\frac{(x+2)(x-3)+1}{(x+3)(x-3)}$$

$$\frac{x^2-x-6+1}{x^2-9}$$

$$\frac{x^2-x-5}{x^2-9}$$

Validation

Let $x=4$

$$\frac{x+2}{x+3}+\frac{1}{x^2-9} \quad \textit{and} \quad \frac{x^2-x-5}{x^2-9}$$

$$\frac{4+2}{4+3}+\frac{1}{(4)^2-9} \qquad \frac{(4)^2-(4)-5}{(4)^2-9}$$

$$\frac{6}{7}+\frac{1}{16-9} \qquad \frac{16-4-5}{16-9}$$

$$\frac{6}{7}+\frac{1}{7} \qquad \frac{7}{7}$$

$$1\ \checkmark \qquad 1\ \checkmark$$

Preparation Inventory 4.4

I can...	OR *Here's my question...*
☐ add and subtract rational expressions.	☐
☐ multiply rational expressions.	☐
☐ divide rational expressions.	☐
☐ determine the LCD of two rational expressions.	☐

Critical Thinking Questions 4.4

1. What are three ways to manipulate rational expressions?

2. What is the purpose of each of these manipulations?

3. What prior knowledge do you use to both simplify and combine rational expressions?

4. How do you assure yourself that a rational expression has been fully simplified?

5. How can you ensure that your simplified form is free of errors?

6. How would you show another student that adding rational expressions is very similar to adding numerical fractions?

7. What is expected from you when you are presenting a simplified rational expression?

8. When dividing rational expressions, what is one thing that you do differently from simplifying and combining rational expressions?

Demonstrate Your Understanding 4.4

1. $\dfrac{2a-b}{a}-\dfrac{a-1}{3a}$

2. $2y + \dfrac{x + 4y}{x - 2}$

3. $\dfrac{2x^2 - 3x - 5}{x - 2} \bullet \dfrac{x^2 - 4x + 4}{x - 1}$

4. $\dfrac{3x-5}{x-y} \div \dfrac{x+y}{x^2+2xy-3y^2}$

5. $\dfrac{\dfrac{2x-3}{x}}{\dfrac{x+3}{x-5}}$

Create and Solve the Hardest Problem 4.4

Your Hardest Problem

Problem Solution

What makes a problem where you must simplify rational expressions a hard problem to solve?

Identify, Describe, & Correct the Errors 4.4

Solve the problem correctly in the appropriate location. Be sure to identify and describe the error you found in the worked solution!

Problem 1

Simplify: $\dfrac{3}{y^2-4}+\dfrac{4}{y-2}$

Worked Solution *(Find the error!)*

$$\frac{3}{(y+2)(y-2)}+\frac{4}{y-2}$$

$3+4(y+2)$

$3+4y+8$

$4y+11$

Correct Process

Describe the Error

Problem 2

Simplify: $\dfrac{5a-1}{2a-3b}\div\dfrac{2a-3b}{4}$

Worked Solution *(Find the error!)*

$$\frac{5a-1}{4}$$

Correct Process

Describe the Error

Problem 3

Simplify: $\frac{s+3}{s+1} + \frac{s^2 - 2s + 5}{2}$

Worked Solution *(Find the error!)*

$$\frac{s+3}{s+1} + \frac{s^2 - 2s + 5}{2}$$

$$\frac{(s+3) + (s^2 - 2s + 5)}{s+3}$$

$$\frac{s^2 - s + 8}{s+3}$$

Correct Process

Describe the Error

What Did You Learn About Learning Math? 4.4

1. What did you learn about simplifying from simplifying rational expressions?

2. How did you use what you know about complying with methodologies in working with rational expressions?

3. How did you learn the differences and similarities between the methodology for adding rational expressions and the methodology for multiplying them?

Section 4.5 Solving Quadratic and Other Polynomial Equations

Why? 4.5

Real world problems, such as calculating the trajectory of an object or modeling certain data, may involve solving quadratic or other polynomial equations.

What Do You Already Know? 4.5

How do you factor a trinomial?

What other kinds of polynomials can you factor?

How do you solve a linear equation?

How do you simplify a number in radical form?

Goals 4.5

When you complete this activity, you should be able to:

1. Solve quadratic equations by factoring
2. Solve quadratic equations using the quadratic formula
3. Use the *Zero Product Rule* to solve other factorable polynomials

Building Mathematical Language 4.5

New Terms:	discriminant	quadratic equation
	quadratic formula	Zero Product Rule

Key Concepts 4.5

A ***quadratic equation*** is an equation that can be written in the form $ax^2 + bx + c = 0$.

The ***Zero Product Rule*** is used to solve polynomial equations that can be factored. It states that the product of two quantities is zero if and only if at least one of the factors of the product is zero. That is, $cd = 0$ if and only if $c = 0$ and/or $d = 0$.

If a quadratic equation $ax^2 + bx + c = 0$ cannot be factored or cannot be factored easily, solutions to the equation can be found using the ***quadratic formula***.

$$x = \frac{-b \pm \sqrt{b^2 - 4ac}}{2a}$$

In the quadratic formula, the expression $b^2 - 4ac$ is called the ***discriminant*** and can be used to determine the nature and number of real number solutions to the equation.

Methodology 1 4.5

Solving Quadratic Equations by Factoring

Limitation/Caution: Few real world polynomials will factor, so this method is of limited use.

	Example 1	Example 2	Your Turn
Solve:	$3a^2 - 2 = 5a$	$4x^3 = 36x$	$6x^3 - 2x^2 - 20x = 0$

Steps	Discussion
1 Change the equation to an equivalent equation with one side equal to zero	**Write the equivalent equation with the exponents in descending order.**

Ex 1

$$3a^2 - 2 = 5a$$
$$3a^2 - 5a - 2 = 0$$

Ex 2

$$4x^3 = 36x$$
$$4x^3 - 36x = 0$$

Your Turn

Steps	Discussion
2 Look for common factors	**Factor out any common monomial factors.**

Ex 1

There are no common factors.

Ex 2

$$4x(x^2 - 9) = 0$$

Your Turn

Steps	Discussion
3 Factor completely	**Use an appropriate methodology from section 4.2 or 4.3 to factor completely.**

Ex 1

$$(3a + 1)(a - 2) = 0$$

Ex 2

$$4x(x + 3)(x - 3) = 0$$

Your Turn

Steps	Discussion
4 Apply the Zero Product Rule	Set each factor equal to zero and solve.

Ex 1

$$3a + 1 = 0 \text{ or } a - 2 = 0$$
$$3a = -1 \quad a = 2$$
$$a = -\frac{1}{3}$$

Ex 2

$$4x = 0 \text{ or } x + 3 = 0 \text{ or } x - 3 = 0$$
$$x = 0 \quad x = -3 \quad x = 3$$

Your Turn

5 Present the solutions	Present the solutions in simplified form.

Ex 1

$$a = -\frac{1}{3} \quad \text{or} \quad a = 2$$

Ex 2

$$x = 0 \text{ or } x = -3 \text{ or } x = 3$$

Your Turn

6 Validate	Validate by substituting solutions back into the original equation.

Ex 1

$$a = -\frac{1}{3}$$
$$3\left(-\frac{1}{3}\right)^2 - 2 \stackrel{?}{=} 5\left(-\frac{1}{3}\right)$$
$$3\left(\frac{1}{9}\right) - 2 \stackrel{?}{=} -\frac{5}{3}$$
$$\frac{1}{3} - \frac{6}{3} \stackrel{?}{=} -\frac{5}{3}$$
$$-\frac{5}{3} = -\frac{5}{3} \checkmark$$

$a = 2$:
$$3(2)^2 - 2 \stackrel{?}{=} 5(2)$$
$$12 - 2 \stackrel{?}{=} 10$$
$$10 = 10 \checkmark$$

Ex 2

$x = 0$
$$4(0)^3 \stackrel{?}{=} 36(0)$$
$$0 = 0 \checkmark$$

$x = -3$:
$$4(-3)^3 \stackrel{?}{=} 36(-3)$$
$$-108 = -108 \checkmark$$

$x = 3$:
$$4(3)^3 \stackrel{?}{=} 36(3)$$
$$108 = 108 \checkmark$$

Steps	Discussion
6 Validate (con't)	Validate by substituting solutions back into the original equation.

Your Turn

Methodology 2 4.5

Solving Quadratic Equations Using the Quadratic Formula

	Example 1	Example 2	Your Turn
Solve:	$x^2 - 8x + 2 = 0$	$2x^3 + 13x^2 = 24x$	$3x^2 = 5x + 4$

Steps	Discussion
1 Change the equation to an equivalent equation with one side equal to zero	Write the equivalent equation with the exponents in descending order.

Ex 1	Ex 2	Your Turn
$x^2 - 8x + 2 = 0$	$2x^3 + 13x^2 - 24x = 0$	

Steps	Discussion
2 Look for common factors	**Factor out any common monomial factors and use the Zero Product Rule.**

Ex 1

There are no common factors.

Ex 2

$$x(2x^2 + 13x - 24) = 0$$

$$x = 0$$

or

$$2x^2 + 13x - 24 = 0$$

Your Turn

3 Identify a, b, and c	**Compare the quadratic expression with the standard form ax^2+bx+c.**

Ex 1

a = 1, b = –8, c = 2

Ex 2

a = 2, b = 13, c = –24

Your Turn

4 Use the quadratic formula	**Recall the quadratic formula and apply it.**

Ex 1

$$x = \frac{-b \pm \sqrt{b^2 - 4ac}}{2a}$$

$$x = \frac{8 \pm \sqrt{64 - 4 \cdot 1 \cdot 2}}{2}$$

$$x = \frac{8 \pm \sqrt{56}}{2}$$

Ex 2

$$x = \frac{-b \pm \sqrt{b^2 - 4ac}}{2a}$$

$$x = \frac{-13 \pm \sqrt{13^2 - 4 \cdot 2 \cdot (-24)}}{2 \cdot 2}$$

$$x = \frac{-13 \pm \sqrt{169 + 192}}{4}$$

$$x = \frac{-13 \pm \sqrt{361}}{4}$$

Your Turn

Steps	Discussion
5 Simplify the solutions	**Simplify the radical and fraction, if possible.**

Ex 1

$$x = \frac{8 \pm \sqrt{56}}{2} = \frac{8 \pm 2\sqrt{14}}{2}$$

$$x = 4 \pm \sqrt{14}$$

Ex 2

$$x = \frac{-13 \pm \sqrt{361}}{4} = \frac{-13 \pm 19}{4}$$

$$x = \frac{3}{2} \quad \text{or} \quad x = -8$$

Your Turn

Steps	Discussion
6 Present the answer	**Include solutions from factors in step 2 and solutions from step 5.**

Ex 1

$$x = 4 + \sqrt{14} \quad \text{or} \quad x = 4 - \sqrt{14}$$

Ex 2

$$x = 0 \quad x = \frac{3}{2} \quad x = -8$$

Your Turn

Steps	Discussion
7 Validate	**Validate by substituting solutions into original equation.**

Ex 1

This is an appropriate use of your calculator:

$4 + \sqrt{14} \rightarrow x$ Enter

$x^2 - 8x + 2$ Enter

0

The left side of the equation is equal to the right side.

$0 = 0$ ✓

Perform a similar check for the other solution:

$4 - \sqrt{14} \rightarrow x$ Enter

$x^2 - 8x + 2$ Enter

0

The left side of the equation is equal to the right side.

$0 = 0$ ✓

Steps	Discussion

Ex 2

$x=0$

$2(0)^3+13(0)^2 \stackrel{?}{=} 24(0)$

$0=0$ ✓

$x=\frac{3}{2}$:

$2\left(\frac{3}{2}\right)^3+13\left(\frac{3}{2}\right)^2 \stackrel{?}{=} 24\left(\frac{3}{2}\right)$

$\frac{27}{4}+\frac{117}{4} \stackrel{?}{=} \frac{72}{2}$

$\frac{144}{4} \stackrel{?}{=} \frac{72}{2}$

$36=36$ ✓

$x=-8$:

$2(-8)^3+13(-8)^2 \stackrel{?}{=} 24(-8)$

$-192=-192$ ✓

Your Turn

Model: Solving Quadratic Equations with Complex Solutions 4.5

Note: If the discriminant $b^2 - 4ac < 0$, the solutions will be a pair of complex numbers, a + bi and a – bi. These are called ***complex conjugates*** (see section 1.3).

Solve: $x^2 - 4x + 8 = 0$

$$x = \frac{4 \pm \sqrt{(-4)^2 - 4(1)(8)}}{2(1)} = \frac{4 \pm \sqrt{-16}}{2} = \frac{4 \pm 4i}{2} = 2 \pm 2i$$

Validation: $2 + 2i \rightarrow x$ Enter

$x^2 - 4x + 8$ Enter

$0 = 0$ ✓

Perform a similar check for $2 - 2i$.

Addressing Common Errors 4.5

The following table shows some of the most common errors that learners tend to make with the kinds of problems covered in this section.

ERROR: Forgetting to include solutions from the monomial factors

Incorrect Process	Resolution
Solve: $3x^3 - 36x^2 + 108x = 0$ $3x(x^2 - 12x + 36) = 0$ $(x - 6)^2 = 0$ or $(x - 6)(x - 6) = 0$ $x - 6 = 0$ $x = 6$	*All factors must be included when listing the solutions.*

Correct Process	Validation
Solve: $3x^3 - 36x^2 + 108x = 0$ $3x(x^2 - 12x + 36) = 0$ $3x = 0$ or $x^2 - 12x + 36 = 0$ $x = 0$ or $(x - 6)(x - 6) = 0$ $x = 0$ or $x - 6 = 0$ $x = 0$ or $x = 6$	$x = 6$ $3(6)^3 - 36(6)^2 + 108(6) \stackrel{?}{=} 0$ $648 - 1296 + 648 \stackrel{?}{=} 0$ $0 = 0$ ✓ $x = 0$ $3(0)^3 - 36(0)^2 + 108(0) \stackrel{?}{=} 0$ $0 = 0$ ✓

ERROR: Not changing the equation into standard form before solving for x

Incorrect Process

$x^2 + 5x = 6$

$x(x + 5) = 6$

$x = 6$ or $x + 5 = 6$

$x = 6$ or $x = 1$

Resolution

Quadratic equations must always be in standard form, ax2 + bx + c = 0, before factoring or using the quadratic formula.

Correct Process

$x^2 + 5x = 6$

$x^2 + 5x - 6 = 0$

$(x + 6)(x - 1) = 0$

$x + 6 = 0$ or $x - 1 = 0$

$x = -6$ or $x = 1$

Validation

$x = -6$	$x = 1$
$(-6)^2 + 5(-6) \stackrel{?}{=} 6$	$(1)^2 + 5(1) \stackrel{?}{=} 6$
$36 - 30 \stackrel{?}{=} 6$	$1 + 5 \stackrel{?}{=} 5$
$6 = 6$ ✓	$6 = 6$ ✓

ERROR: Incorrectly simplifying the fraction in the quadratic formula

Incorrect Process

Solve: $x^2 - 4x = 5$

$x^2 - 4x - 5 = 0$

$$x = \frac{4 \pm \sqrt{(-4)^2 - 4(1)(-5)}}{2}$$

$$= \frac{4 \pm \sqrt{16 + 20}}{2} = \frac{4 \pm \sqrt{36}}{2}$$

$$= \frac{4 \pm 6}{2} = \frac{\cancelto{2}{4} \pm 6}{\cancelto{1}{2}}$$

$x = 2 + 6 = 8$ or $x = 2 - 6 = -4$

Resolution

Write each rational expression with a common denominator.

Correct Process

$$x = \frac{4 \pm \sqrt{(-4)^2 - 4(1)(-5)}}{2}$$

$$= \frac{4 \pm \sqrt{16 + 20}}{2} = \frac{4 \pm \sqrt{36}}{2} = \frac{4 \pm 6}{2}$$

$$x = \frac{4 + 6}{2} = 5 \text{ or } x = \frac{4 - 6}{2} = -1$$

Validation

$x = 5$:	$x = -1$:
$5^2 - 4(5) \stackrel{?}{=} 5$	$(-1)^2 - 4(-1) \stackrel{?}{=} 5$
$25 - 20 \stackrel{?}{=} 5$	$1 + 4 \stackrel{?}{=} 5$
$5 = 5$ ✓	$5 = 5$ ✓

ERROR: Incorrectly simplifying the square root in the quadratic formula

Incorrect Process

Solve: $2x^2 - 5x + 2 = 0$

Using the quadratic formula,

$$x = \frac{5 \pm \sqrt{(-5)^2 - 4 \cdot 2 \cdot 2}}{2 \cdot 2}$$

$$= \frac{5 \pm \sqrt{25 - 16}}{4} = \frac{5 \pm (5 - 4)}{4}$$

$$x = \frac{5 + 1}{4} = \frac{6}{4} = \frac{3}{2}$$

or

$$x = \frac{5 - 1}{4} = \frac{4}{4} = 1$$

Resolution

Perform all operations under the radical before simplifying the square root.

Correct Process

$$x = \frac{5 \pm \sqrt{(-5)^2 - 4 \cdot 2 \cdot 2}}{2 \cdot 2}$$

$$= \frac{5 \pm \sqrt{25 - 16}}{4} = \frac{5 \pm \sqrt{9}}{4}$$

$$x = \frac{5 + 3}{4} = \frac{8}{4} = 2$$

or

$$x = \frac{5 - 3}{4} = \frac{2}{4} = \frac{1}{2}$$

Validation

$x = 2:$

$$2(2)^2 - 5(2) + 2 \stackrel{?}{=} 0$$

$$8 - 10 + 2 \stackrel{?}{=} 0$$

$$0 = 0 \checkmark$$

$x = \frac{1}{2}:$

$$2\left(\frac{1}{2}\right)^2 - 5\left(\frac{1}{2}\right) + 2 \stackrel{?}{=} 0$$

$$\left(\frac{1}{2}\right) - \left(\frac{5}{2}\right) + \left(\frac{4}{2}\right) \stackrel{?}{=} 0$$

$$0 = 0 \checkmark$$

Preparation Inventory 4.5

I can...	OR *Here's my question...*
☐ solve a quadratic equation by factoring.	☐
☐ use the quadratic formula to solve a quadratic equation.	☐
☐ simplify solutions found with the quadratic formula.	☐
☐ validate the solutions, either by hand or with a calculator.	☐

Critical Thinking Questions 4.5

1. What is the first thing you do when solving a quadratic equation of the form $ax^2 + bx + c = 0$?

2. What does the symbol $\pm$ mean?

3. How many real number solutions can a quadratic equation have? Explain.

4. If the discriminant of a quadratic equation is zero, what does that imply about the solutions of the quadratic equation? What if the discriminant is greater than 0? What if it is less than 0?

5. If a polynomial $p(x)$ factors into three linear factors, how many solutions will the equation $p(x) = 0$ have?

6. Why do both the quadratic formula method and the factoring method for solving quadratic equations require that a zero appear on one side of the equation?

Demonstrate Your Understanding 4.5

Solve each of the following, using the method indicated.

1. $3x^2 - 16x = 12$ (quadratic formula)

2. $3x^2 - 16x = 12$ (factoring)

3. $2a^2 = 5a + 8$ (quadratic formula)

4. $a^4 - 7a^2 + 10 = 0$ (factoring)

5. $12x^4 - 2x^3 = 70x^2$ (factoring)

Create and Solve the Hardest Problem 4.5

Your Hardest Problem

Problem Solution

What makes a problem where you must solve quadratics and other polynomial equations a hard problem to solve?

Identify, Describe, & Correct the Errors 4.5

Solve the problem correctly in the appropriate location. Be sure to identify and describe the error you found in the worked solution!

Problem 1

Solve for x:

$$x(3x-1)(x+2)=0$$

Worked Solution *(Find the error!)*

$$x(3x-1)(x+2)=0$$

$$3x-1=0 \text{ or } x+2=0$$

$$x=\frac{1}{3} \text{ or } x=-2$$

Correct Process

Describe the Error

Problem 2

Solve for y: $2y^2 - y = 5$

Worked Solution *(Find the error!)*

$$2y(y-1)=5$$

$$2y=0 \text{ or } y-1=0$$

$$y=0 \text{ or } y=1$$

Correct Process

Describe the Error

Problem 3

Solve for y: $2y^2 - 8y = 1$

Worked Solution *(Find the error!)*

a = 2, b = –8, c = 1

$$y = \frac{8 \pm \sqrt{64-8}}{4}$$

$$y = \frac{8 \pm \sqrt{56}}{4}$$

$$y = \frac{8 \pm 2\sqrt{14}}{4}$$

$$y = \frac{4+\sqrt{14}}{2} \text{ or } y = \frac{4-\sqrt{14}}{2}$$

Correct Process

Describe the Error

Problem 4

Solve for x: $2x^2 - 8x = 1$

Worked Solution *(Find the error!)*

$$2x^2 - 8x - 1 = 0$$

a = 2, b = –8, c = –1

$$\frac{8 \pm \sqrt{64-4(2)(-1)}}{4}$$

$$x = \frac{8 \pm \sqrt{64+8}}{4}$$

$$x = \frac{8 \pm \sqrt{72}}{4} = \frac{8 \pm 6\sqrt{2}}{4}$$

$$x = \frac{{}^{2}\cancel{8} \pm 6\sqrt{2}}{{}^{1}\cancel{4}} = 2 \pm 6\sqrt{2}$$

$$x = 2 + 6\sqrt{2} \text{ or } x = 2 - 6\sqrt{2}$$

Correct Process

Describe the Error

What Did You Learn About Learning Math? 4.5

1. How does pattern recognition help you solve quadratic equations?

2. Why is it important to validate your solutions?

3. How did you make use of prior knowledge to help you solve quadratic equations?

Concept Map — Chapter 4

Terminology from the chapter
check the terms used in your map

__completely factored

__complex fraction

__discriminant

__division algorithm

__factor (verb)

__factor by grouping

__factoring

__missing term(s) in a polynomial

__quadratic equation

__quadratic form

__quadratic formula

__quadrinomial

__rational expression

__Zero Product Rule

Your instructor will specify how many terms from the chapter you should use in your concept map.

CHAPTER 5

This chapter introduces the four representations of functions: verbal, tabular (numeric), graphical, and symbolic. You will learn the connections between and among these representations and how to describe the characteristics and appearance of functions using function cards. In addition, you will learn about functions that have inverse functions and how to represent these inverse functions. Finally, you will learn to analyze, compare and manipulate the representations of functions.

Section 5.1 Basics of Functions

Why? 5.1

Many variables or quantities that appear in real life are functions of other variables: your cell phone bill may be a function of the minutes you have used; your electric bill is a function of the kilowatts-hours you have used; the distance you travel is a function of your speed and the time you spend driving.

What Do You Already Know? 5.1

How do you graph a linear equation?

How do you find the intercepts of an equation?

How do you make a T-table for a function from a graph of the function?

Goals 5.1

When you complete this activity, you should be able to:

Use function notation to define linear relationships

Building Mathematical Language 5.1

New Terms:	domain	one-to-one function
	function	range
	function notation	relation

Key Concepts 5.1

A ***function*** is a relationship between two quantities or, for our purposes, two sets of numbers. A function associates each input number with a unique output number. A function can be represented symbolically in several ways. A function can be represented as an equation in two variables, x (input) and y (output), as long as for each x value there is just one y value. For example, the equations $-2x + y = 5$ and $y = x^2 + 3x$ can be representations of functions. The function given by $-2x + y = 5$ can be represented as the function in what is called ***function notation*** by $f(x) = 2x + 5$ where $y = f(x)$. Similarly, $y = x^2 + 3x$ can be represented as the function g by $g(x) = x^2 + 3x$. This new function notation focuses your attention on the input variable.

Limitation/Caution: Some equations in two variables cannot be representations of functions. For example, in the equation $y^2 = 4 - x^2$ when the input value for x is 0, the output value for y can be either -2 or 2.

A ***one-to-one function*** is a function that has only one input for each output and only one output for each input.. The function $f(x) = x^2$ is not a one-to-one function because the output 4 is associated with two inputs—namely, 2 and -2. A relationship represented by equations such as $y^2 = 4 - x^2$ where there are two or more outputs for some inputs is called a ***relation***. For functions and relations, the set of inputs is called the ***domain*** and the set of outputs is called the ***range***.

Note: As with equations, we can represent functions in three ways: symbolically, numerically, and graphically.

Methodology 5.1

Graphing a Function

Displaying the graph of a function is a standard way to examine the behavior of a function.

Limitation/Caution: A single graph of a function in a specific graphing window may not show all the important aspects of the function.

	Example 1	**Example 2**	**Your Turn**
Graph the function given by the equation:	$f(x) = -2x + 5$	$h(t) = \frac{1}{2}t - 3$	$g(x) = -3x + 2$

Steps	**Discussion**
1 Identify the domain	**Identify the domain variable and its set of possible values.**

Ex 1

The domain variable is x and the domain is all real numbers.

Ex 2

The domain variable is t and the domain is all real numbers.

Your Turn

Steps	**Discussion**
2 Identify the maximum or minimum	**Determine if the function has a minimum or maximum value.**

Ex 1

The graph is a straight line with no minimum or maximum.

Ex 2

The graph is a straight line with no minimum or maximum.

Your Turn

Steps	Discussion
3 Identify some x-values for a T-table	Determine which x-values to put in your T-table by determining the function's maximum and minimum values, if they exist, and values where there are special behaviors, like intercepts.

Ex 1

x	$f(x)$
$\frac{5}{2}$	
0	

→

x	$f(x)$
$\frac{5}{2}$	0
0	5

The x-intercept is $\left(\frac{5}{2}, 0\right)$.

The y-intercept is (0, 5).

Ex 2

t	$h(t)$
0	–3
6	0

The x-intercept is (6, 0).

The y-intercept is (0, –3).

Your Turn

Steps	Discussion
4 Identify the range of the function	Determine what the minimum and maximum y-values will be.

Ex 1

The range is all real numbers since the maximum will be $+\infty$ (positive infinity) and the minimum will be $-\infty$ (negative infinity).

Ex 2

The range is all real numbers since the maximum will be $+\infty$ and the minimum will be $-\infty$.

Your Turn

Steps	Discussion
5 Determine the length of the table of ordered pairs	**Determine how many ordered pairs are needed to effectively graph the specific function.**

Ex 1

Determine a third point for the T-table to check.

x	$f(x)$
$\frac{5}{2}$	0
0	5
1	3

The three points on the graph of the straight line are: (0, 5), $\left(\frac{5}{2}, 0\right)$ and (1, 3).

Ex 2

Determine a third point for the T-table to check.

t	$h(t)$
0	–3
6	0
2	–2

Your Turn

6 Connect the points	**Plot all the points and then draw a line through all the plotted points.**

Ex 1

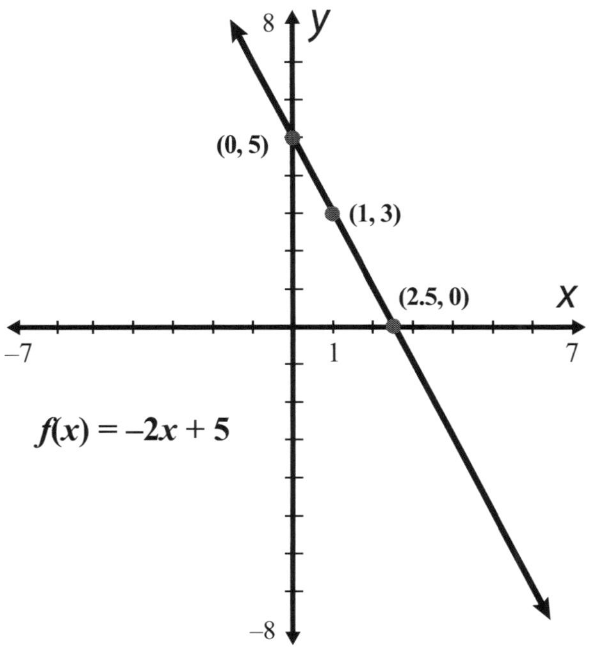

Ex 2

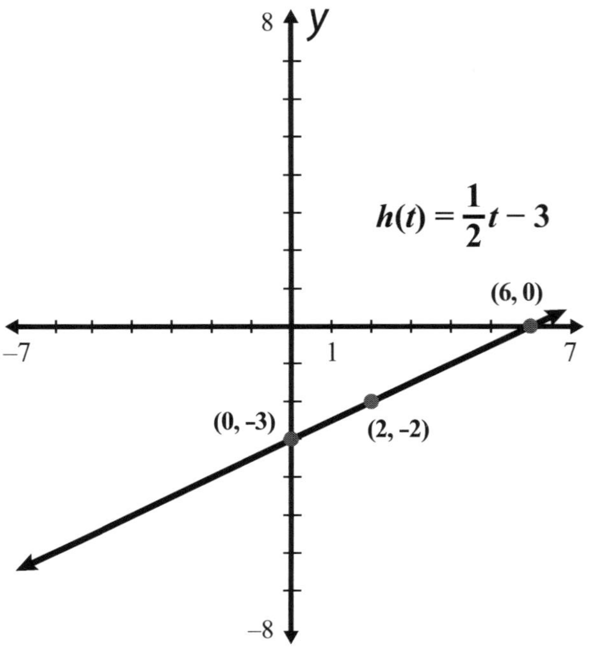

Steps	Discussion

Your Turn

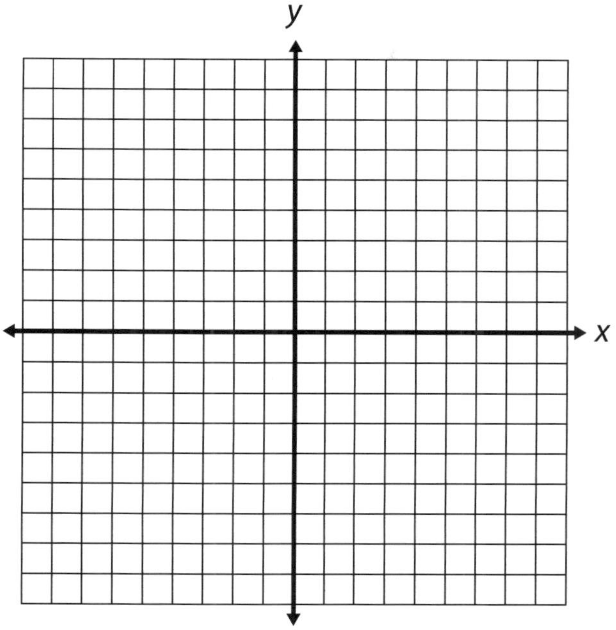

7 Validate the graph	Check that the points are on a straight line and that a different point on the graph satisfies the function.

Ex 1

Use the point (2, 1) to validate:

$f(x) = -2x + 5$

$f(2) = -2(2) + 5 = 1$

The point (2, 1) on the graph satisfies the function. ✓

Ex 2

Use the point (4, –1) to validate:

$h(t) = \frac{1}{2}t - 3$

$h(4) = \frac{1}{2}(4) - 3 = 2 - 3 = -1$

The point (4, –1) on the graph satisfies the function ✓

Your Turn

Model 5.1

Graph the function given by: $f(x) = 2x - 4$

Step 1 Identify the domain

The domain variable is x and the domain is all real numbers.

Step 2 Identify the minimum or maximum

The graph is a straight line with no minimum or maximum.

Step 3 Identify some x-values for a T-table

The x-intercept is (2, 0).
The y-intercept is (0, –4).

Step 4 Identify the range

The range is all real numbers since the maximum will be positive infinity and the minimum will be negative infinity.

Step 5 Determine the length of the table of ordered pairs

Determine three points on the graph of the straight line: (0, –4), (2, 0), and (3, 2)

Step 6 Connect the points

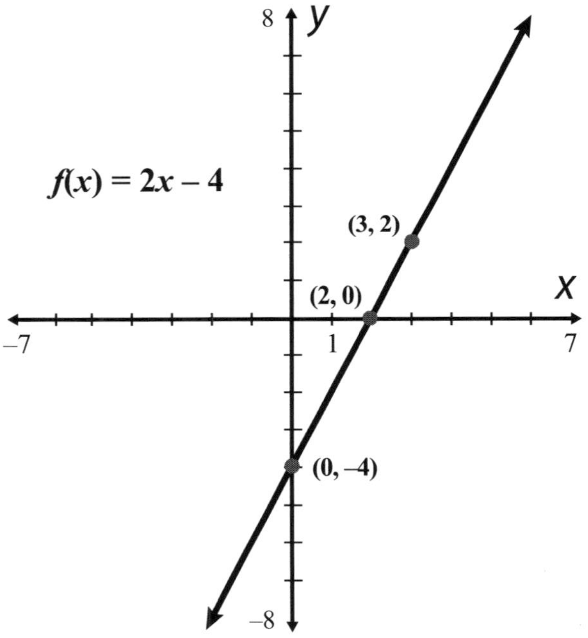

Step 7 Validate the graph

The three points are on a straight line.

$$f(x) = 2x - 4$$
$$f(1) = 2(1) - 4$$
$$f(1) = 2 - 4 = -2$$

The point (1, –2) on the graph satisfies the function. ✓

Addressing Common Errors 5.1

The following table shows some of the most common errors that learners tend to make with the kinds of problems covered in this section.

ERROR: **Creating a line segment rather than a line**

Incorrect Process

Graph $f(x) = -3x + 5$

x	$f(x)$
0	−3
6	0
2	−2

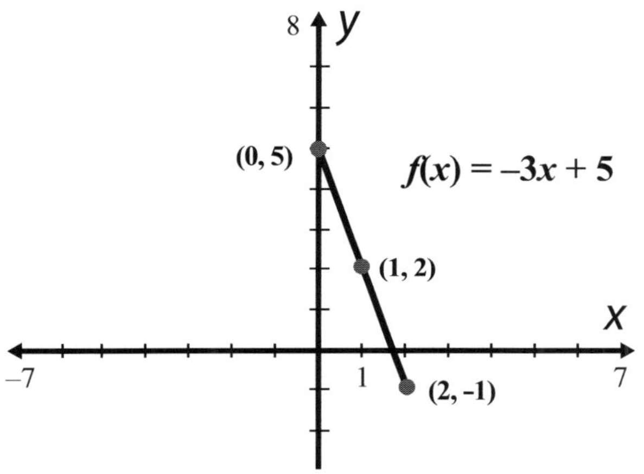

Correct Process

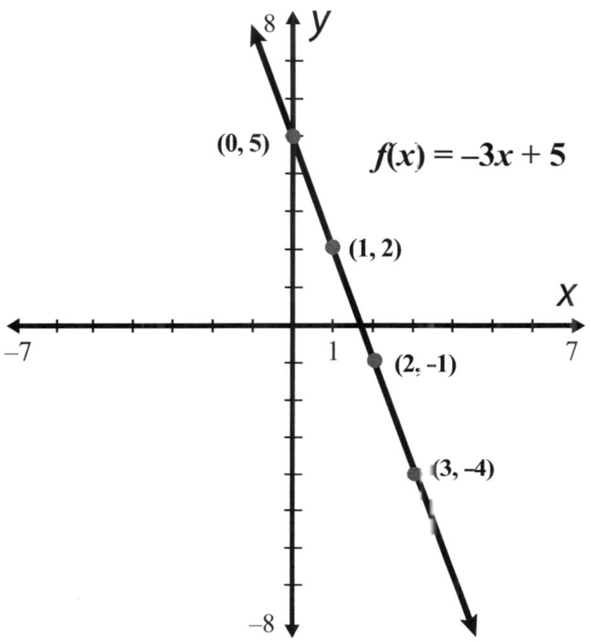

Resolution

Draw a line through the points that extends beyond them. Adding arrows to the ends of the line you draw shows that the line keeps going beyond what you've drawn.

Validation

The line goes through the three points and extends in both directions. The point (3, −4) that satisfies the equation is on the line, but not on the original line segment. You can keep checking points to verify that the line extends through the original three points, in both directions.

ERROR: Incorrect computation of a function value

Incorrect Process

Calculate $f(2)$ if $f(x) = 4x + 6$

$f(2) = (4x + 6)(2) = 8x + 12$

Resolution

$f(2)$ indicates the output of the function when $x = 2$. Parentheses do not mean multiplication here.

Correct Process

$f(2) = 4(2) + 6 = 8 + 6 = 14$

Validation

Validate by substituting 14 for the value of $f(x)$ to see if we recover a value of 2 for x:

$$14 = 2x + 6$$

$$8 = 4x$$

$$x = 2 \checkmark$$

Preparation Inventory 5.1

I can...	OR *Here's my question...*
☐ identify the domain of a function.	☐
☐ identify the intercepts of a function.	☐
☐ create a T-table of points to graph.	☐
☐ compute $f(2)$ from the function defined by $f(x) = 3x + 5$.	☐
☐ use function notation to express a linear relationship.	☐

Critical Thinking Questions 5.1

1. What defines the domain of a function?

2. What determines the range of a function?

3. How do you convert an algebraic representation of a function to a numerical representation?

4. Is there an easy way to go from a graphical to an algebraic representation of a function?

5. What strategies do you use to find special points to graph?

6. How do you determine if you have enough points plotted to graph a function correctly?

7. What is the strategy behind validating that a graph is drawn correctly?

8. What is the difference between $a(3)$ and $f(3)$ where a is a real number and f is a function?

Demonstrate Your Understanding 5.1

Graph each of the following functions. Identify the domain and range of each.

1. $x - 3y = 4$

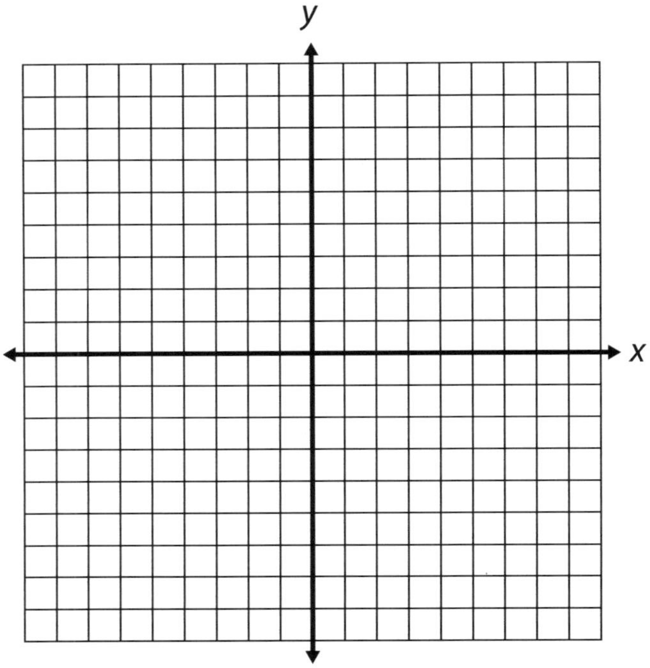

Domain

Range

2. $y = \frac{1}{2}x - 3$

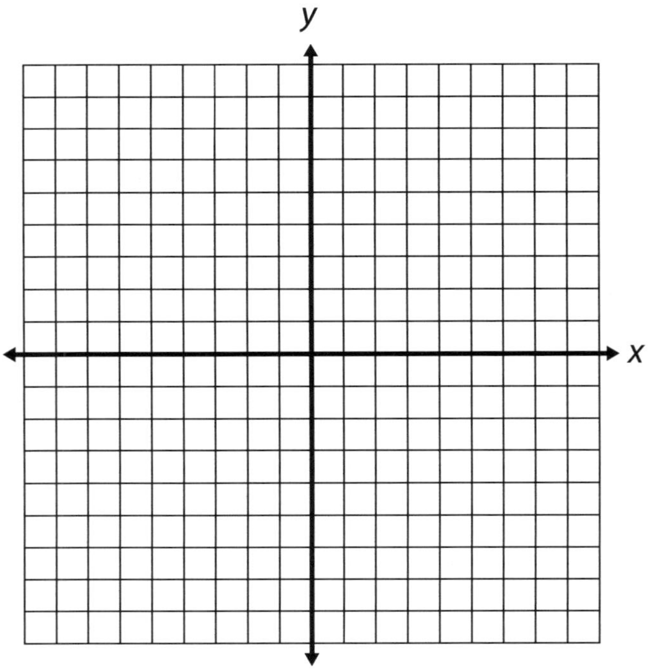

Domain

Range

3. $f(x) = 2x + 3$

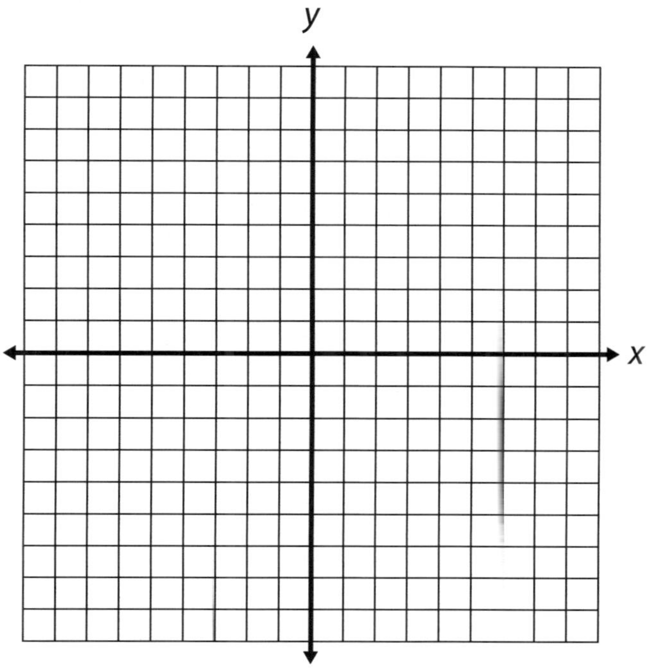

Domain

Range

4. $g(x) = -2x - 4$

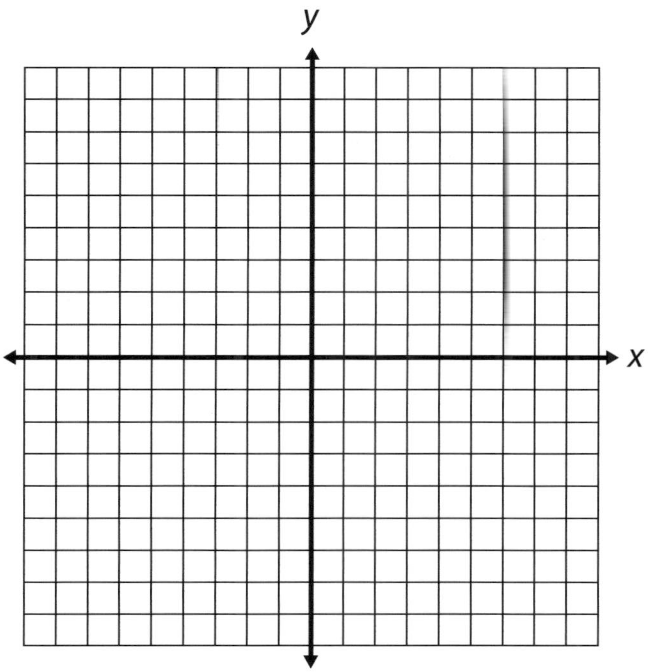

Domain

Range

Create and Solve the Hardest Problem 5.1

Your Hardest Problem

Problem Solution
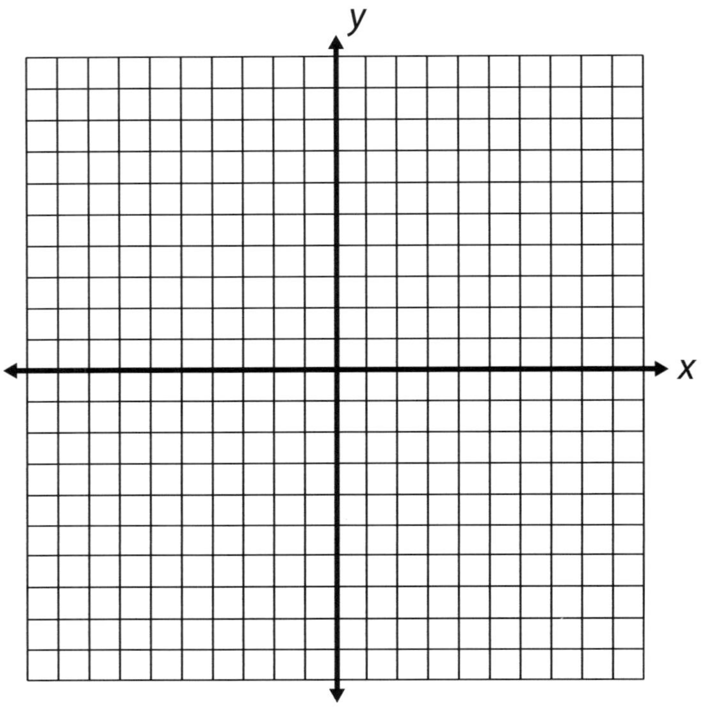

What makes a problem where you must graph a function a hard problem to solve?

Identify, Describe, & Correct the Errors 5.1

Solve the problem correctly in the appropriate location. Be sure to identify and describe the error you found in the worked solution!

Problem 1

Graph: $f(x) = 2x - 1$

Choose ordered pairs, find the domain and range, and choose a validation point.

Worked Solution *(Find the error!)*

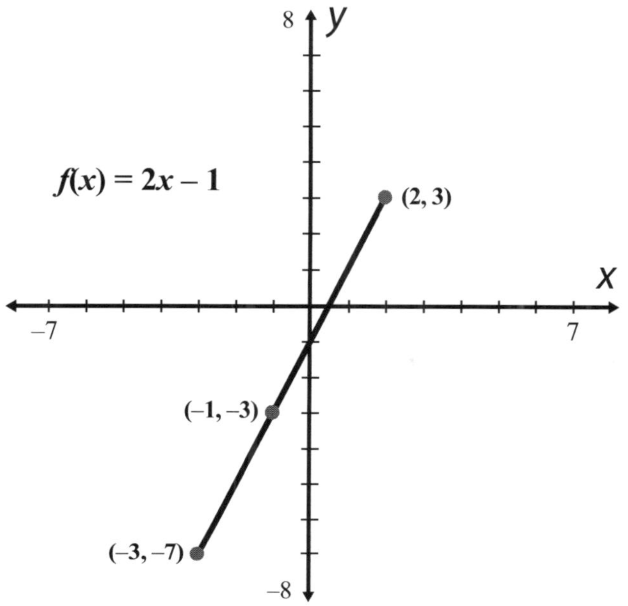

The domain is all real numbers.
The range is all real numbers.
Validation point is (–3, –7).

Describe the Error

Correct Process

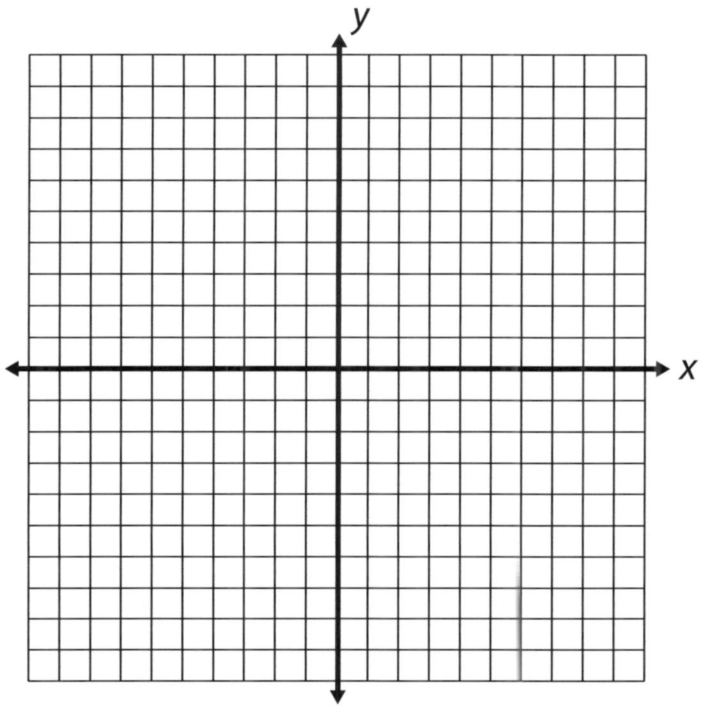

Problem 2

Graph: $g(x) = -2x + 5$

Choose ordered pairs, find the domain and range, and choose a validation point.

Worked Solution *(Find the error!)*

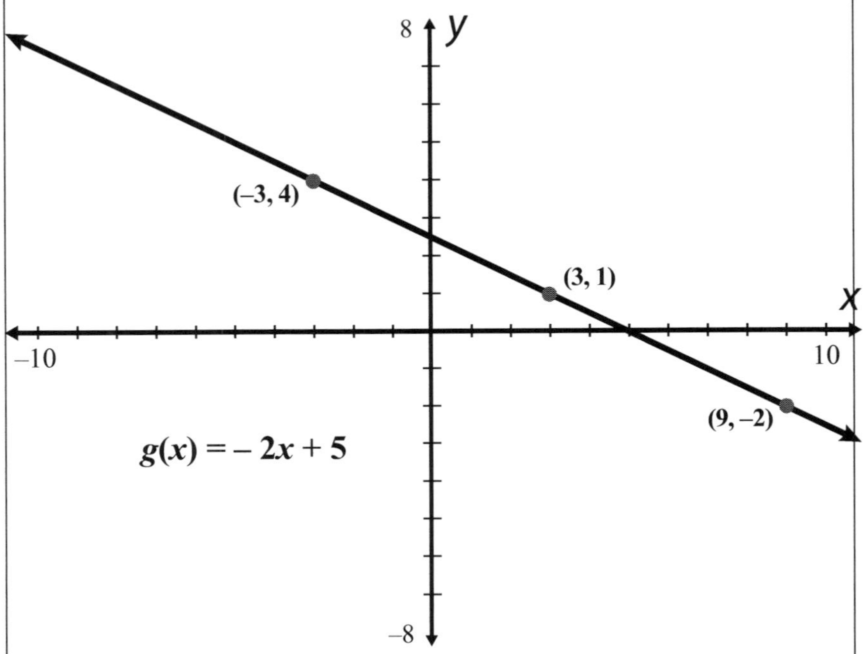

The domain is all real numbers. The range is all real numbers.

Validation point is (3, 1).

Describe the Error

Correct Process

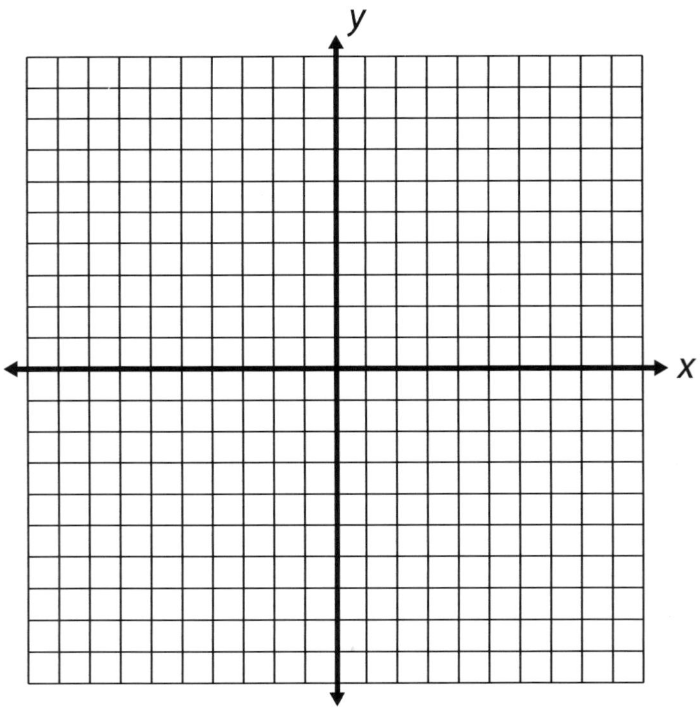

Problem 3

Graph: $y = -x + 2$

Choose ordered pairs, find the domain and range, and choose a validation point.

Describe the Error

Worked Solution *(Find the error!)*

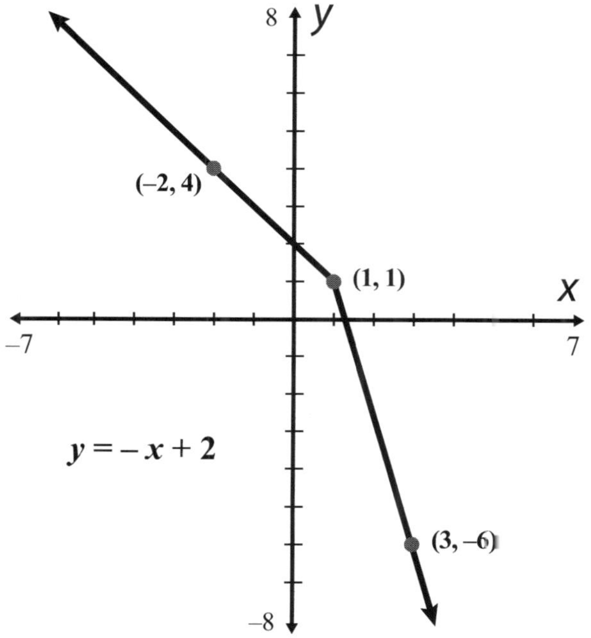

The domain is all real numbers. The range is all real numbers.

Validation point is (1, 1).

Correct Process

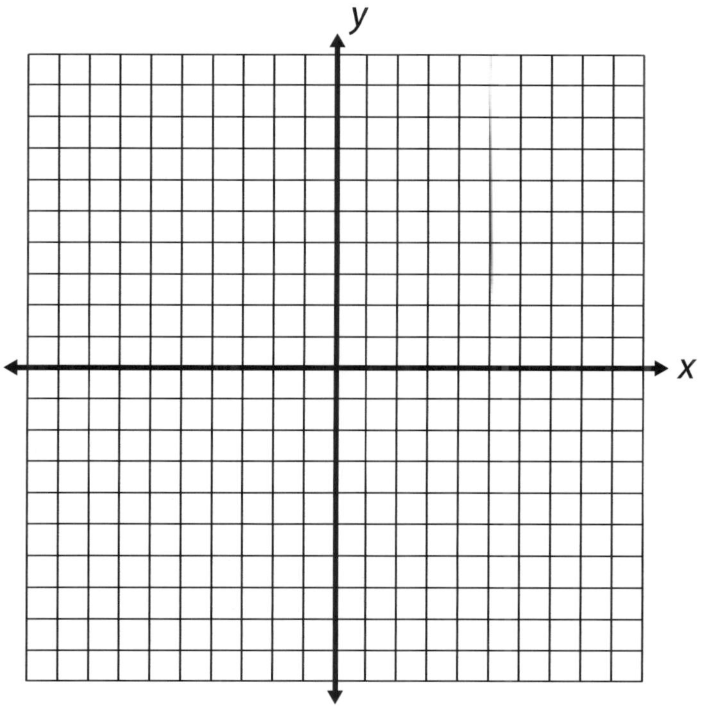

Problem 4

Graph: $f(x) = 2x - 1$

Choose ordered pairs, find the domain and range, and choose a validation point.

Describe the Error

Worked Solution *(Find the error!)*

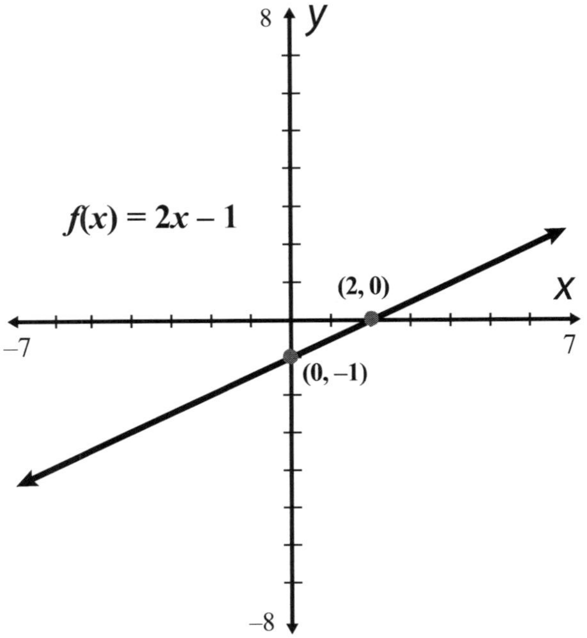

The domain is all real numbers. The range is all real numbers.

Correct Process

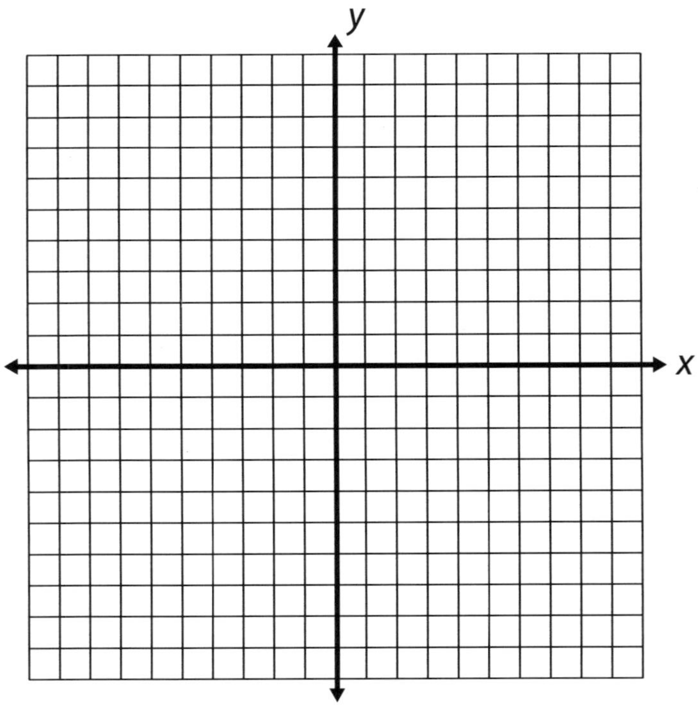

What Did You Learn About Learning Math? 5.1

1. What pattern did you use to describe linear relationships using function notation?

2. How do you visualize the graph of a linear function?

3. How do linear equations relate to linear functions?

Section 5.2 Families of Functions

Why? 5.2

Functions play a critical role in describing physical phenomena, but they are also important in everyday life. For example, the cost of the gasoline you buy at a service station is determined by the function that associates the number of gallons you pump with the cost (per gallon of gas). You can see the associated table flash by as you look at the spinning numbers on the pump.

What Do You Already Know? 5.2

How do you compute the cost of a given quantity of salad from a salad bar in a grocery store?

How do you compute $f(3)$ if $f(x) = 2x^2 - 5x - 1$?

What is the difference between $a(x)$ and $f(x)$ where a is a real number and f is a function?

Goals 5.2

When you complete this activity, you should be able to:

1. Use function notation for a variety of functions
2. Relate a graph of a function to its behavior
3. Interpret a family of functions card

Building Mathematical Language 5.2

New Terms:

asymptotes (horizontal and vertical)

function behavior

families of functions:

absolute value	logarithmic	quadratic
exponential	logistic	radical
linear	polynomial	rational

Key Concepts 5.2

Families of functions provide a convenient way to analyze various types of functions that occur frequently in the real world. This book presents a comprehensive analysis laid out like baseball cards with key components for each of 12 different families of functions. We call each card a family of functions card. The purpose of the cards is to provide a concise way of analyzing a specific family of functions. These cards are available in this section of your book.

Limitation/Caution: Although comprehensive, these cards do not profile every aspect of each family of functions.

When analyzing a function, you identify key characteristics of that function, including its domain and range, trends (increasing or decreasing behavior), intercepts, and asymptotes.

Limitation/Caution: Care must be taken to ensure that you identify all the important behaviors of the function.

By graphing functions using the same domain for each, you can more easily compare and contrast several functions' behavior.

Limitation/Caution: You need not use the complete range of each function. Also, the range of a function can be difficult to determine before you graph it.

You can observe the trends of functions (when they are increasing and decreasing) from their graphs.

Limitation/Caution: Functions may have "kinks" where they break from a trend briefly, decreasing before continuing to increase or vice versa, that are difficult to spot.

Intercepts (where functions cross the x- or y-axes) are important features of graphs of functions as they are points of reference for the graphs and give you information about when the function value is 0 and what the function value is when the domain value is 0.

Limitation/Caution: x-intercepts can be difficult to determine exactly.

If you know the general shape of a family of functions, then you know the general shape of any function in that family. Such knowledge can guide you in graphing the function and in determining its behavior.

Limitation/Caution: As with trends, a function's general shape may have slight variations that are hard to detect.

The word ***asymptote*** is derived from the Greek words *symptotos* meaning to meet and *a* meaning not. Thus, *asymptote* means *not meeting*.

The graph of $f(x) = 1/x$ is asymptotic to both the x-axis and y-axis as shown in the graph at right. You use the *horizontal asymptotes* in comparing the long-term behaviors of functions and the *vertical asymptotes* to determine at what points function behavior becomes arbitrarily large (in absolute value). Usually asymptotes are found algebraically.

Limitation/Caution: Asymptotes can be inappropriately graphed by graphing calculators.

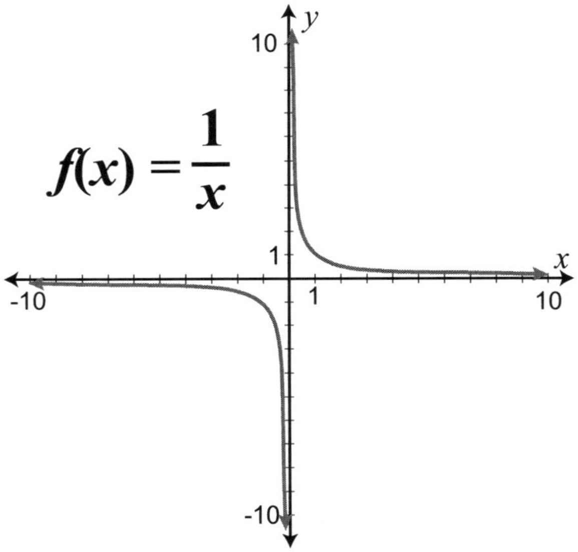

Functions that are built from members of different families of functions can have *interesting features* or *characteristics.* Such features can be combinations of features from the different families.

Limitation/Caution: Interesting features are sometimes in the eye of the beholder.

Linear $f(x) = mx + b$

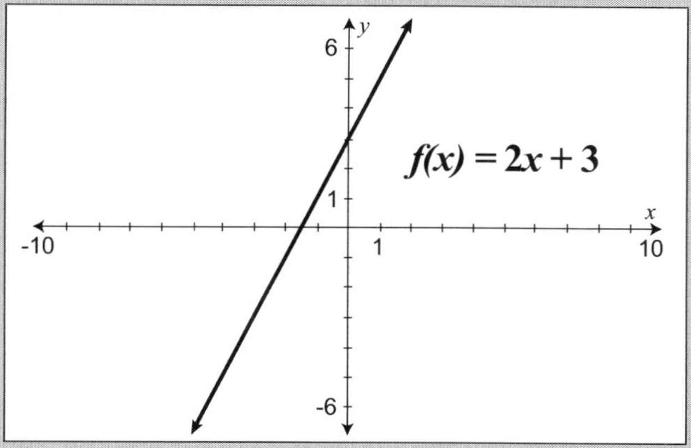

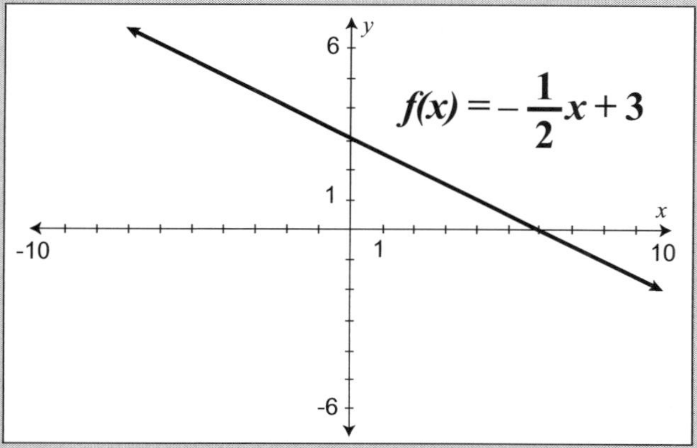

Linear $f(x) = mx + b$

Simplest Form	$y = x$
Domain	All real numbers
Range	All real numbers except when $m = 0$ and $f(x) = b$ (then the range is b).
Characteristic 1	*(0, b)* is the *y*-intercept
Characteristic 2	*m* is the slope

Appearance

A straight line that is increasing, decreasing, or constant depending on whether m is positive, negative or zero, respectively.

Interesting Features

These functions have a constant rate of change and the algebraic definition of these functions can be determined from two ordered pairs.

Real-World Application

The manufacturing cost of a cell phone is a linear function of the number of cell phones produced.

Logarithm $f(x) = \log_b x$

(and $b > 0$, $b \neq 1$)

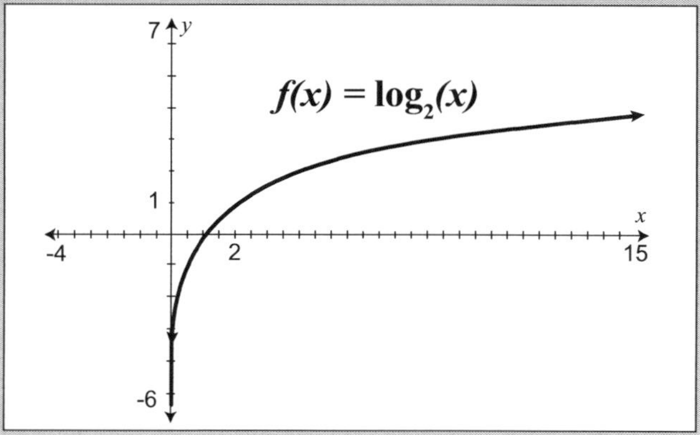

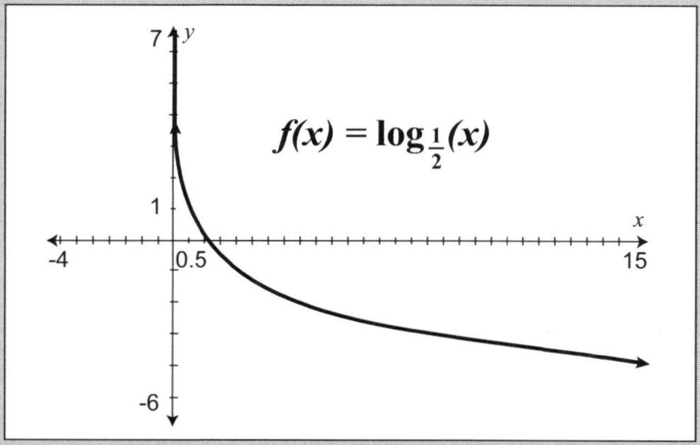

Logarithm $f(x) = \log_b x$

(and $b > 0$, $b \neq 1$)

Simplest Form	$y = \log_2 x$
Domain	All positive real numbers
Range	All real numbers
Characteristic 1	*There is a vertical asymptote at y = 0*
Characteristic 2	*There is no horizontal asymptote*

Appearance

Always increasing if b > 1 and always decreasing if b < 1.

Interesting Features

These function values increase very slowly when b > 5 and x > 100 but do not have horizontal asymptotes.

Real-World Application

The loudness of sound is measured with a logarithmic function.

Polynomial $p(x)=a_nx^n+...+a_1x^1+a_0$

(and n a positive integer)

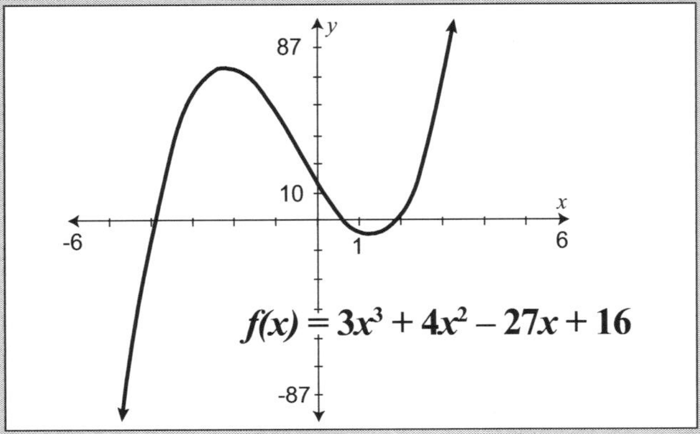

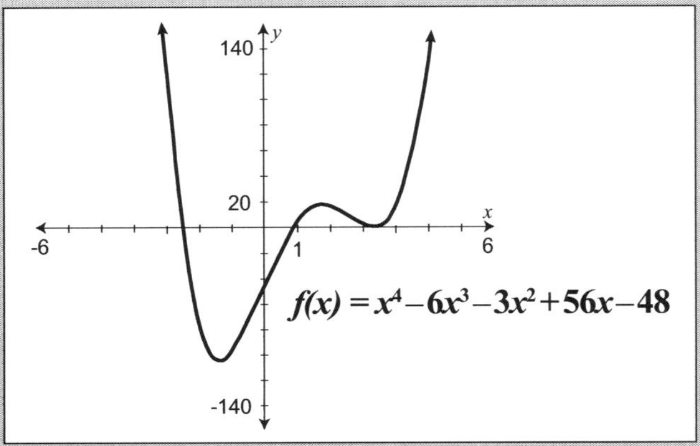

Polynomial $p(x)=a_nx^n+...+a_1x^1+a_0$

(and n a positive integer)

Simplest Form	$y = x^3$
Domain	All real numbers
Range	If n is odd, the range is all real numbers. If n is even and $a_n < 0$, the range is all real numbers less than the maximum value of the function. If n is even and $a_n > 0$, then all real numbers are greater than the minimum function value.
Characteristic 1	*It crosses the x-axis no more than n times.*
Characteristic 2	*If n is odd, it will cross the x-axis at least once.*

Appearance

As x gets farther from 0, the function approximates the behavior of its term of highest degree (a_nx^n).

Real-World Application

To handicap races between dragsters and "funny cars," the National Hot Rod Association uses a polynomial: $f(x) = 71.682x - 60.427x^2 + 84.710x^3 - 27.769x^4 + 4.296x^5 - 0.262x^6$

Exponential $f(x)=b \cdot g^x$

($g > 0$ and $b > 0$)

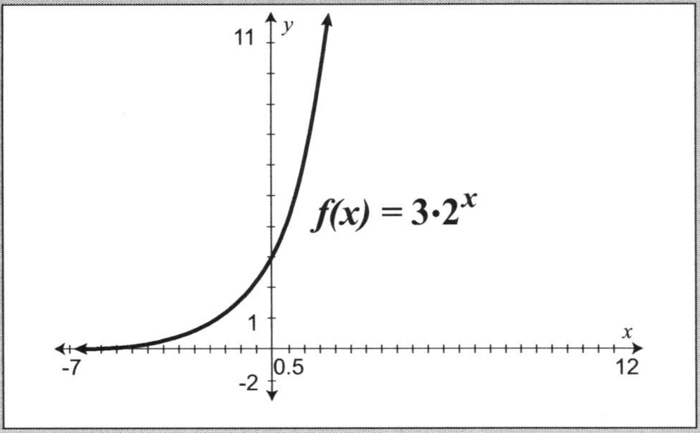

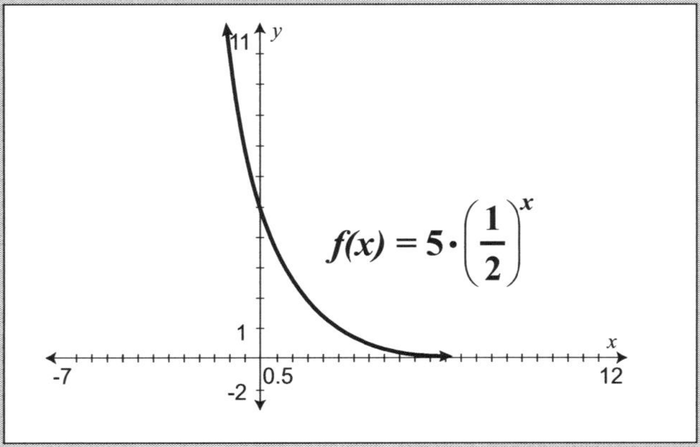

Exponential $f(x)=b \cdot g^x$

($g > 0$ and $b > 0$)

Simplest Form	$y = 2^x$
Domain	All real numbers
Range	All positive real numbers
Characteristic 1	*(0, b) is the y-intercept*
Characteristic 2	*g is the growth factor*

Appearance

Always increasing if g > 1 and always decreasing if g < 1.

Interesting Features

The algebraic definition of these functions can be determined from two ordered pairs.

Real-World Application

The value of an investment as it grows at a constant, compounded interest rate is an exponential function of time. As it decays, the amount of radioactive carbon-14 in an animal bone is an exponential function of time.

Quadratic $f(x) = ax^2 + bx + c$

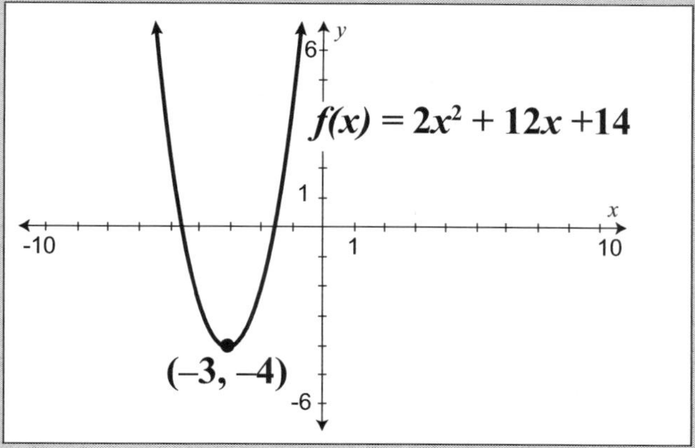

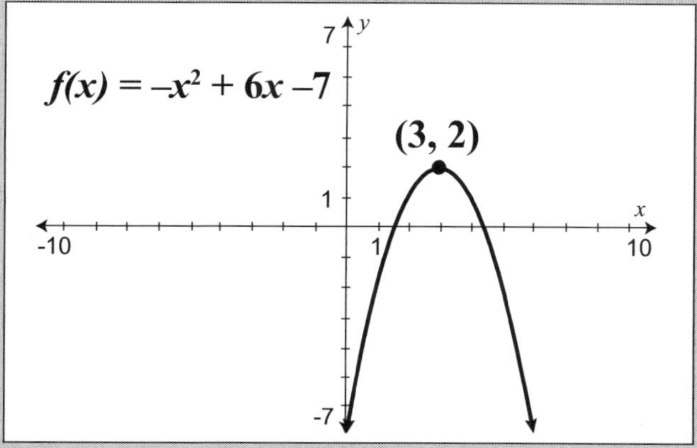

Quadratic $f(x) = ax^2 + bx + c$

Simplest Form	$y = x^2$

Domain	All real numbers
Range	Given that (h, k) is the vertex: If $a > 0$, then the range is all real numbers *greater* than k. If $a < 0$, then the range is all real numbers *less* than k. If a = 0, it is not a quadratic function.

Characteristic 1	*The vertex (h, k) is on the line of symmetry*
Characteristic 2	*The line of symmetry is x = h*

Appearance

Looks like ∪ if a > 0 or ∩ if a < 0

Interesting Features

$y - k = a(x - h)^2$ is another symbolic representation of the function.

Real-World Application

The rate at which a ball falls in Earth's gravitational field is a quadratic function of time.

Rational $f(x) = \frac{p(x)}{q(x)}$ (in reduced form)

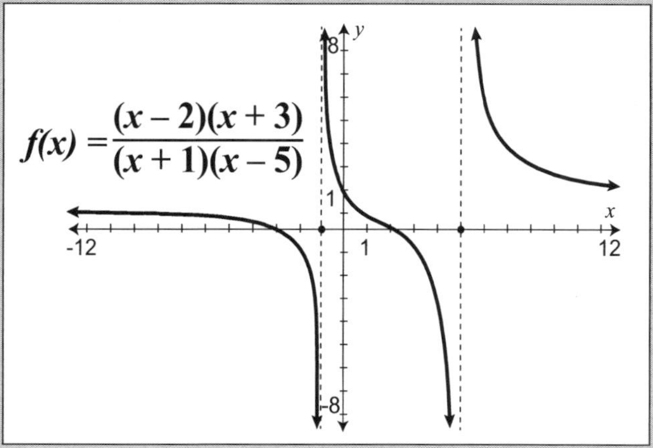

Rational $f(x) = \frac{p(x)}{q(x)}$ (in reduced form)

Simplest Form	$y = \frac{1}{x}$

Domain	All numbers for which $q(x) \neq 0$
Range	Can be all real numbers

Characteristic	*The graph of the function has a vertical asymptote wherever q(x) = 0.*

Appearance

Crosses the x-axis the same number of times as the number of solutions to p(x) = 0

Interesting Features

These functions have both vertical and horizontal asymptotes when q is not a constant.

Real-World Application

A rational function can be used to determine how much pure water needs to be added to dilute a 20% saline solution to a 5% saline solution.

Power (q is odd) $f(x)=a \cdot x^{p/q}$

(and p, q positive integers, reduced form, $q \neq 1$)

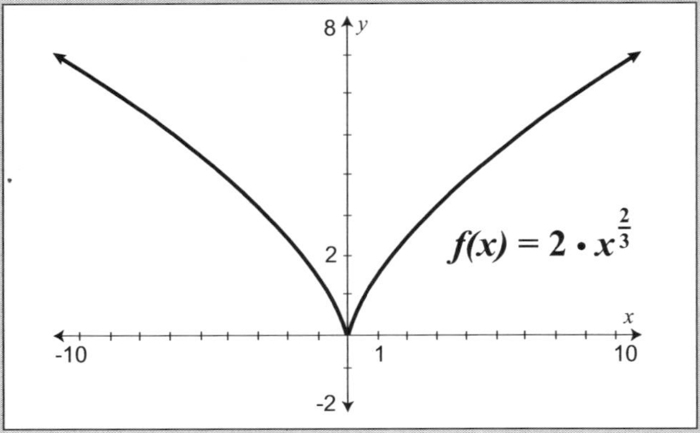

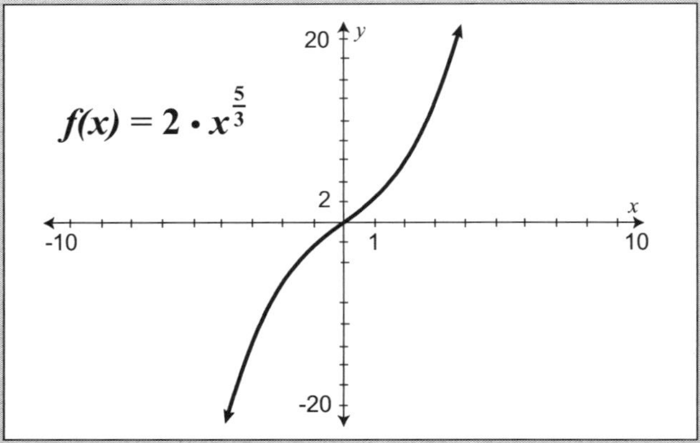

Power (a > 0, q is odd) $f(x)=a \cdot x^{p/q}$

(and p, q positive integers, reduced form, $q \neq 1$)

Simplest Form	$y = x^{\frac{1}{3}}$
Domain	All real numbers
Range	All non-negative real numbers (p even); all real numbers (p odd).
Characteristic	*The point (0, 0) is on the graph.*

Appearance

Touches the x-axis at (0, 0).

Interesting Features

These functions are either symmetric to the origin or symmetric in the y-axis.

Real-World Application

The surface area of a human being can be modeled by the power function
$S(h) = 327h^{17/40}$
where h is height in inches.

Power (q is even) $f(x)=a \cdot x^{p/q}$

(and p, q positive integers, reduced form, $q \neq 1$)

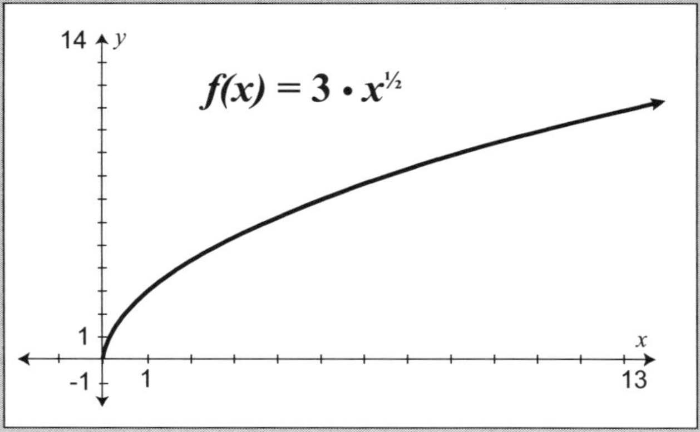

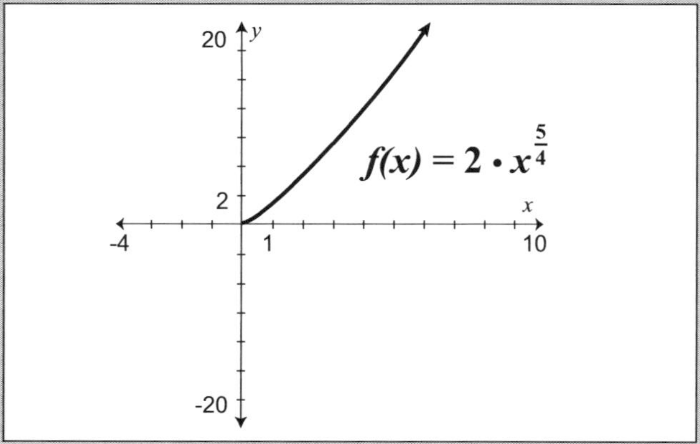

Power (a > 0, q is even) $f(x)=a \cdot x^{p/q}$

(and p and q positive integers, reduced form, $q \neq 1$)

Simplest Form	$y = x^{\frac{1}{2}}$
Domain	All non-negative real numbers
Range	All non-negative real numbers
Characteristic	*The point (0, 0) is on the graph.*

Appearance

The graph starts at the point (0, 0).

Interesting Features

Always positive and increasing

Real-World Application

The time required for a planet to make one complete revolution about the sun is approximated by a power function of the planet's distance from the sun:
$T(d) = k \cdot d^{3/2}$

Logistic

$$f(x) = \frac{a}{1 + c \cdot b^{-x}}$$

(where $b > 1$, a and c greater than 0)

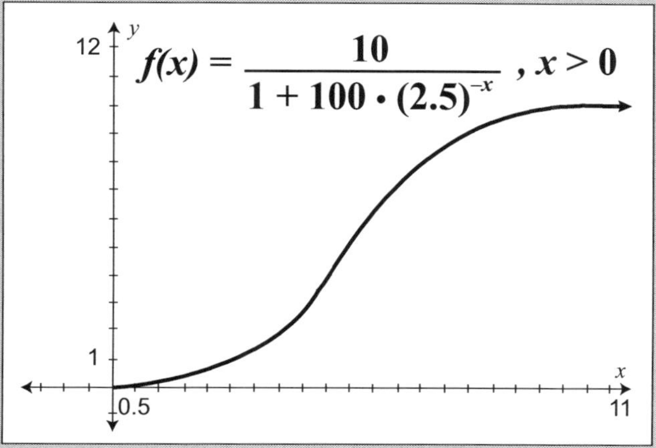

Logistic

$$f(x) = \frac{a}{1 + c \cdot b^{-x}}$$

(where $b > 1$, a and c greater than 0)

Simplest Form	$y = \frac{1}{1 + 2^{-x}}$
Domain	All real numbers
Range	$0 < y < a$

Appearance

Looks like an S and is sometimes referred to as a sigmoid

Interesting Features

Real-World Application

Logistic functions are used in determining limits on catching fish so that fish populations will be maintained.

Absolute Value

$$f(x) = a \cdot |x - h| + k$$

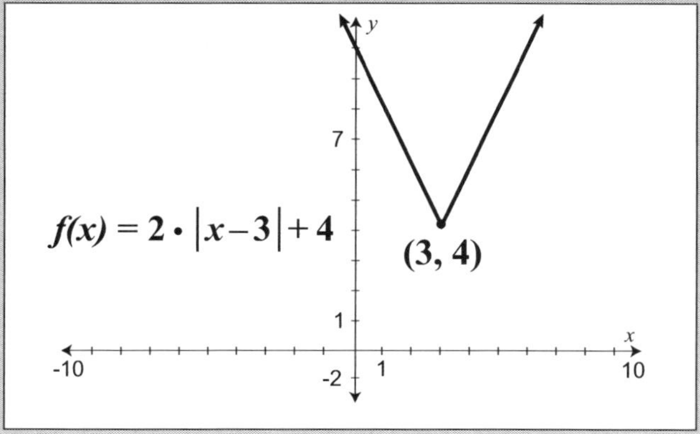

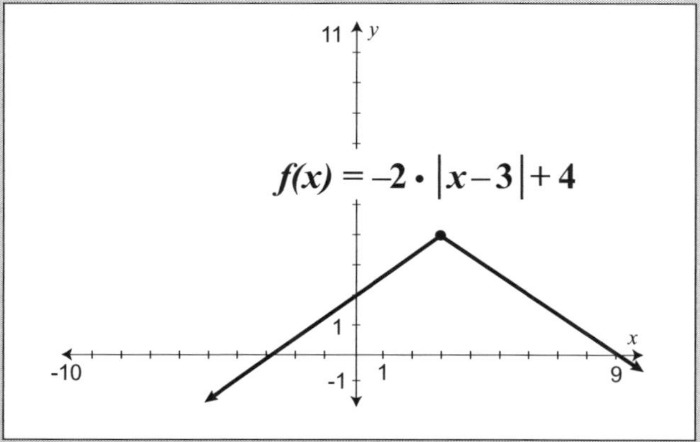

Absolute Value

$$f(x) = a \cdot |x - h| + k$$

Simplest Form	$y = \lvert x \rvert$
Domain	All real numbers
Range	$x \geq k$ or $x \leq k$, depending on whether $a > 0$ or $a < 0$

Appearance

Looks like V *or* Λ.

Interesting Features

Has a vertex at (h, k).

Real-World Application

The absolute value of your velocity is your speed.

Radical $f(x) = a \cdot \sqrt[n]{x-h} + k$ (and $a \neq 0$)

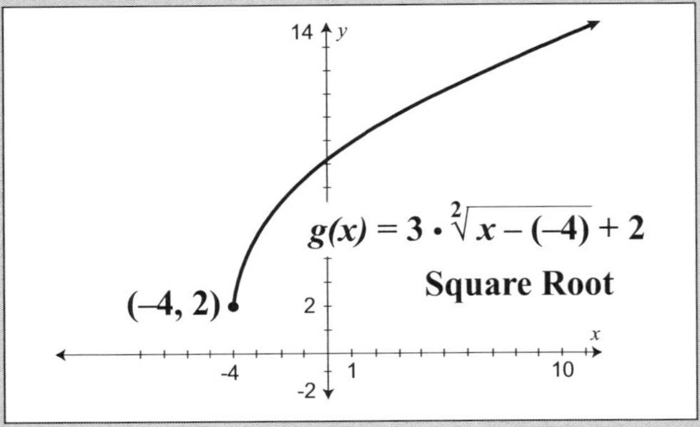

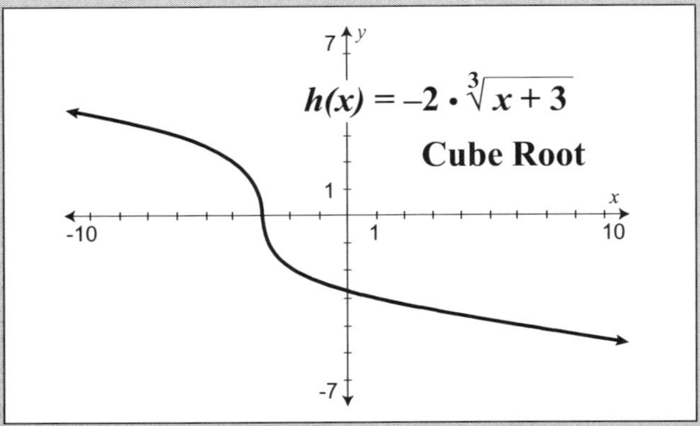

Radical $f(x) = a \cdot \sqrt[n]{x-h} + k$ (and $a \neq 0$)

Simplest Form	$y = \sqrt{x}$

Domain	Index even: All numbers for which $x \geq h$
	Index odd: All real numbers
Range	Index even: $a > 0$, $y \geq k$ and $a < 0$, $y \leq k$
	Index odd: All real numbers

Characteristic 1	*Any even index (n) will result in a graph similar to the square root graph.*
Characteristic 2	*Any odd index (n) will result in a graph similar to the cube root graph.*

Real-World Application

The hang time in football (the time elapsed between the time a punt is kicked and when it is caught) is modeled by the function at right, where T is in seconds and h is in feet.

$$T(h) = \frac{\sqrt{h}}{2}$$

Cyclical $f(x) = a \cdot \sin(x-h) + k$

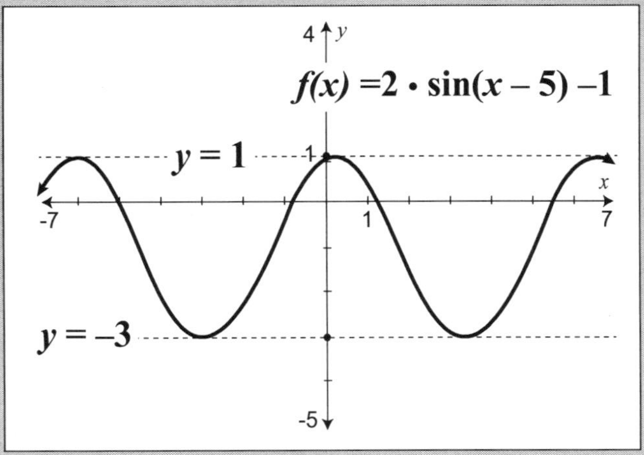

Cyclical $f(x) = a \cdot \sin(x-h) + k$

Simplest Form	$y = \sin x$

Domain	All real numbers
Range	$-\lvert a\rvert + k < y < \lvert a\rvert + k$

Appearance

Looks like a repeating wave.

Interesting Features

Maximum and minimum values are infinitely recurring.

Real-World Application

Analysis of alternating current electrical circuits requires cyclical functions.

Methodology 5.2

Interpreting a Function Using a Family of Functions Card

Graphing a function is very helpful when attempting to understand its behavior.

Limitation/Caution: There are many functions that do not fall into any family.

	Example 1	**Example 2**	**Your Turn**
Interpret the function given by the equation:	$y+6=2x^2-4x$	$y=\dfrac{x-1}{x^2-9}$	$3x^2=y-4x+12$

Steps	**Discussion**
1 Rewrite in functional notation	**If the function is given in equation form, solve for the dependent variable and present it in functional notation.**

Ex 1

$$y+6=2x^2-4x$$
$$y=2x^2-4x-6$$
$$f(x)=2x^2-4x-6$$

Ex 2

$$g(x)=\frac{x-1}{x^2-9}$$

Your Turn

Steps	**Discussion**
2 Determine which family the functions belongs to	**Compare the symbolic structure of the function to each family of functions to determine which is the appropriate family.**

Ex 1 Quadratic Function Family

Ex 2 Rational Function Family

Your Turn

Steps	**Discussion**
3 Rewrite the function in the standard form	**Rewrite the function in the standard form that appears on its family of functions card.**

Ex 1

$$f(x)=2x^2-4x-6$$

This is the standard form displayed on the card.

Ex 2

$$g(x)=\frac{x-1}{(x+3)(x-3)}$$

This is the standard form displayed on the card.

Your Turn

Steps	Discussion
4 Graph the function based on the key characteristics of its family of functions	Use the key characteristics of this family to determine the function's intercepts, key ordered pairs, and its basic behaviors.

Ex 1

Vertex:

$$x = -\frac{b}{2a} = -\frac{-4}{2(2)} = 1$$

$$y = 2(1)^2 - 4(1) - 6$$

$$y = 2 - 4 - 6$$

$$y = -8$$

x-intercepts (zeros):

$$2x^2 - 4x - 6 = 0$$

$$x^2 - 2x - 3 = 0$$

$$(x+1)(x-3) = 0$$

$$x = -1,\ 3$$

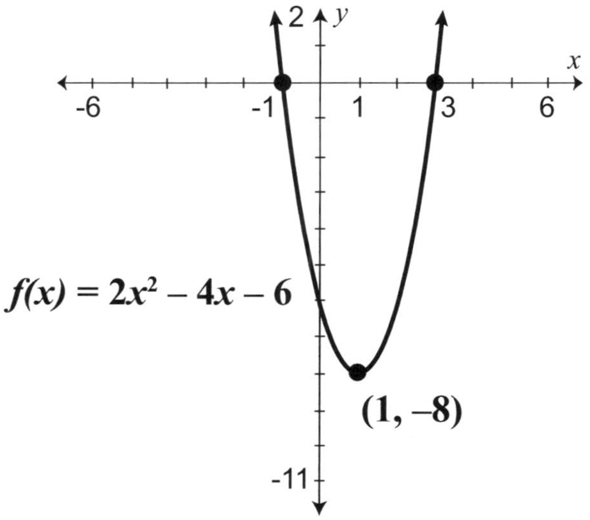

Ex 2

Vertical asymptotes at $x = 3$ and $x = -3$
Horizontal asymptote at $y = 0$

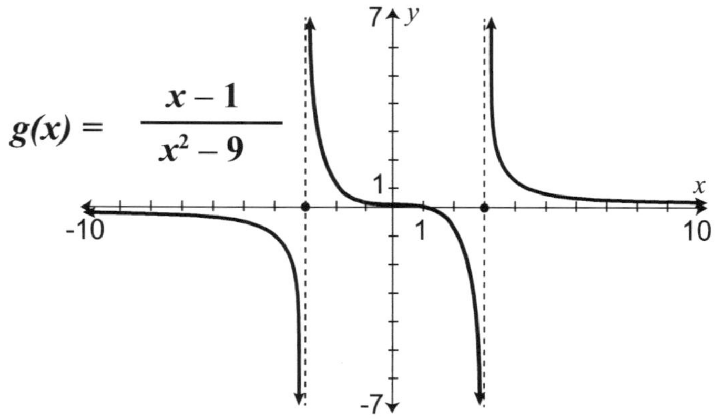

Steps	Discussion

Your Turn

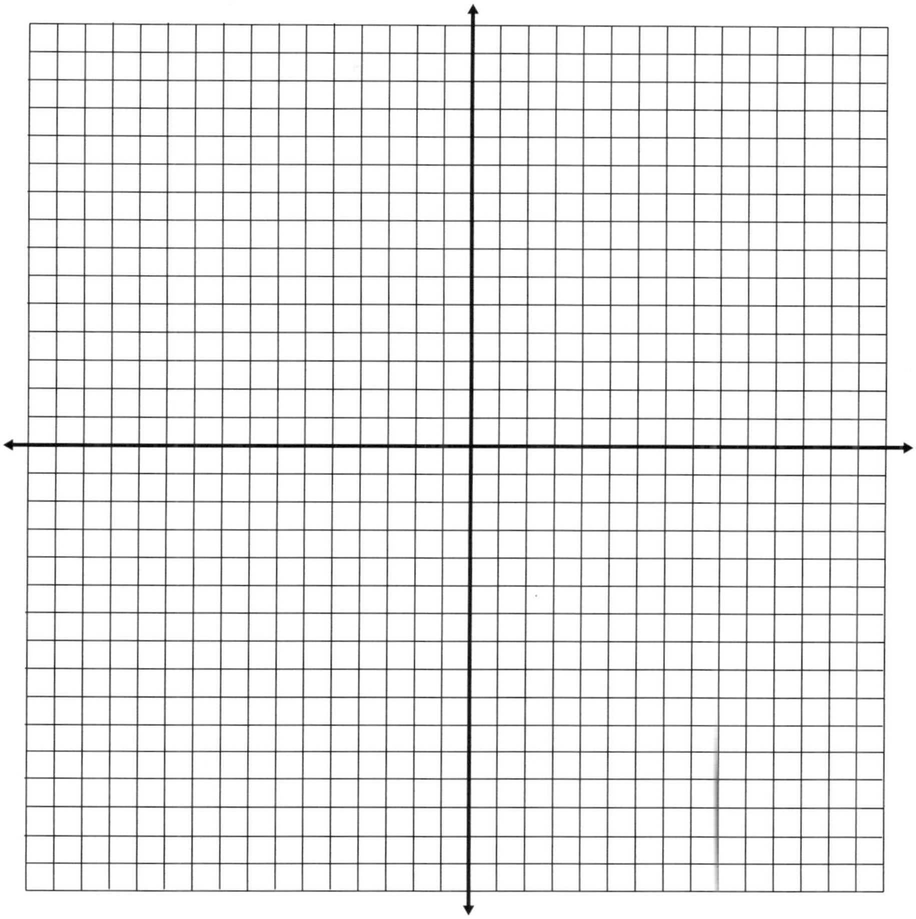

Steps	Discussion
5 Validate your graph	**Choose 5 key ordered pairs from the graph to make sure those ordered pairs satisfy the symbolic representation of the function.**

Ex 1

Choose 5 key points.

x	y
–1	0
3	0
1	–8
0	–6
2	–6

$f(-1) = 2(-1)^2 - 4(-1) - 6 = 0$ ✓

$f(3) = 2(3)^2 - 4(3) - 6 = 0$ ✓

$f(1) = 2(1)^2 - 4(1) - 6 = -8$ ✓

$f(0) = 2(0)^2 - 4(0) - 6 = -6$ ✓

$f(2) = 2(2)^2 - 4(2) - 6 = -6$ ✓

Ex 2

Choose 5 key points.

x	y
1	0
3	undefined
–3	undefined
0	0.1
–1	0.3

$g(x) = \dfrac{(1)-1}{(1)^2-9} = 0$ ✓

$g(3) = \dfrac{(3)-1}{(3)^2-9} = undefined$ ✓

$g(-3) = \dfrac{(-3)-1}{(-3)^2-9} = undefined$ ✓

$g(0) = \dfrac{(0)-1}{(0)^2-9} = \dfrac{-1}{-9} = 0.11$ ✓

$g(-1) = \dfrac{(-1)-1}{(-1)^2-9} = \dfrac{-2}{-8} = 0.25$ ✓

Your Turn

You can complete a function card for a given function after you have analyzed it in terms of a family of functions card.

Addressing Common Errors 5.2

The following table shows some of the most common errors that learners tend to make with the kinds of problems covered in this section.

ERROR: Misidentifying a function

Incorrect Process

Identify the family of functions for the following function: $f(x)=\frac{1}{2}x+3$

This function is a member of the rational function family.

Correct Process

This function is a member of the linear function family.

Resolution

Recall the difference between a rational expression and the coefficient of a linear function.

Validation

Graph the function.

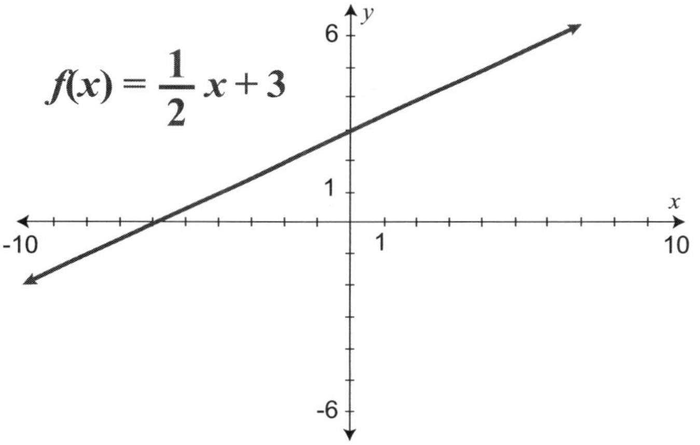

The graph of the function is a straight line.

ERROR: Plotting too few points

Incorrect Process

Graph the following member of the family of rational functions:

$$g(x) = \frac{x^2 + 2x - 6}{x + 4}$$

x	$g(x)$
–2	–3
0	$\frac{-3}{2}$
2	$\frac{1}{3}$
4	$\frac{9}{4}$
6	$\frac{21}{5}$

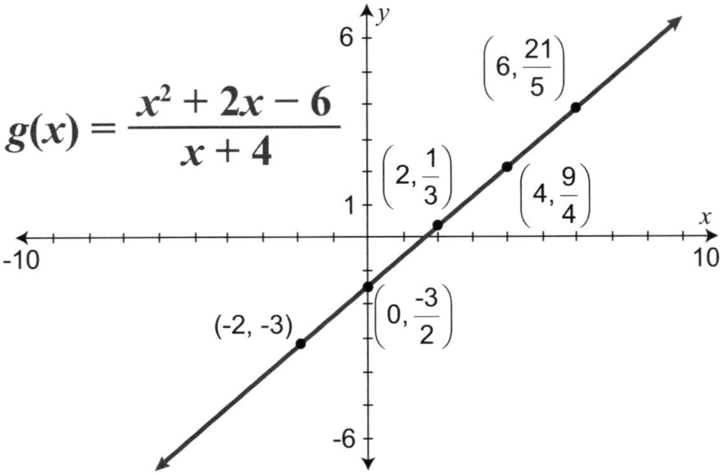

Correct Process

x	$g(x)$
–8	$\frac{-21}{2}$
–6	–9
–4	undef.
–2	–3
0	$\frac{-3}{2}$
2	$\frac{1}{3}$
4	$\frac{9}{4}$
6	$\frac{21}{5}$

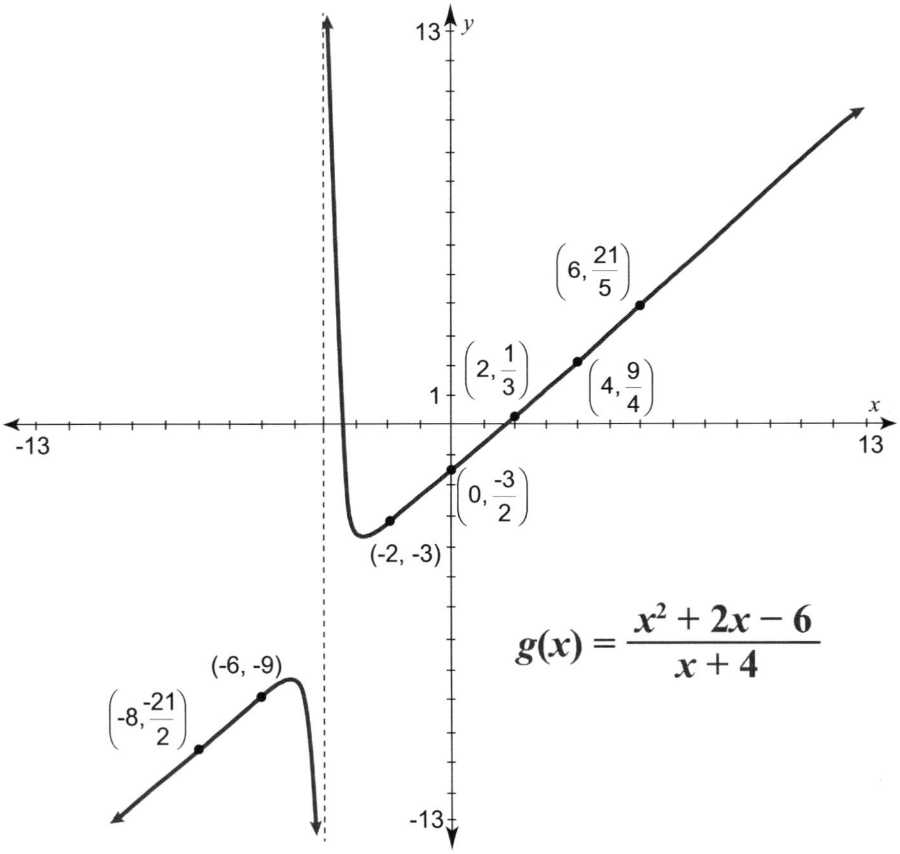

Resolution

Graph the rational function with a calculator knowing that there may be a vertical asymptote probably at $x = -4$ (the function may be undefined at $x = -4$).

Validation

The function is undefined at $x = -4$ and the point on the graph with approximate coordinates (–3, –3) should satisfy the function:

$$\frac{x^2 + 2x - 6}{x + 4} = \frac{(-3)^2 + 2(-3) - 6}{-3 + 4} = \frac{9 - 6 - 6}{1} = \frac{-3}{1} = -3$$

ERROR: **Choosing the wrong domain**

Incorrect Process

Find the domain of the function

$$f(x)=\frac{x^2-4}{x-2}$$

The domain of the function is $x \neq -2$.

Resolution

Correctly determine the domain of the function by solving the equation $x-2=0$.

Correct Process

$$x-2=0$$
$$x-2+2=0+2$$
$$x-0=2$$
$$x=2$$

The domain is all numbers except $x = 2$.

Validation

The function is undefined at $x = 2$.

$$f(x)=\frac{x^2-4}{x-2}$$

$$f(0)=\frac{2^2-4}{2-2}=\frac{0}{0}$$ is undefined.

Preparation Inventory 5.2

I can...	OR *Here's my question...*
☐ determine the domain of a function.	☐
☐ determine the range of a function.	☐
☐ determine the asymptotes of a function.	☐
☐ find the zeros of a function.	☐
☐ graph a function by hand and with technology.	☐

Critical Thinking Questions 5.2

1. Why do we graph a function to analyze it?

2. What are four common key components of a family of function card to consider when graphing?

3. Once you have identified which family a function belongs to, how do you write it in function form?

4. What patterns do you look for when choosing a family of functions card?

5. What are three examples of asymptotes on the family of functions cards?

6. Why are you developing the skill of analyzing a function?

Demonstrate Your Understanding 5.2

1. Identify the family of functions to which each of the following functions belongs.

a. $f(x) = 3x - 4$

b. $g(x) = \dfrac{3x - 5}{2x^4 - 5x^2 - 4x + 7}$

c. $h(x) = -3x^{8/3}$

2. Complete a function card for each of the given equations.

a. $2x - y = x^2 + 4$

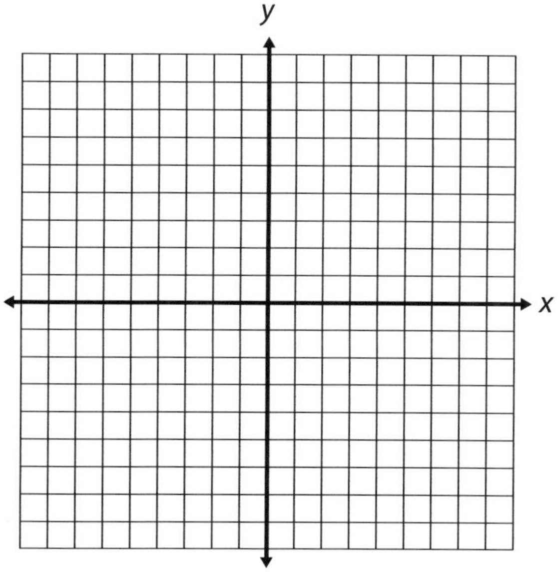

Simplest Form	
Domain	
Range	
Characteristic 1	
Characteristic 2	
Appearance	

c. $y = 3 \cdot 2^x$

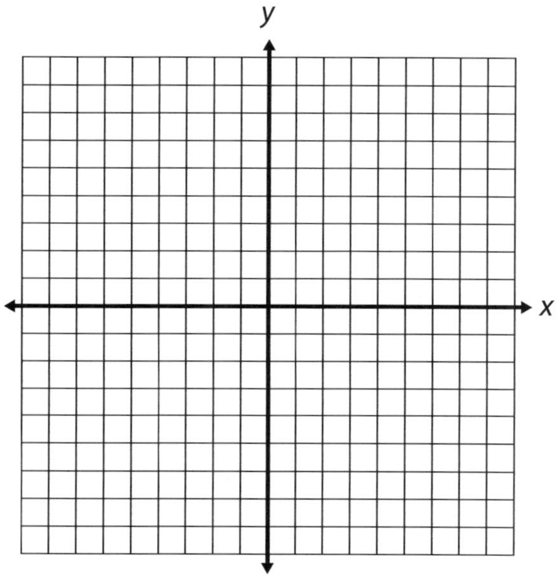

Simplest Form	
Domain	
Range	
Characteristic 1	
Characteristic 2	
Appearance	

b. $g(x) = \dfrac{3x+6}{(x^2-4)(x+5)}$

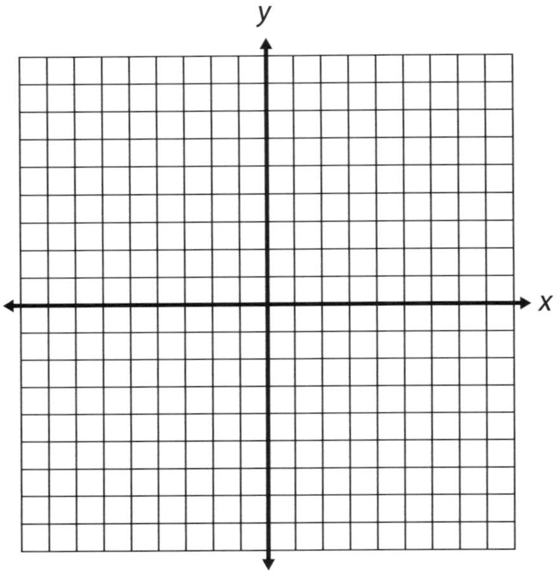

Simplest Form	
Domain	
Range	
Characteristic 1	
Characteristic 2	

Appearance

Create and Solve the Hardest Problem 5.2

Your Hardest Problem

Problem Solution

What makes a problem where you must identify a function as belonging to a family of functions a hard problem to solve?

Identify, Describe, & Correct the Errors 5.2

Solve the problem correctly in the appropriate location. Be sure to identify and describe the error you found in the worked solution!

Problem 1

Identify this function's family:

$$g(x) = -5x^{2/3}$$

Worked Solution *(Find the error!)*

This function is a member of the **polynomial family of functions**.

Correct Process

Describe the Error

Problem 2

Identify this function's family:

$$h(x) = \frac{7}{4}x^2 - \frac{1}{2}x - \frac{3}{5}$$

Worked Solution *(Find the error!)*

This function is a member of the **rational family of functions**.

Correct Process

Describe the Error

Problem 3

Identify the vertical asymptote of the following function:

$$h(x) = \frac{x-3}{x+5}$$

Worked Solution *(Find the error!)*

The vertical asymptote is at $x = 3$

Correct Process

Describe the Error

Problem 4	Correct Process	Describe the Error
Identify the domain of the following function: $g(x) = 3\sqrt[3]{x-4} + 7$		
Worked Solution *(Find the error!)*		
The domain is $x \geq 4$		

What Did You Learn About Learning Math? 5.2

1. What patterns did you learn to recognize when working with the family of functions cards?

2. What did you learn about visualization when working with family of functions cards?

3. What interesting contexts did you learn about in the real-world applications on the family of functions cards?

Section 5.3 Analyzing a Function

Why? 5.3

Many variables or quantities in life are functions of other variables. Being able to analyze the behavior of such functions, both algebraically and graphically, can explain a lot about phenomenon and may be used to predict future behavior as well.

What Do You Already Know? 5.3

How do you determine the domain of a function?

What are asymptotes and how do you find them?

Name five families of functions.

Goals 5.3

When you complete this activity, you should be able to:

1. Analyze a function from its graph
2. Analyze a function from its equation

Building Mathematical Language 5.3

New Terms: behavior of a function   end behavior of a function

Key Concepts 5.3

You describe the ***behavior of a function*** by stating its trends, x-intercepts, y-intercepts, maximums and minimums, asymptotes, and undefined numbers moving left to right through domain values. When a function models a real world phenomenon, its behavior can reveal a great deal about said phenomenon.

Limitation/Caution: We have not encountered all the possible behaviors of a function yet.

The ***end behavior of a function*** is the behavior exhibited by the function as domain values approach infinity ($x \to \infty$) or as domain values approach negative infinity ($-\infty \leftarrow x$).

Limitation/Caution: A function that is defined by a finite interval of real numbers will not have end behavior.

Methodology 5.3

Analyzing Functions

You can analyze a function using both its algebraic and graphical representations. Connections between these representations can provide important information about the function's behavior.

Limitation/Caution: A function's graphical representation may not make obvious important information.

	Example 1	**Example 2**	**Your Turn**
Analyze the function:	$f(x) = \dfrac{3}{1+x}, x > 0.$	$h(t) = t^2 - 5t - 6$	$g(x) = 3 \cdot 2^x \ x > 0.$

Steps	Discussion
1 Determine the function's family	**Families of functions have standard features.**

Ex 1 This is a rational function.

Ex 2 This is a quadratic function.

Your Turn

Steps	Discussion
2 Determine the domain of the function	**If not specified, the domain is the set of values for which the defining expression exists. Knowing the function's domain can help in determining the appropriate graphing window.**

Ex 1 The domain is specified as $x > 0$.

Ex 2 The domain is all real numbers.

Your Turn

Steps	Discussion
3 Graph the function	Many of the features of a function can be determined from its graph.

Ex 1

$f(x) = \frac{3}{1+x}, x > 0$

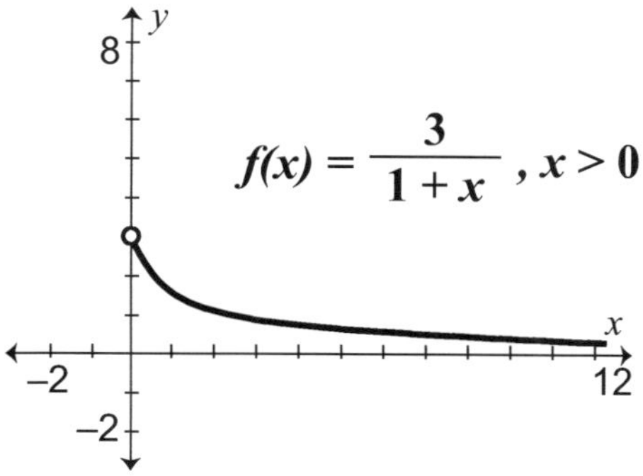

Ex 2

$h(t) = t^2 - 5t - 6$

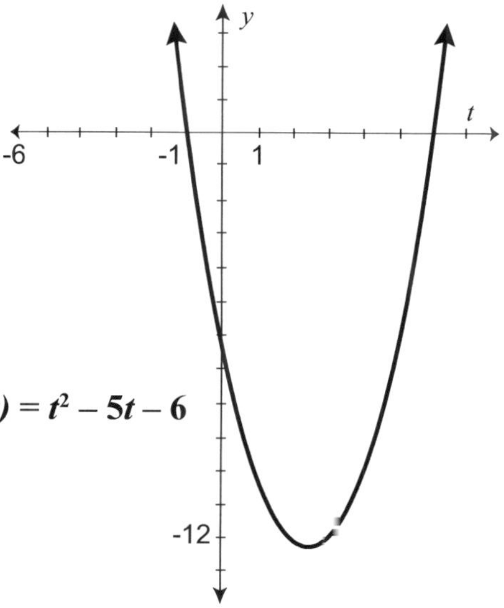

Your Turn

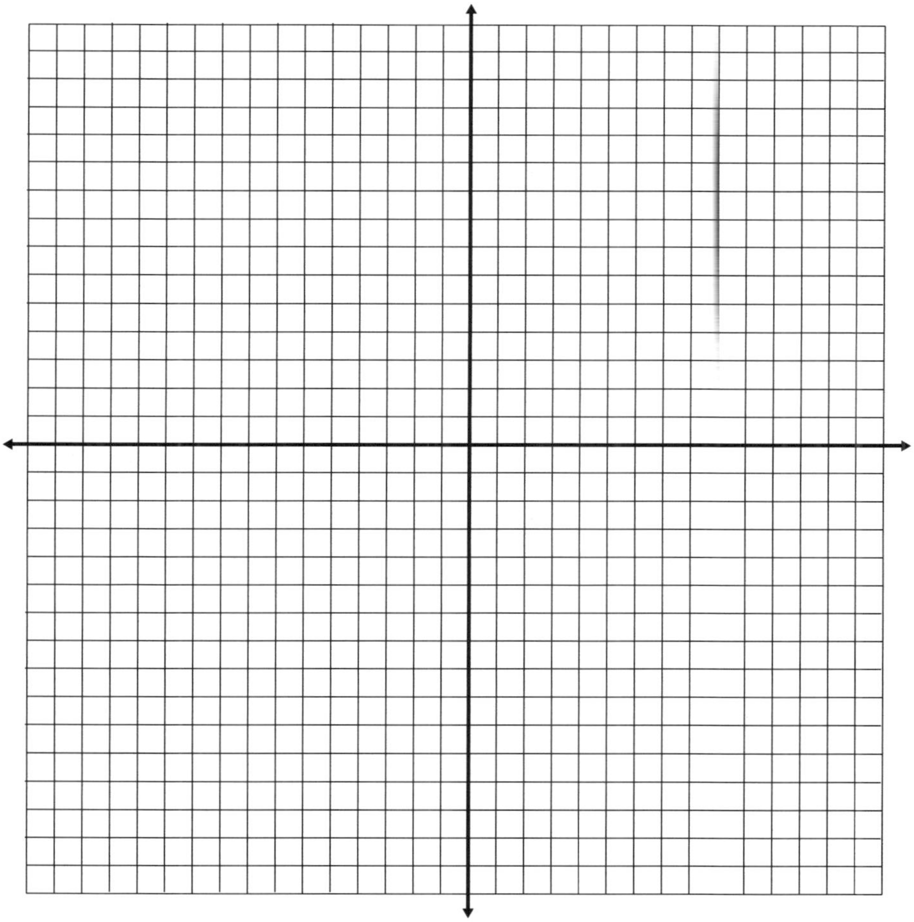

Steps	Discussion
4 Asymptotes	**A function's horizontal and vertical asymptotes can be determined from its symbolic representation and can be verified by its graph.**

Ex 1

There is a no vertical asymptote in this domain.

The horizontal asymptote is the x-axis ($y = 0$).

Ex 2

Quadratic functions have no asymptotes.

Your Turn

Steps	Discussion
5 Determine the range of the function	**The range of a function can be determined from its graph and its symbolic representation. A function's graph also gives information about its maximum and minimum.**

Ex 1

From the graph, the range is $0 < x < 3$ or the interval (0, 3) since $\frac{3}{1+0} = 3$ and as x gets large the denominator $1 + x$ gets large and the defining expression gets closer and closer to zero. The function has no maximum or minimum value.

Ex 2

From the graph, the range is

$[-12.25, \infty)$

Note that the minimum occurs when

$$t = \frac{-b}{2a} = 2.5$$

$h(2.5) = -12.25$

Your Turn

Steps	Discussion
6 Determine the function's intercepts	**The intercepts of a function can be determined from its graph or algebraic representation.**

Ex 1

There is no x-intercept or y-intercept.

Ex 2

The y-intercept (0, –6) occurs when $t = 0$. Solve $0 = t^2 - 5t - 6$ to find the t-intercepts.

$0 = (t - 6)(t + 1)$, so $t = 6$ or $t = -1$. The t-intercepts are (6, 0) and (–1, 0).

Your Turn

Steps	Discussion
7 Determine if the function is increasing or decreasing	**An important feature of a function is when it is increasing and when it is decreasing.**

Ex 1: The function is always decreasing.

Ex 2: The function is decreasing on the interval $(-\infty, 2.5)$ and increasing on the interval $(2.5, \infty)$.

Your Turn

Steps	Discussion
8 Determine function's end behavior	**Determine what the behavior of the function is when its domain values approach $-\infty$ or $+\infty$.**

Ex 1: The function approaches 0 as x gets large.

Ex 2: The function goes to ∞ as x gets large (positive) and as x gets very small (negative).

Your Turn

Steps	Discussion
9 Validate your analysis	**Compare your analysis against the family of function cards in Section 5.2 of this book. You can validate your graph through the use of a graphing calculator.**

Ex 1: The analysis fits the Rational family of functions card, within the constraints of $x > 0$. ✓

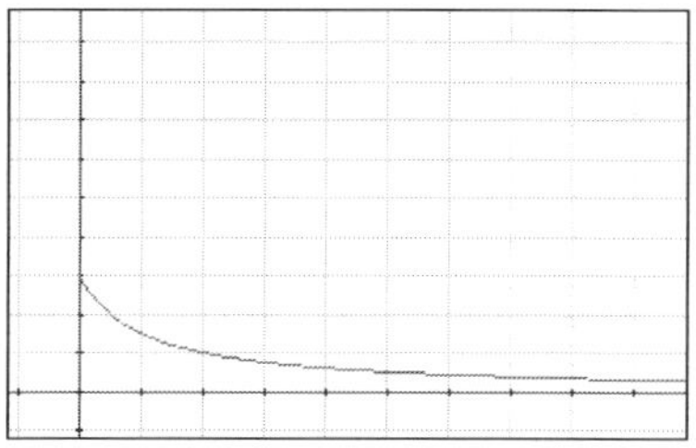

✓

Ex 2: The analysis fits the Quadratic family of functions card, with a range of $[-12.25, \infty)$. ✓

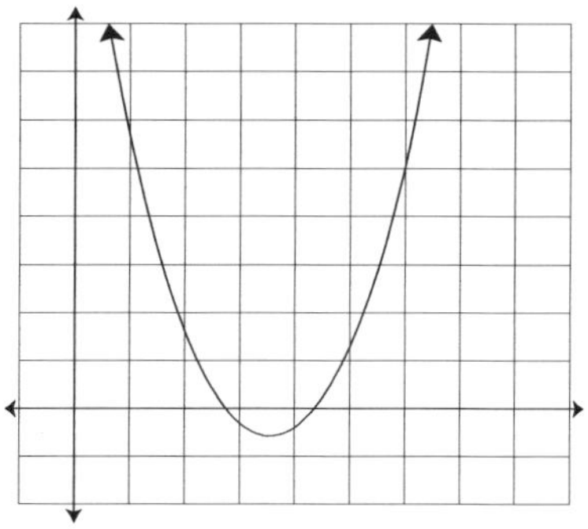

✓

Your Turn

(Do not graph your function again here. Simply analyze based on use of a graphing calculator or comparison with the family of function cards.)

Model 5.3

Analyze the function: $f(x) = x^{2/3}$

Step 1 Family of functions Power function

Step 2 Domain All real numbers

Step 3 Graph

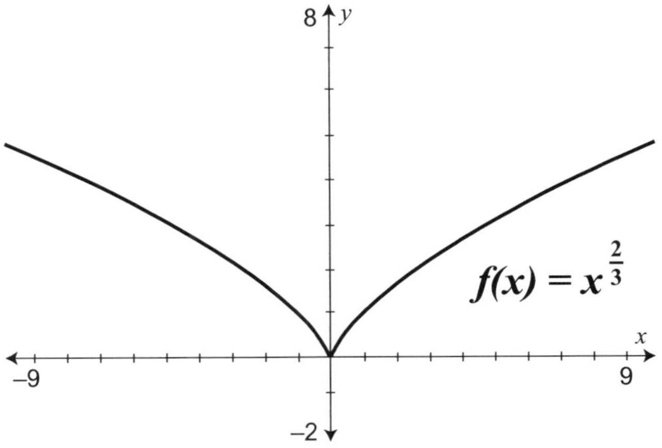

Step 4 Asymptotes No asymptotes

Step 5 Range Non-negative real numbers

Step 6 Intercepts x- and y-intercepts at (0,0)

Step 7 Increasing/decreasing Decreasing from $-\infty$ to 0 Increasing from 0 to ∞

Step 8 End behavior The graph is always positive, decreasing from $-\infty$ to 0 and increasing from 0 to ∞. Unbounded.

Step 9 Validate analysis Should behave like $x^{1/3}$ for $x > 0$. Check against the function card for the family of power functions. ✓

Addressing Common Errors 5.3

The following table shows some of the most common errors that learners tend to make with the kinds of problems covered in this section.

ERROR: Incorrectly identifying the family of a function

Incorrect Process

Identify the family of the function $f(x) = 3x^{-1} + 2$.

This function is from the linear family of functions.

Resolution

Rewrite the function with no negative exponents.

Correct Process

$$f(x) = 3x^{-1} + 2$$
$$= \frac{3}{x} + 2 = \frac{3+2x}{x}$$

This function belongs to the rational family of functions.

Validation

Compare this function with the rational family of functions card.

ERROR: **Incorrectly describing the behavior of a function.**

Incorrect Process

Describe the behavior of the function $g(x) = -2x^2 + 4x + 3$.

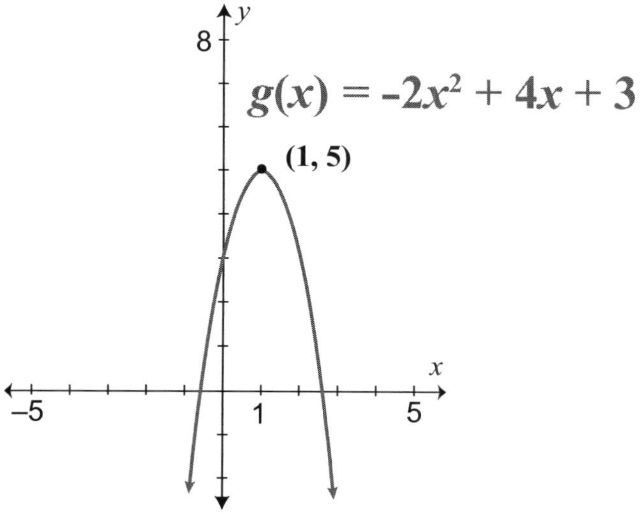

The function is a decreasing function.

Resolution

Describe the behavior of the function from left to right rather than from its minimum or maximum point.

Correct Process

The function is increasing from left to right to its maximum at (1,5) and decreasing from there.

Validation

The description of the behavior of this function matches the graph as the domain values move from left to right.

Preparation Inventory 5.3

I can...	OR ***Here's my question...***
☐ identify the domain of a function.	☐
☐ identify the intercepts of a function.	☐
☐ analyze the graph of a function.	☐
☐ analyze a function from its equation.	☐

Critical Thinking Questions 5.3

1. Why is it important to graph the function early in the process of analyzing it?

2. What are three families of functions which possess asymptotes?

3. What are four common key characteristics of families of functions to consider when graphing?

4. What impact does the exponent have on the behavior of a function?

5. What are three similarities that linear and exponential functions share?

6. What are three differences between linear and exponential functions?

7. Why are the domains of the square root and cube root functions different?

8. In what ways are the symbolic representations of polynomial and power functions the same?

Demonstrate Your Understanding 5.3

1. Create a function card for each of the following functions.

   a) $f(x) = 2 \cdot 3^x$

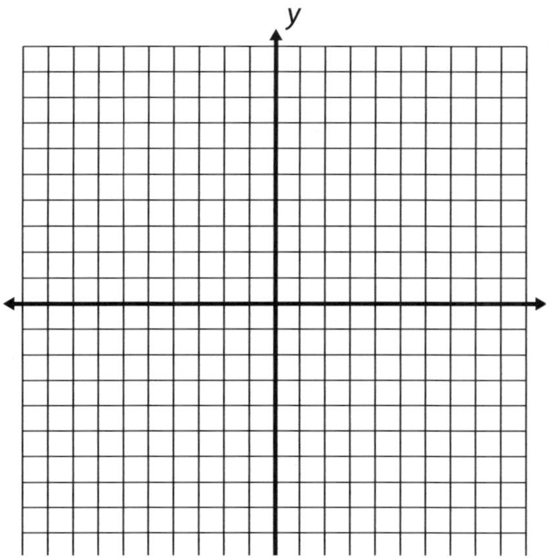

Simplest Form	
Domain	
Range	
Characteristic 1	
Characteristic 2	

Appearance

c) $r(x) = 2\sqrt{x-4}$

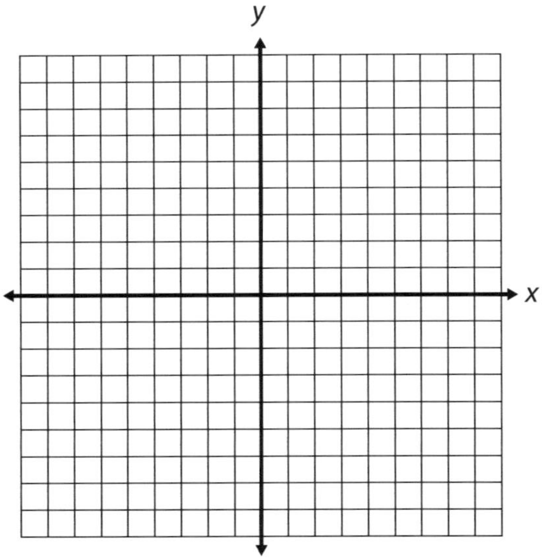

SIMPLEST FORM	
DOMAIN	
RANGE	
CHARACTERISTIC 1	
CHARACTERISTIC 2	

APPEARANCE

b) $g(x) = \dfrac{x}{2} - \dfrac{2}{x}$

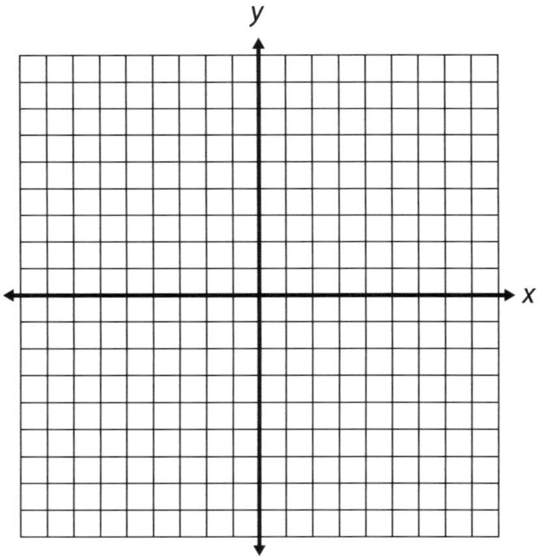

SIMPLEST FORM	
DOMAIN	
RANGE	
CHARACTERISTIC 1	
CHARACTERISTIC 2	

APPEARANCE

d) $h(x) = -2^{-x}$ Hint: $-2^{-x} = -\left(\frac{1}{2}\right)^{x}$

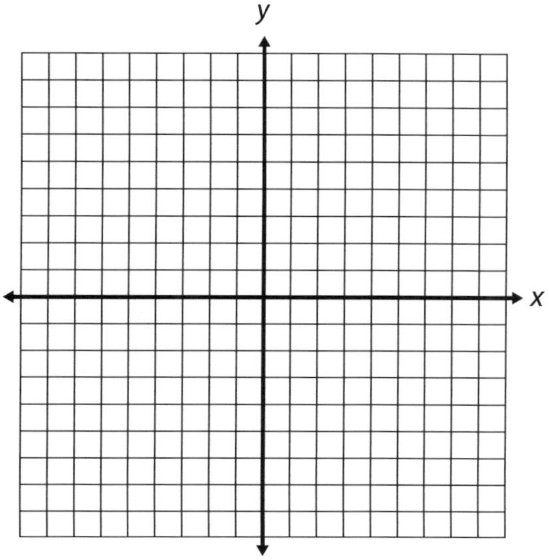

SIMPLEST FORM	
DOMAIN	
RANGE	
CHARACTERISTIC 1	
CHARACTERISTIC 2	
APPEARANCE	

2. What effect will changing the coefficient of x in $h(x) = 5 - 2^{-x}$ from -1 to 2 have on the graph of $h(x)$?

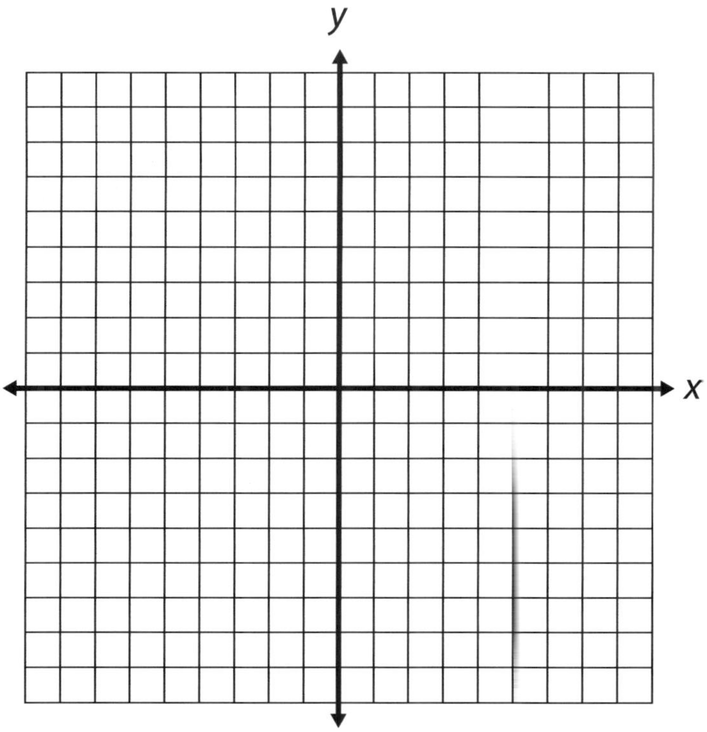

Create and Solve the Hardest Problem 5.3

Your Hardest Problem

Problem Solution

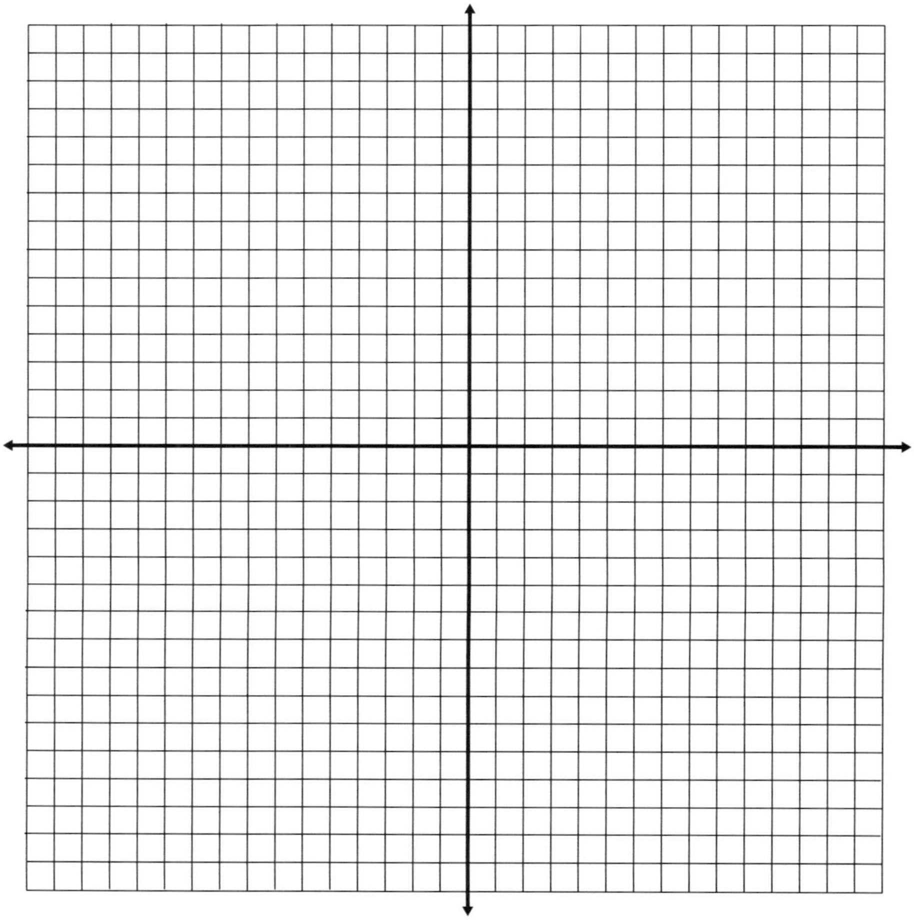

What makes a problem where you must analyze a function a hard problem to solve?

Identify, Describe, & Correct the Errors 5.3

Solve the problem correctly in the appropriate location. Be sure to identify and describe the error you found in the worked solution!

Problem 1

Identify the family this function belongs to:

$$f(x) = 5x^{0.34}$$

Worked Solution *(Find the error!)*

This function is a member of the **polynomial family of functions**.

Correct Process

Describe the Error

Problem 2

Identify the family this function belongs to:

$$g(x) = 3x^{-2}$$

Worked Solution *(Find the error!)*

This function is a member of the **polynomial family of functions**.

Correct Process

Describe the Error

Problem 3

Identify the domain of the following function:

$$f(x) = \frac{x^2 - 1}{x + 1}$$

Worked Solution *(Find the error!)*

$$f(x) = \frac{x^2 - 1}{x + 1} = \frac{(x+1)(x-1)}{(x+1)} = x - 1$$

The domain is **all real numbers**.

Correct Process

Describe the Error

What Did You Learn About Learning Math? 5.3

1. What similarities do you look for first when trying to identify the family of a function?

2. What limitation did you experience when trying to match a function with a family of functions card?

3. What assumptions can you make about the graph of your function compared with the graph on a family of functions card?

Section 5.4 Inverse Relations and Functions

Why? 5.4

Many variables or quantities in life are functions of other variables, and sometimes it is helpful to be able to look at those relationships from multiple perspectives. For example, one might want to know how long it would take for an investment to grow to $10,000, given the starting amount and the interest (growth) rate. This would involve solving an exponential function, where the variable is the exponent. The inverse of the exponential function would be needed to solve this problem.

What Do You Already Know? 5.4

How do you know if a relation is a function?

How do you solve $3x - 2y = 6$ for y?

How do you use a vertical line test to determine if a relation is a function?

How do you make a table of values for a function?

How do you graph a function?

Goals 5.4

When you complete this activity, you should be able to:

1. Find the inverse of a function numerically (from a table)
2. Find the inverse of a function from its graph
3. Find the inverse of some functions from their equations

Building Mathematical Language 5.4

New Terms: inverse function, one-to-one function, inverse relation, reflection

Key Concepts 5.4

If it exists, the function that is the inverse of a function (called, logically enough, the ***inverse function***) can be determined numerically, graphically, or algebraically.

Limitation/Caution: Not all functions have inverse functions.

A ***one-to-one function*** is a function that has just one output value for every input value. For example, the function *f*, shown at right is one-to-one because each input value has just one output value. However, the function *g* (below) is not one-to-one because the output value 4 appears twice in the output column and thus has two input values, 2 and 8.

$f(x)$	
x	y
2	4
5	7
8	1
9	11
10	5

$g(x)$	
x	y
2	4
5	7
8	4
9	11
10	5

Limitation/Caution: It is often difficult to determine if an algebraically defined function is one-to-one. It is much easier to determine if a graphically defined function is one-to-one.

Recall that the vertical line test can be used to determine if a relation is a function. Similarly a horizontal line test can be used to determine if a function is one-to-one. If any horizontal line crosses the graph more than once, then there are at least two distinct points whose second coordinates are the same. This means the function is not one-to-one.

An ***inverse function*** of a function undoes what the function does. For example, the inverse of the doubling function is the halving function. A specific numerical example: the doubling function turns 2 into 4, while its inverse, the halving function, turns 4 into 2. In function notation, double(2) = 4 and half(4) = 2. Using standard function notation, if f takes a to b or $f(a) = b$, then the inverse function (labeled f^{-1}) takes b back to a or $f^{-1}(b) = a$. In this case, (a, b) is a point on the graph of f and (b, a) is a point on the graph of f^{-1}.

The graph of the linear equation $y = x$ is the perpendicular bisector of the line segment joining (a, b) and (b, a) because the slope of the line is 1 while the slope of the line segment is –1:

$$\left(m = \frac{b-a}{a-b} = -1\right)$$

and the midpoint of the line segment is

$$\left(\frac{a+b}{2}, \frac{a+b}{2}\right)$$

which is on the line $y = x$.

This can be seen on the graph at right.

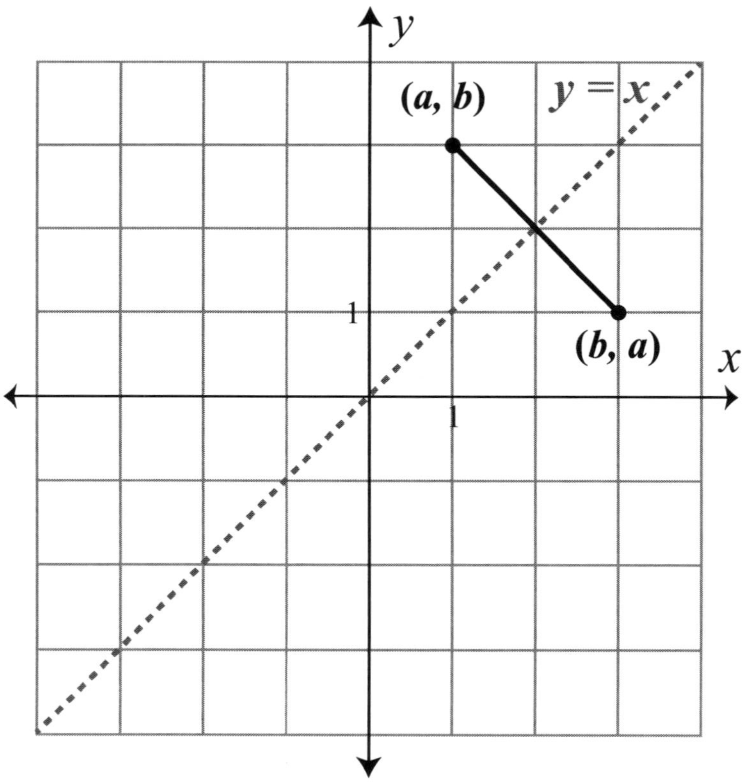

A ***reflection*** in or across a line results in a mirror image of the original figure. In particular, a reflection of a graph in the line $y = x$ will be a collection of points whose coordinates are reversed from those in the original graph. For example, if the point (2, 4) is on the original graph, then (4, 2) will be a point on the reflected graph.

Methodology 1 — 5.4

Finding Values of an Inverse Function Numerically (Using a Table)

All functions (even those that are not one-to-one) have ***inverse relations*** which can be determined by building a table of inverse values.

Limitation/Caution: Again, not every function has an ***inverse function***.

	Example 1	Example 2	Your Turn
Find the inverse of the function:	$f(x) = 2x$	$g(x) = 2^x$	$h(x) = 3 - x$

Steps	Discussion
1 Create a T-table for the function	**Choose appropriate values of x and calculate their corresponding y values.**

Ex 1

$f(x) = y = 2x$

f	
x	y
0	0
1	2
2	4
3	6
4	8

Ex 2

$f(x) = y = 2^x$

f	
x	y
0	1
1	2
2	4
3	8
4	16

Your Turn

Steps	Discussion
2 Interchange the columns of the T-table	**Interchange the columns of the T-table to create the table for the inverse function.**

Ex 1

The inverse function

f^{-1}	
x	y
0	0
2	1
4	2
6	3
8	4

Ex 2

The inverse function

f^{-1}	
x	y
1	0
2	1
4	2
8	3
16	4

Your Turn

Steps	Discussion
3 Check that the function is one-to-one	**Check both tables to make sure that they represent functions**

Ex 1	Ex 2	Your Turn
OK	OK	

Steps	Discussion
4 If the function is linear, write the equation of its	**Use the table to find the slope and y-intercept.**

Ex 1

$y = \text{m}x + \text{b}$ From the table,

$m = \frac{1}{2}$ and $b = 0.$

$y = \frac{1}{2}x + 0$

$f^{-1}(x) = \frac{1}{2}x$

Ex 2

This function is not linear.

Your Turn

Steps	Discussion
5 Validate	**Find the output value for a certain input value. This output value should be the input value for the inverse function and the input value should be the inverse function's output. That is, if $f(a) = b$, then $f^{-1}(b) = a$.**

Ex 1

$f(4) = 2(4) = 8$

$f^{-1}(8) = \frac{1}{2}(8) = 4$ ✓

Ex 2

Nothing to validate.

Your Turn

Note: The function f^{-1} in Example 1 reverses the effect of f. Denoting f as the "doubling" function, and f^{-1} as the "halving" function, you reverse the effect of doubling a number by halving the result. For example, $f(5) = 10$ (doubling) and then $f^{-1}(10) = 5$ (halving) or $5 \xrightarrow{f} 10 \xrightarrow{f^{-1}} 5$

Example 2 does not have an immediately apparent formula, as does Example 1, where doubling and halving were obvious. This function rewrites the input number as a power of 2 and then takes that power as the output value. For example $8 = 2^3$ and $f^{-1}(8) = 3$. The symbol for this inverse function is $\log_2(x)$ by convention. For example, $\log_2(16) = \log_2(2^4) = 4$.

This is the second time (at least) that you have encountered a symbol that is more a description of a number rather than the number itself. (For example, $\sqrt{2}$.)

For this function and its inverse, $3 \xrightarrow{f} 8 \xrightarrow{f^{-1}} 3$ or, in general, $x \xrightarrow{f} f(x) \xrightarrow{f^{-1}} x$.

Methodology 2 5.4

Finding Values of an Inverse Function Graphically

Graphing the inverse of a function can help you determine whether said inverse is a function.

Limitation/Caution: Inverse functions where the variables are measurements with units cannot be graphed in this way. Also, accurately graphing the original function may be difficult.

	Example 1	**Example 2**	**Your Turn**
Graph the function and its inverse:	$f(x) = 2^x$	$g(x) = x^3 - 4$	$h(x) = -2x + 8$

Steps	**Discussion**
1 Graph the function	**Refer to its family of functions card for help, if needed.**

Ex 1

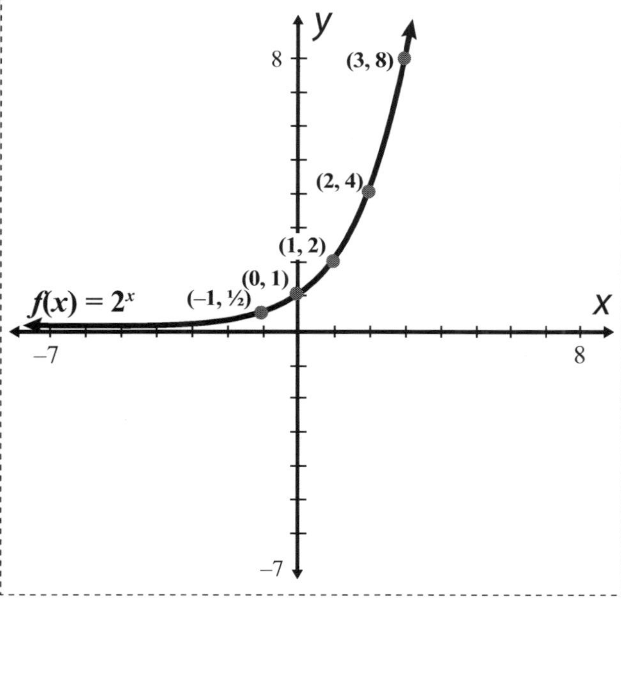

Ex 2

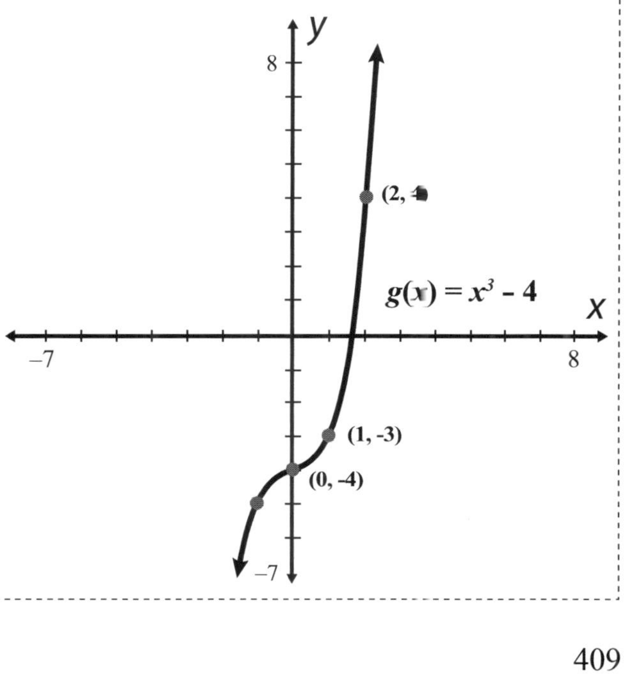

Steps	Discussion

Your Turn

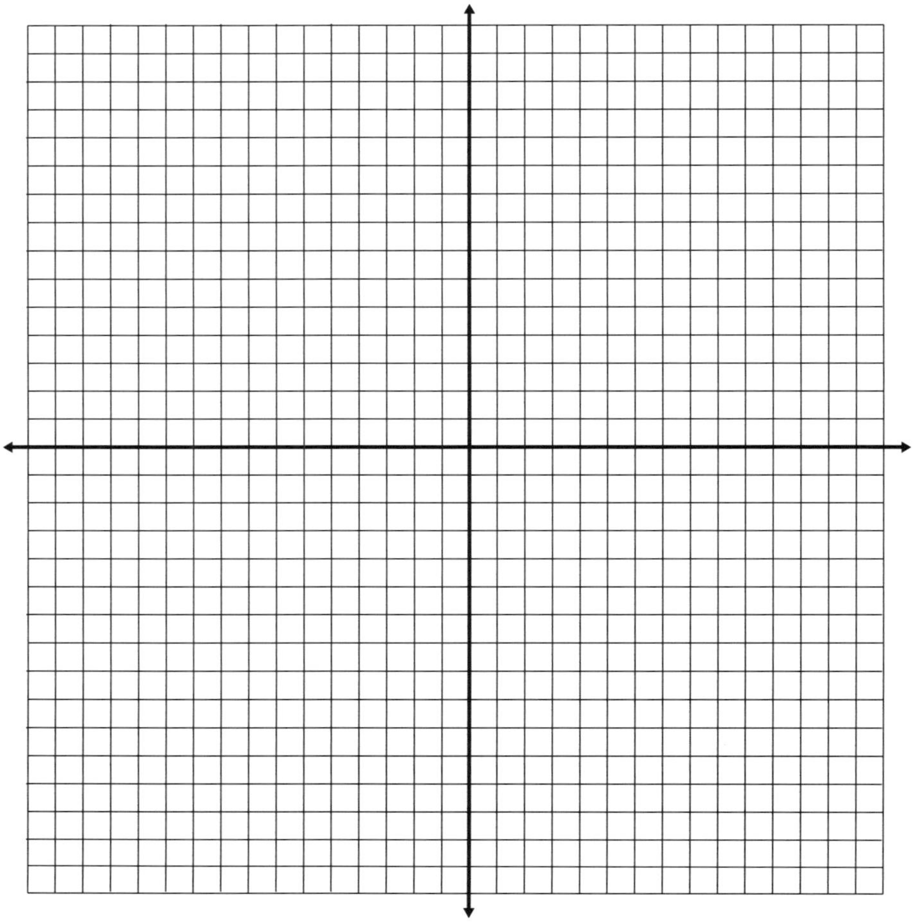

2 Check that the function is one-to-one	Use the horizontal line test.

Ex 1	Ex 2	Your Turn
OK	OK	

Steps	Discussion
3 Graph the line $y = x$	This will be the *line of reflection.*

Ex 1

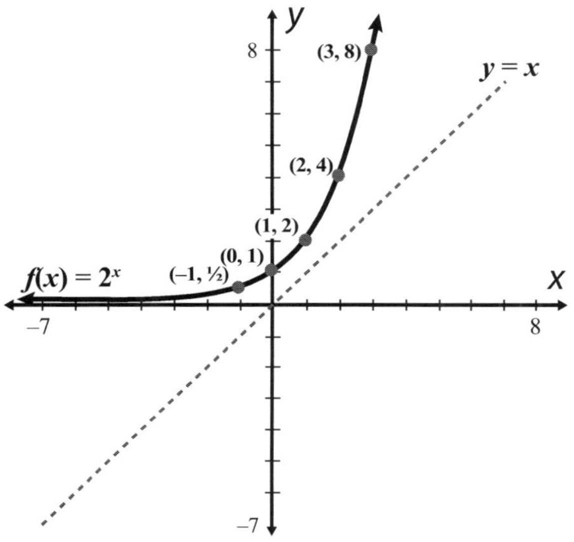

Ex 2

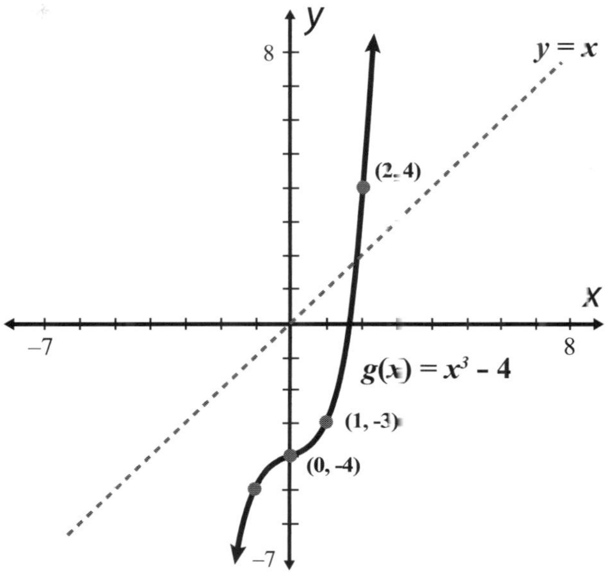

Your Turn

Use your previous graph.

4 Draw the reflection	Plot points that are the reflection (in this line) of points on the graph of the function. Draw the inverse function graph through these points of reflection.

Ex 1

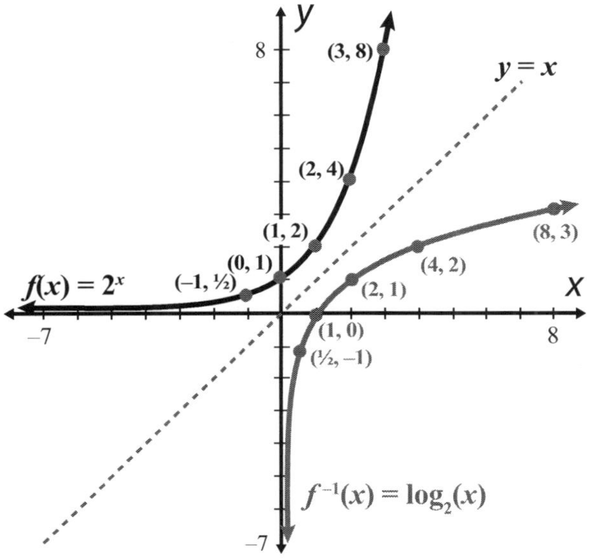

Ex 2

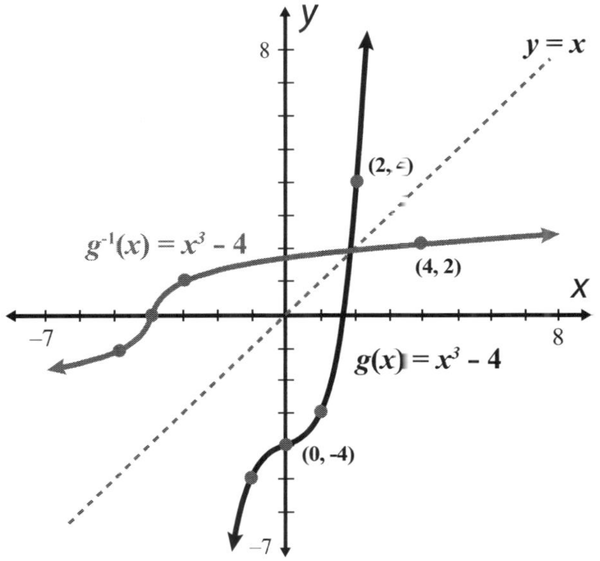

Your Turn

Use your previous graph.

Steps	Discussion
5 Validate	**Draw a line perpendicular to the line $y = x$. The line $y = x$ should be the perpendicular bisector of the line segment between the two points where this line intersects the graphs of f and f^{-1}.**

Ex 1

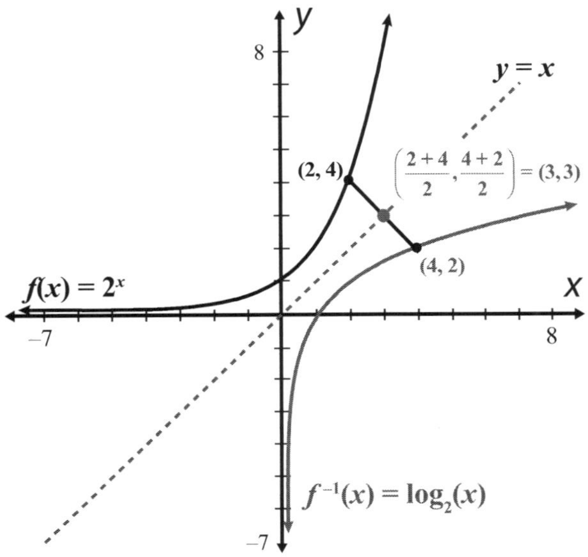

The line drawn between (2, 4) and (4, 2) has slope –1:

$$m = \frac{4-2}{2-4} = \frac{2}{-2} = -1$$

Therefore the line $y = x$ with slope 1 is perpendicular to it.

The mid-point of the line segment between these two points has coordinates (3, 3), by calculation, which is on the line $y = x$. ✓

All line segments between pairs of points like this (with the coordinates reversed) will have the line $y = x$ as their perpendicular bisector.

Ex 2

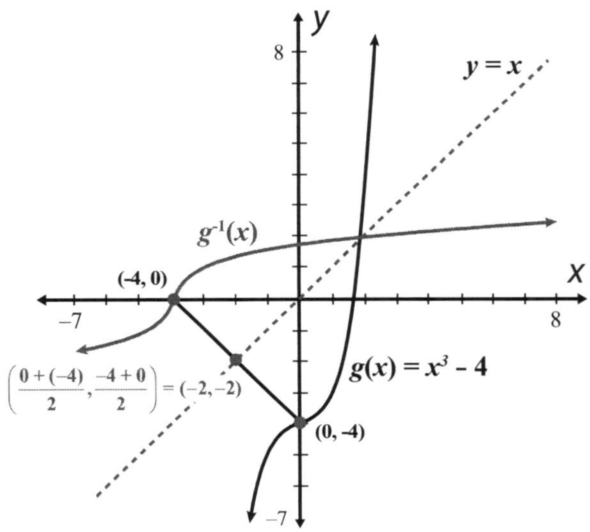

The line drawn between the points of (–4, 0) and (0, –4) has slope –1:

$$m = \frac{-4-0}{0-(-4)} = \frac{-4}{4} = -1$$

Thus the line $y = x$ with slope 1 is perpendicular to it.

The mid-point of the line segment between these two points has coordinates (–2, –2), by calculation, which is on the line $y = x$. ✓

Your Turn

Use your previous graph.

Methodology 3 — 5.4

Determining the Inverse of a Linear Function Algebraically

Limitation/Caution: Not all inverse functions can be found algebraically, because it is not always possible to solve for y after the x and y variables are interchanged in the original function.

	Example 1	Example 2	Your Turn
Find the inverse of the given function:	$f(x) = 2x - 5$	$f(x) = 5 - 3x$	$g(x) = 3 - 4x$

Steps	Discussion
1 State the function	Write the function in function notation.

Ex 1

$f(x) = 2x - 5$

Ex 2

$f(x) = 5 - 3x$

Your Turn

Steps	Discussion
2 Graph the function	Determine if the function is one-to-one (every element of the range corresponds to only one element of the domain). Do this by graphing the function.

Ex 1

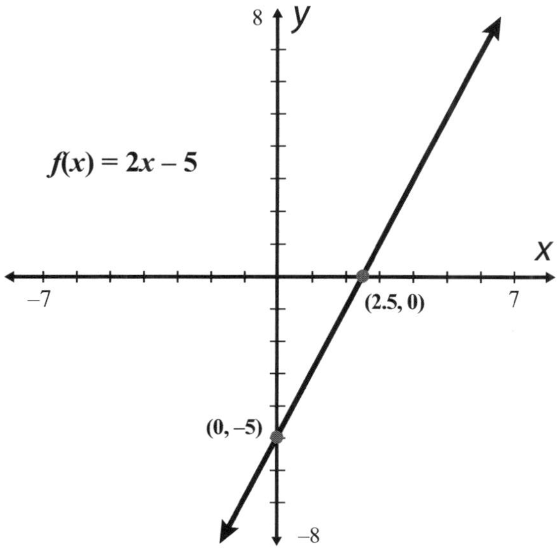

The function is one-to-one.

Ex 2

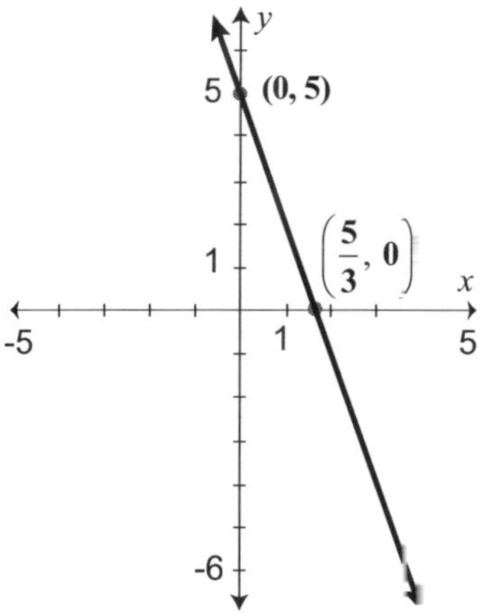

The function is one-to-one.

Steps	Discussion

Your Turn

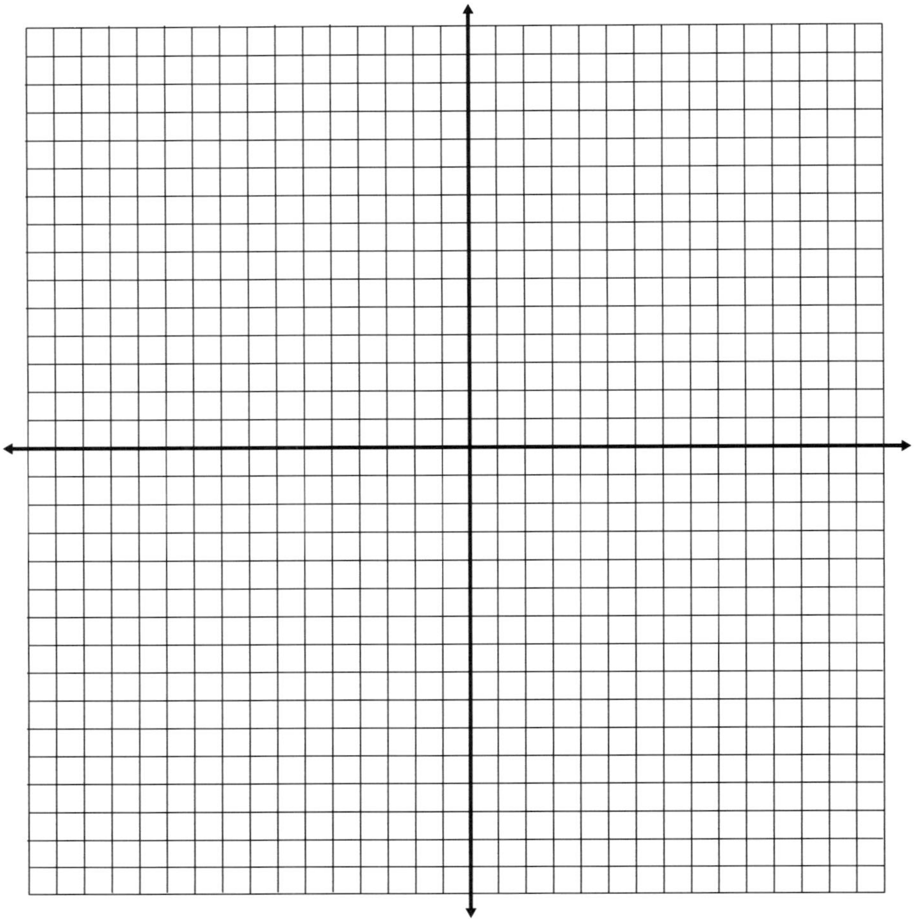

Steps	Discussion
3 Substitute *y* for *f*(*x*)	**Write the equation as an equation with two variables.**

Ex 1

$$y = 2x - 5$$

Ex 2

$$y = 5 - 3x$$

Your Turn

4 Interchange variables *x* and *y*.	**Replace *y* with *x* and *x* with *y*.**

Ex 1

$$x = 2y - 5$$

Ex 2

$$x = 5 - 3y$$

Your Turn

5 Solve for *y*	**Use the Methodology for Solving Basic Equations from Section 2.2.**

Ex 1

$$y = \frac{x+5}{2}$$

Ex 2

$$x - 5 = -3y$$

$$-3y = x - 5$$

$$y = \frac{x-5}{-3} = -\frac{x}{3} + \frac{5}{3}$$

Your Turn

6 Write the inverse in function notation	**Present the inverse function in function notation.**

Ex 1

$$f^{-1}(x) = \frac{x+5}{2}$$

Ex 2

$$f^{-1}(x) = -\frac{x}{3} + \frac{5}{3}$$

Your Turn

Steps	Discussion
7 Validate	Show that one pair of values from the table for *f* appears switched in the table for f^{-1}. You can also graph the two functions to see if they are inverses.

Ex 1

$f(x) = 2x - 5$	
x	y
0	–5
3	**1**
1	–3
$\frac{5}{2}$	0

$f^{-1}(x) = \frac{x+5}{2}$	
x	y
–5	0
1	**3**
–3	1
0	$\frac{5}{2}$

$f(3) = 2(3) - 5 = 1$

$$f^{-1}(1) = \frac{(1)+5}{2} = 3$$

The pair (3, 1) is in the table for $f(x)$.
The pair (1, 3) is in the table for $f^{-1}(x)$. ✓

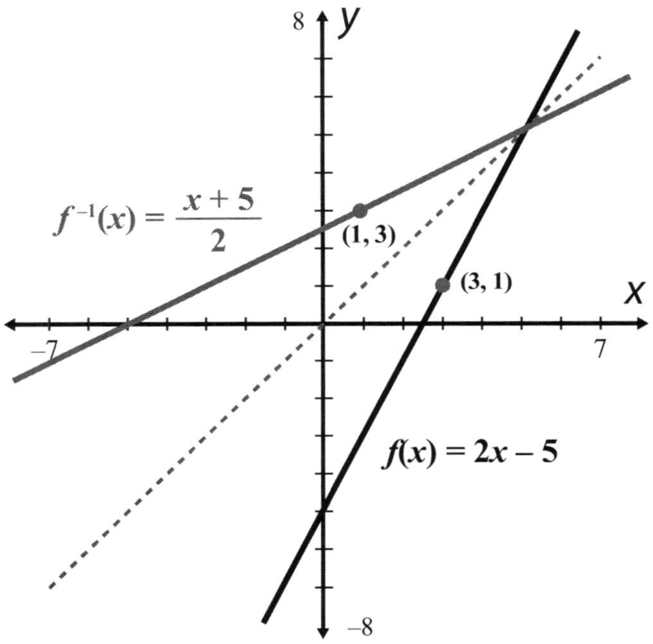

Ex 2

$f(x) = 5 - 3x$	
x	y
0	5
1	2
2	**–1**
$\frac{5}{3}$	0

$f^{-1}(x) = -\frac{x}{3} + \frac{5}{3}$	
x	y
5	0
2	1
–1	**2**
0	$\frac{5}{3}$

$f(x) = 5 - 3x$

$f(2) = 5 - 3(2) = -1$

$$f^{-1}(x) = -\frac{x}{3} + \frac{5}{3}$$

$$f^{-1}(-1) = -\frac{(-1)}{3} + \frac{5}{3}$$

$$f^{-1}(-1) = \frac{1}{3} + \frac{5}{3} = 2$$

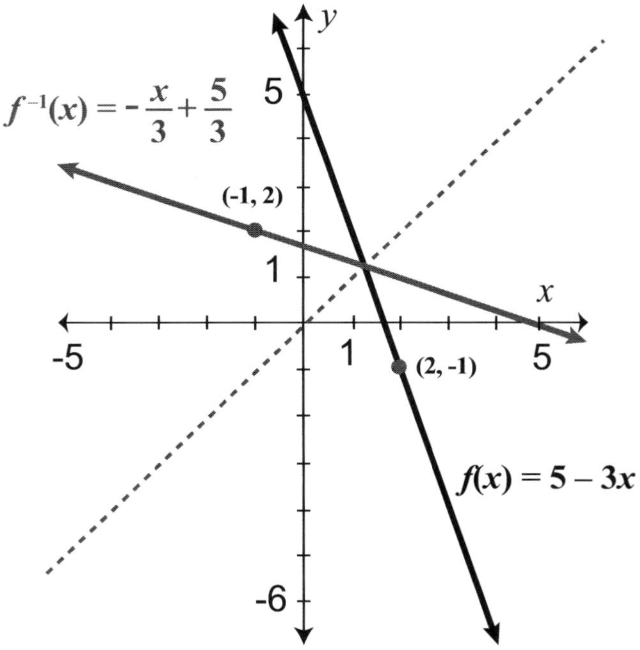

The pair (2, –1) is in the table for f and the pair (–1, 2) is in the table for f^{-1}. ✓

Your Turn

Use your previous graph and the space provided there.

Addressing Common Errors 5.4

The following table shows some of the most common errors that learners tend to make with the kinds of problems covered in this section.

ERROR: Graphing too little of the graph

Incorrect Process

Graphically find the inverse: $f(x) = x^2 + 2$

x	$f(x)$
0	2
1	3
2	6
3	11
4	18

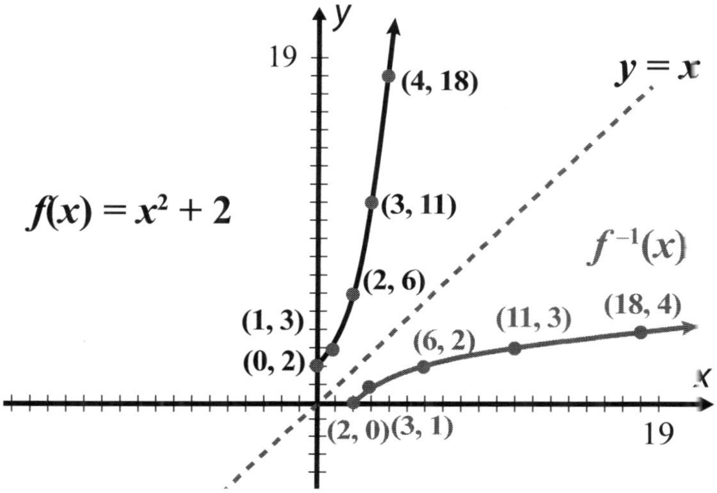

Correct Process

This function is not one-to-one and therefore has no inverse function.

Resolution

Refer to the family of functions cards when needed, to remind yourself of what the representative portion of the graph should look like.

Validation

This graph fails the horizontal line test. ✓

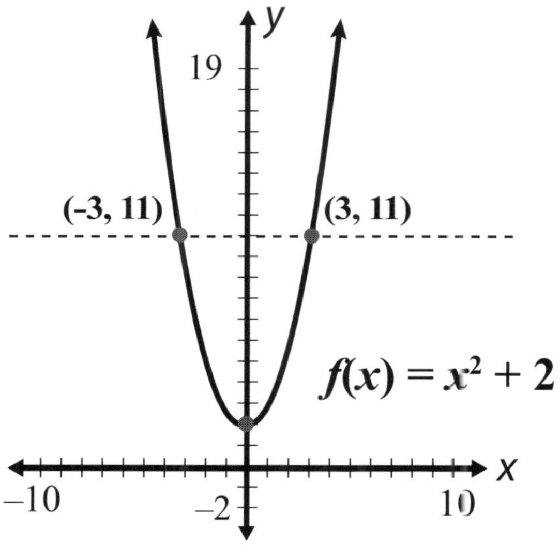

ERROR: Confusing inverse function notation with an exponent of –1

Incorrect Process

Find $f^{-1}(x)$ if $f(x) = x^5$

$$f^{-1}(x) = \frac{1}{f(x)} = \frac{1}{x^5}$$

Resolution

Remember that $f^{-1}(x)$ means the inverse function of $f(x)$, not the reciprocal of $f(x)$.

Correct Process

Find $f^{-1}(x)$ if $f(x) = x^5$

$y = x^5$

Interchange x and y: $x = y^5$

Solve for y: $y = \sqrt[5]{x}$

$f^{-1}(x) = \sqrt[5]{x}$

Validation

Reverse the process. Find the inverse function of $f^{-1}(x)$; it should be the same as $f(x)$.

$y = \sqrt[5]{x}$

Interchange x and y: $x = \sqrt[5]{y}$

Solve for y by raising each side to the 5th power: $y = x^5$

$f(x) = x^5$ ✓

Preparation Inventory 5.4

I can...	OR *Here's my question...*
☐ make a T-table for a given function.	☐
☐ create a graph of a given function.	☐
☐ find an inverse function algebraically.	☐
☐ find an inverse function graphically.	☐

Critical Thinking Questions 5.4

1. Why do we graph a function to analyze it?

2. Why is the inverse of a function important?

3. How do you validate that an inverse relation is a function graphically?

4. What is the relationship between the x-values of a function and the y-values of its inverse function?

5. Why can you replace x with y to find the inverse relation of a function?

6. Why do one-to-one functions always have inverse functions?

7. Why can't functions that are *not* one-to-one have inverse functions?

8. What are four examples of functions and their inverse functions?

9. What is true about the graph of the inverse function of $f(x)$?

10. How do you validate that an inverse relation is a function graphically?

11. How do the domain and range of a one-to-one function relate to the domain and range of its inverse function?

12. In the first methodology, Finding Values of an Inverse Function Numerically (Using a Table), you found that the inverse function of $h(x) = 3 - x$ was $h^{-1}(x) = 3 - x$. In other words, the function was its own inverse. Explain why this happened, and for what other functions will it happen again. (Hint: Finding the inverse graphically or algebraically might shed some light on this situation.)

Demonstrate Your Understanding 5.4

1. Graph the function of $f(x) = 2^x$. Also graph its inverse, $f^{-1}(x)$, on the same graph.

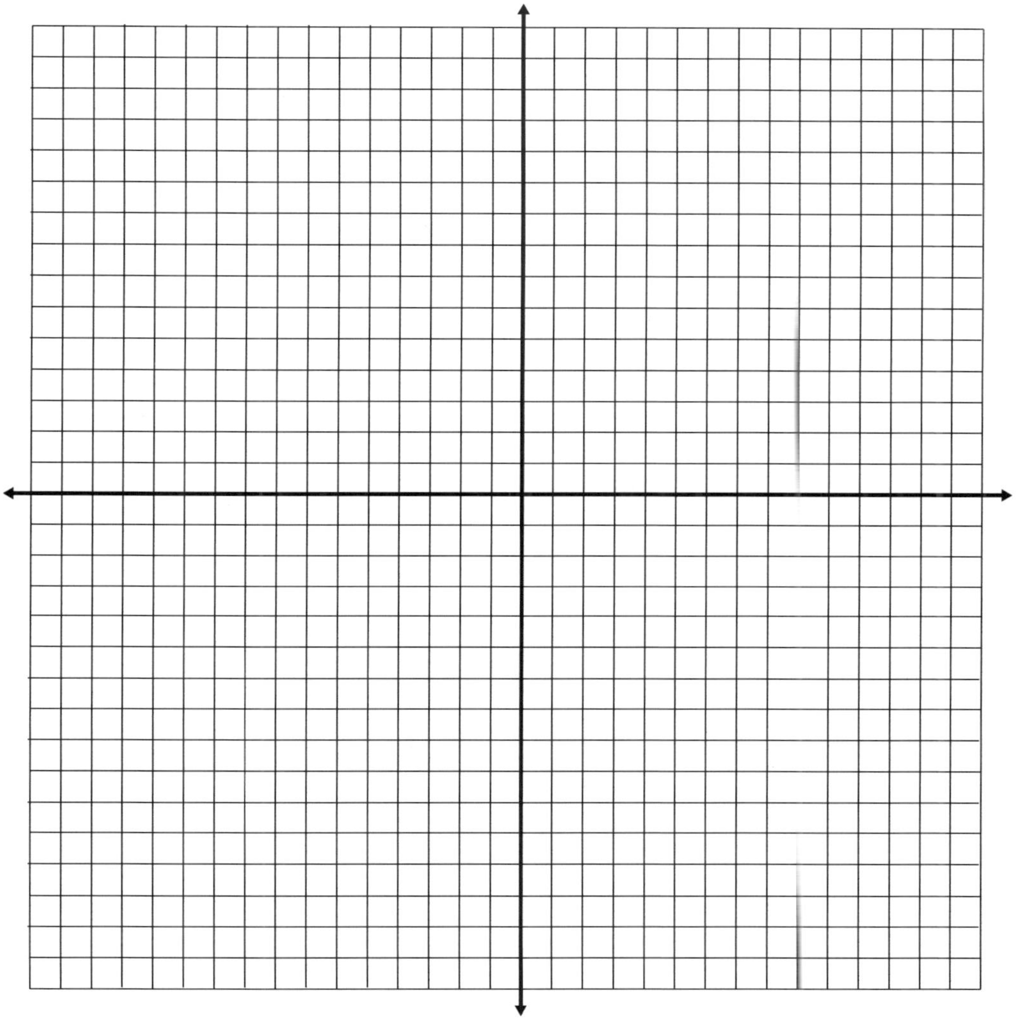

2. Find the inverse of the function $f(x) = 3 - 2x$, algebraically. Validate your answer by graphing the function and its inverse on the same coordinate plane.

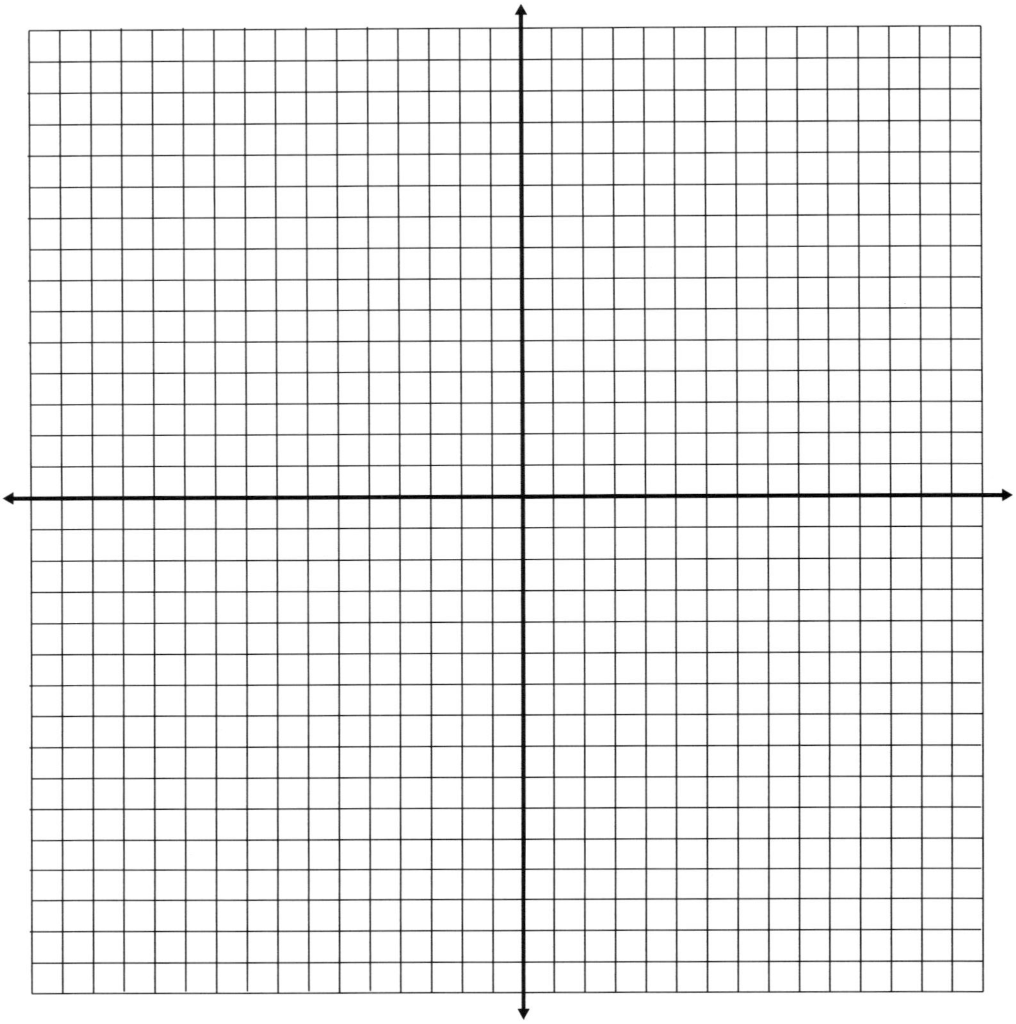

3. Determine if the function $f(x) = 2 - x^2$ has an inverse function. Validate your answer.

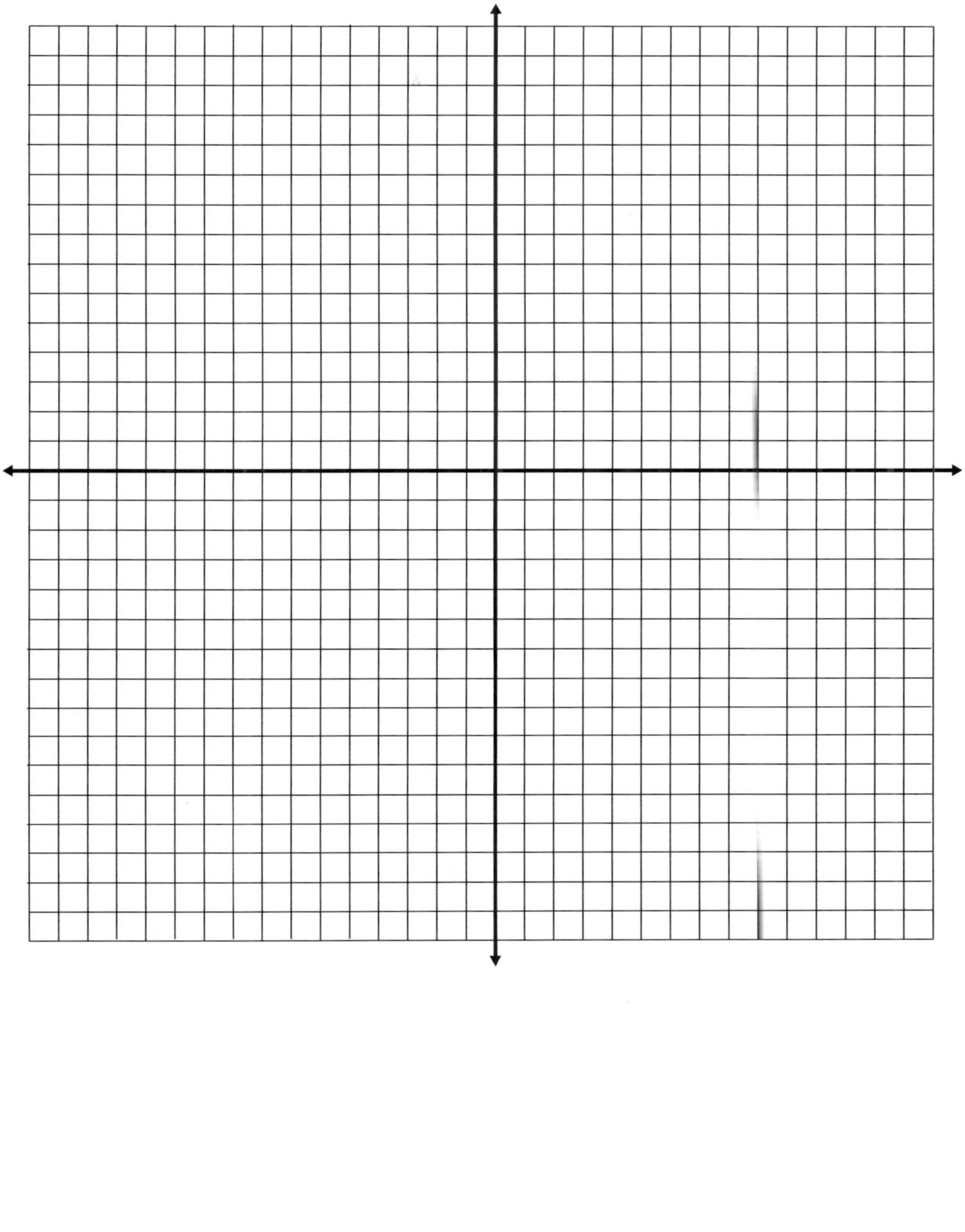

4. Find the inverse function of $f(x) = x^3$ algebraically and graphically.

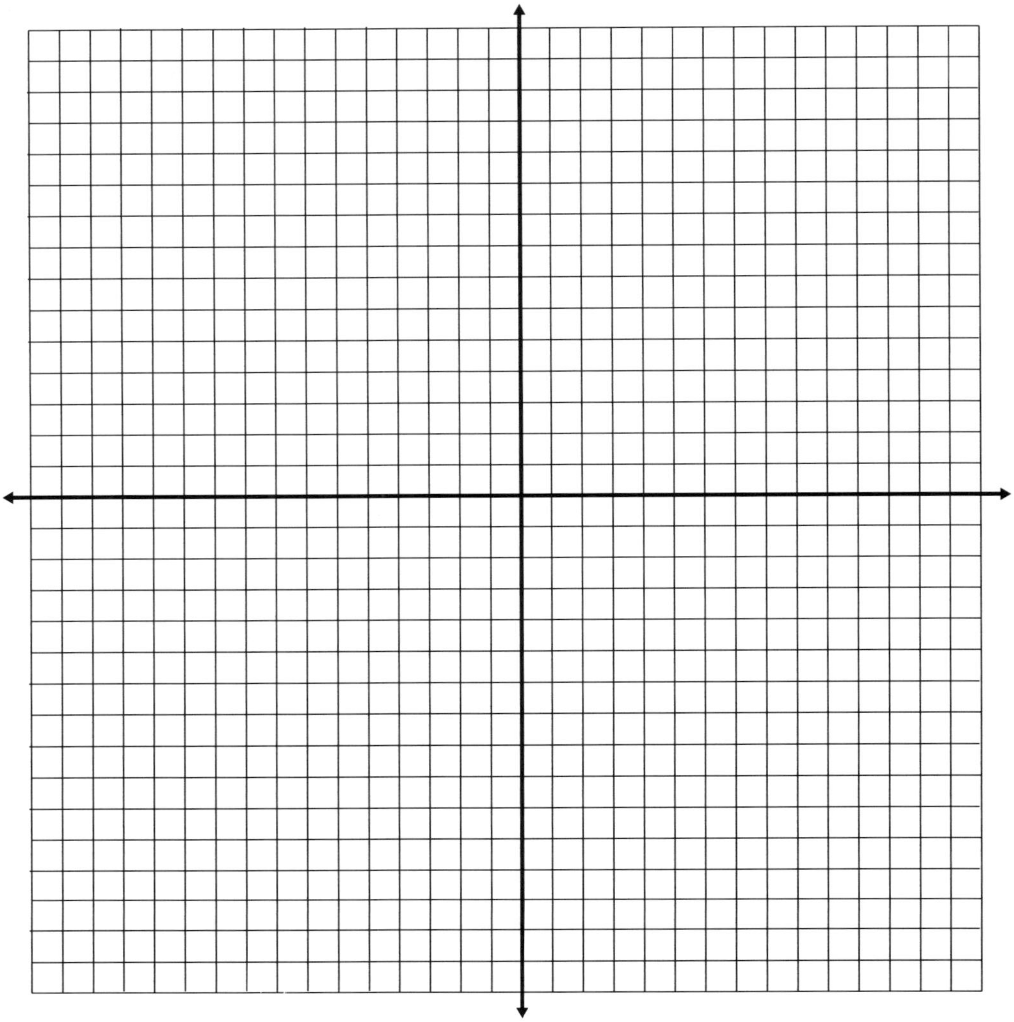

Create and Solve the Hardest Problem 5.4

Your Hardest Problem

Problem Solution

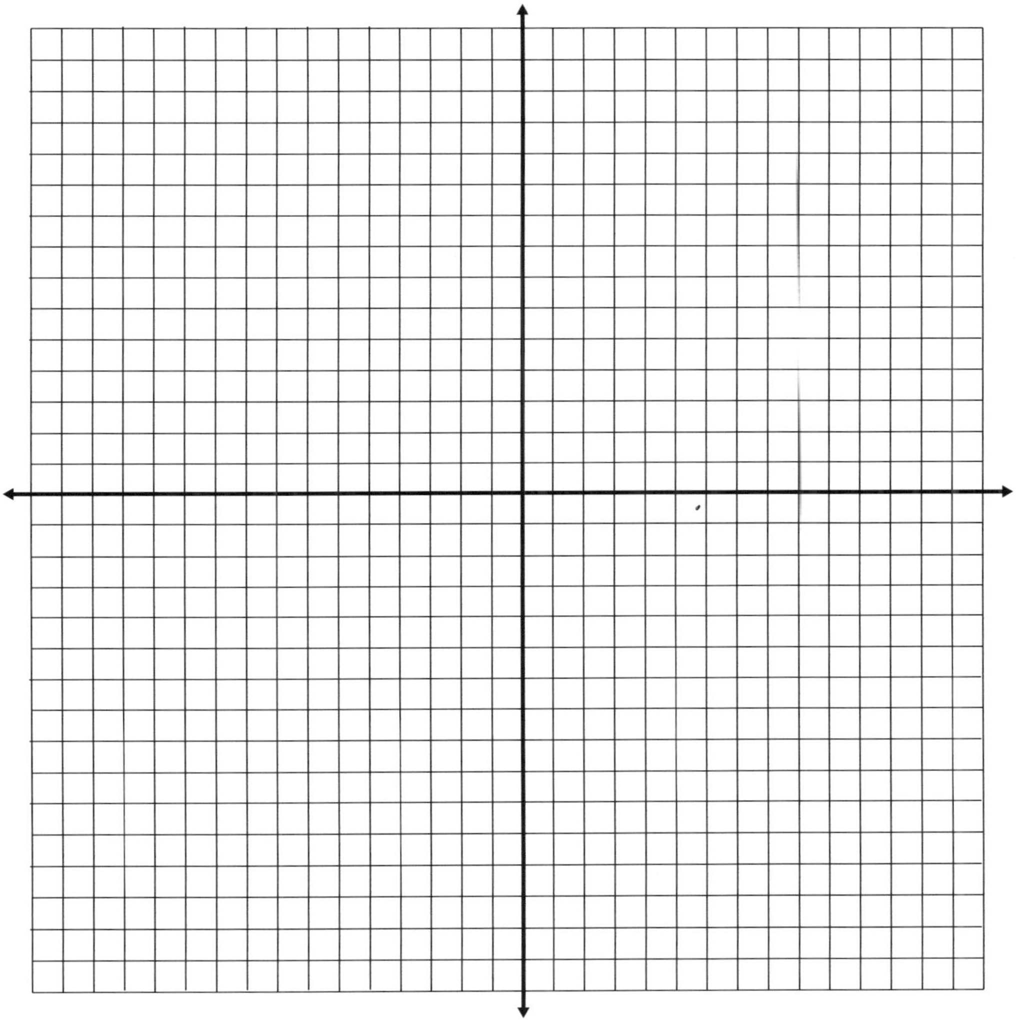

What makes a problem involving inverse relations and functions a hard problem to solve?

Identify, Describe, & Correct the Errors 5.4

Solve the problem correctly in the appropriate location. Be sure to identify and describe the error you found in the worked solution!

Problem 1

Given $f(x) = 2 - 3x$, find $f^{-1}(3)$.

Worked Solution *(Find the error!)*

$$f(x) = 2 - 3x$$

$$f^{-1}(x) = \frac{1}{2 - 3x}$$

$$f^{-1}(3) = \frac{1}{2 - 3(3)}$$

$$f^{-1}(3) = -\frac{1}{7}$$

Correct Process

Describe the Error

Problem 2

Find $f^{-1}(x)$ for the function

$f(x) = 3x - 1$

Worked Solution *(Find the error!)*

$$f(x) = 3x - 1$$

$$y = 3x - 1$$

$$x = 3y - 1$$

$$x + 1 = 3y$$

$$\frac{x+1}{3} = y$$

$$y = \frac{x+1}{3}$$

Correct Process

Describe the Error

Problem 3

Does the function $g(x) = \dfrac{1}{x^2}$ have an inverse function?

Describe the Error

Worked Solution *(Find the error!)*

$$y = \frac{1}{x^2}$$

$$x = \frac{1}{y^2}$$

$$y^2 = \frac{1}{x}$$

$$y = \frac{1}{\sqrt{x}}, \quad g^{-1}(x) = \frac{1}{\sqrt{x}}$$

$g(x) = \dfrac{1}{x^2}$	
x	y
1	1
2	$\frac{1}{4}$
3	$\frac{1}{9}$

$g^{-1}(x) = \dfrac{1}{\sqrt{x}}$	
x	y
1	1
$\frac{1}{4}$	2
$\frac{1}{9}$	**3**

Yes, it does have an inverse function.

Correct Process

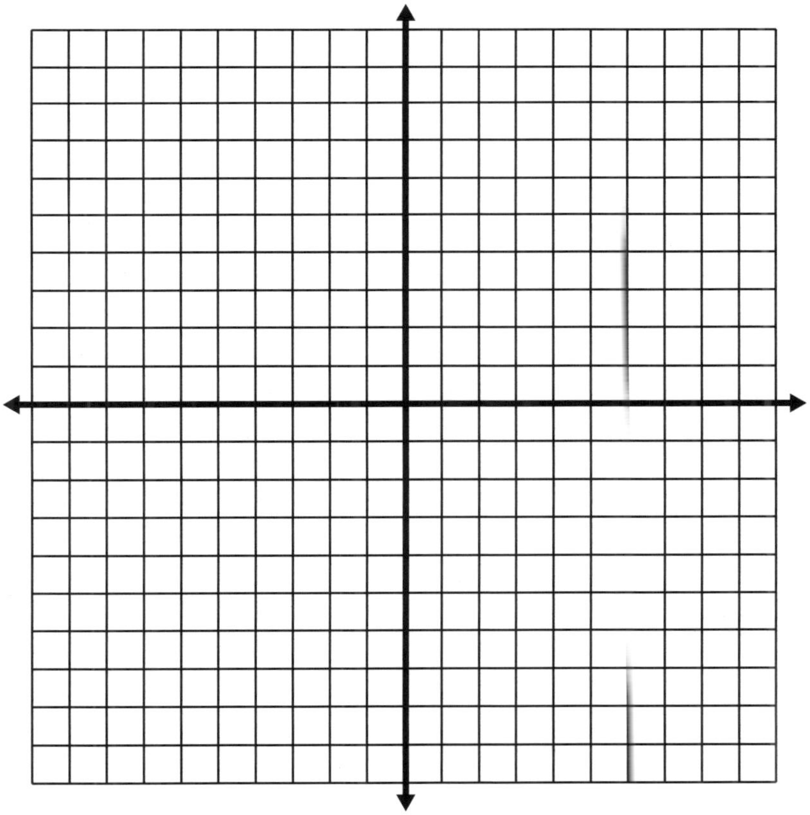

What Did You Learn About Learning Math? 5.4

1. How do you visualize the graph of a function and its inverse?

2. What patterns do you look for in the T-table of a function in order to determine if it has an inverse function and what the inverse function might look like?

3. What methods did you learn for validating that an inverse function is correct?

Section 5.5 Creating New Functions from Old

Why? 5.5

We use functions to model real-world situations. For example, the flight of a softball can be modeled with a quadratic function (on a windless day). We can use family of functions cards to decide which family has a graph whose shape matches the data; then we can use function manipulations to determine the best member of the family.

What Do You Already Know? 5.5

What do you think is the most important step when graphing a non-linear function?

How do you determine the appropriate graphing window for the graph of a function?

How do you determine if a function has an inverse function?

If $f^{-1}(5) = 7$, what is $f(7)$?

Goals 5.5

When you complete this activity, you should be able to:

1. Compose functions
2. Find the inverse of some non-linear functions
3. Translate and scale the graphs of functions

Building Mathematical Language 5.5

New Terms:	function composition	scaling	translation

Key Concepts 5.5

To create the **function composition** of two functions (**compose two functions**), you apply one function to the output of the other function. The composition of two functions is represented by $f \circ g(x)$ by which we mean $f(g(x))$.

Limitation/Caution: It can be difficult to know which function to apply first.

The ***output of the composition of a function and its inverse***, depending on which function is applied first, is written in several ways: $f \circ f^{-1}(x) = f(f^{-1}(x))$ or $f^{-1} \circ f(x) = f^{-1}(f(x))$. The result of this composition (with f or f^{-1} applied first) is given by:

$$f \circ f^{-1}(x) = f(f^{-1}(x)) = x \text{ or } f^{-1} \circ f(x) = f^{-1}(f(x)) = x.$$

In other words, f^{-1} reverses f and f reverses f^{-1}.

Limitation/Caution: The function f may not have an inverse, but may *appear* to have one. For example, $f(x) = x^2$ is not one-to-one, but it appears that $g(x) = \sqrt{x}$ is its inverse.

To ***translate*** a function graph, you move the function graph horizontally (left or right) or vertically (up or down) by adding a constant to each occurrence of the input variable or the output variable, respectively. For example, the graph of $y = f(x - 2)$ will be the graph of $y = f(x)$ moved to the right 2 units. The graph of $y = f(x) + 2$ will be the graph of $y = f(x)$ moved up 2 units.

Limitation/Caution: It is difficult to know which way the function moves when you add a negative number to the independent variable. You should check your graph with a graphing calculator.

To ***scale*** a function graph, you expand or compress it horizontally or vertically by multiplying the input or output variable by a constant, respectively. For example, the graph of $y = f(2x)$ will be the graph of $y = f(x)$ compressed by a factor of 2. The graph of $y = 2(f(x))$ will be the graph of $y = f(x)$ expanded by a factor of 2.

Limitation/Caution: It is difficult to know whether you are expanding or compressing the graph of a function simply by looking at the function's equation. You should check your graph with a graphing calculator.

Methodology 1 — 5.5

Finding the Composition of Two Functions

Functions can be composed using the $f \circ g(x)$ notation, where g is applied first, followed by f.

Limitation/Caution: Remember that the function closest to the variable (on the right) is the function that is applied first to the variable.

	Example 1	Example 2	Your Turn
Find the composition specified:	$f \circ g$ where $g(x) = x^2$ and $f(x) = x + 4$	$h \circ g$ where $g(x) = \sqrt{x}$ and $h(x) = 2x - 5$	$g \circ f$ where $f(x) = 3^x$ and $g(x) = 2 - 3x$

Steps	Discussion
1 Write the composition	Write in function composition notation.

Ex 1	Ex 2	Your Turn
$f \circ g(x) = f(g(x))$	$h \circ g(x) = h(g(x))$	

Steps	Discussion
2 Insert the first function	Substitute the value of the first function in.

Ex 1	Ex 2	Your Turn
$f(x^2)$	$h(\sqrt{x})$	

Steps	Discussion
3 Evaluate the second function	Apply the second function to the expression.

Ex 1

$x^2 + 4$

Ex 2

$2\left(\sqrt{x}\right) - 5$

Your Turn

Steps	Discussion
4 Present the result	Write the result in composition notation.

Ex 1

$f \circ g(x) = x^2 + 4$

Ex 2

$h \circ g(x) = 2\sqrt{x} - 5$

Your Turn

Steps	Discussion
5 Validate	Choose a number (not 0 or 1). Substitute that number into the composition of the functions and into the functions in order.

Ex 1

$f \circ g(x) = x^2 + 4$

$f \circ g(2) = 2^2 + 4 = 4 + 4 = 8$

$f(g(2)) = f(4) = 8$ ✓

Ex 2

$h \circ g(x) = 2\sqrt{x} - 5$

$h \circ g(9) = 2\sqrt{9} - 5 = 2(3) - 5 = 1$

$h(g(9)) = h(3) = 2(3) - 5 = 1$ ✓

Your Turn

Methodology 2 5.5

Finding the Inverse of a Non-Linear Function

Some non-linear functions have inverse functions.

Limitation/Caution: Algebraically determining the inverse of a non-linear function can be difficult or impossible. A function whose inverse function you wish to find can have constraints on its domain that must be carried through the computation of the inverse.

	Example 1	**Example 2**	**Your Turn**
Find the inverse of the given function:	$f(x) = x^2 + 2$ with $x \ge 0$	$g(x) = \sqrt[3]{x} + 2$	$h(x) = x^2 - 2x + 1$ with $x \ge 1$

Steps	Discussion
1 State the function	**Write the function in function notation.**

Ex 1 $f(x) = x^2 + 2$ with $x \ge 0$

Ex 2 $g(x) = \sqrt[3]{x} + 2$

Your Turn

2 Graph the function	**Determine if the function is one-to-one on the given domain. Do this by graphing it and performing the horizontal line test.**

Ex 1

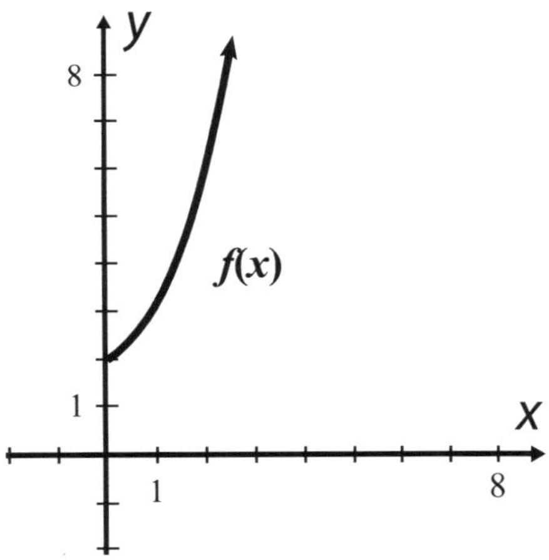

one-to-one

Ex 2

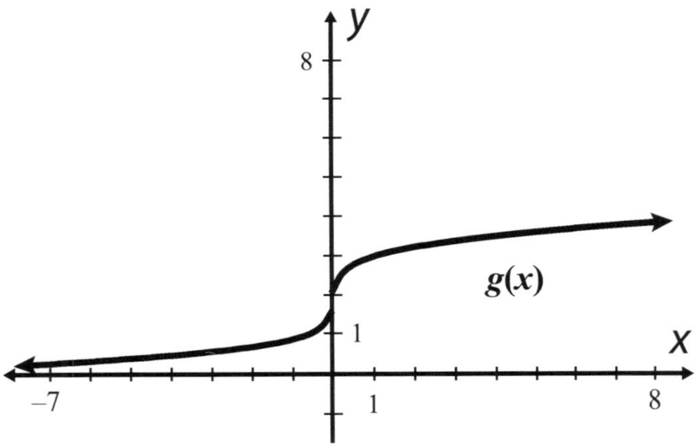

one-to-one

Steps	Discussion

Your Turn

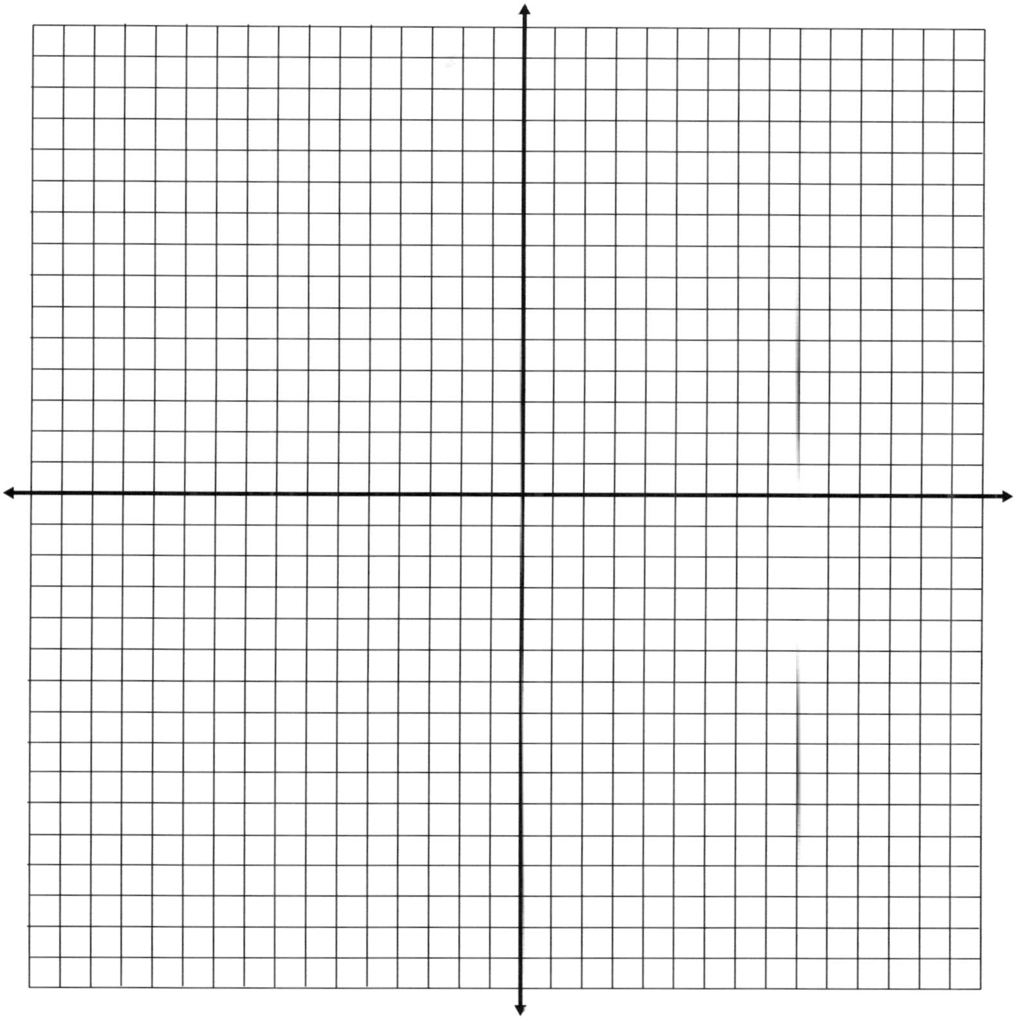

Your Turn

Steps	Discussion
3 Substitute *y* for *f*(*x*)	Write the equation as an equation with two variables.

Ex 1

$$y = x^2 + 2 \text{ with } x \geq 0$$

Ex 2

$$y = \sqrt[3]{x} + 2$$

Your Turn

Steps	Discussion
4 Interchange the names of the variables	Replace y with x and x with y.

Ex 1

$$x = y^2 + 2 \text{ with } y \geq 0$$

Ex 2

$$x = \sqrt[3]{y} + 2$$

Your Turn

Steps	Discussion
5 Solve for *y*	Use the methodologies for solving equations, noting the constraints at each step (domains) of the variables.

Ex 1

$x = y^2 + 2$ with $y \geq 0$

Note that $x \geq 2$.

$x - 2 = y^2$ with $y \geq 0$ and $x \geq 2$

$y^2 = x - 2$ with $y \geq 0$ and $x \geq 2$

$y = \pm\sqrt{x-2}$ with $y \geq 0$ and $x \geq 2$

$y = \sqrt{x-2}$ with $x \geq 2$

Ex 2

There are no constraints because the variables can be any real number.

$$x = \sqrt[3]{y} + 2$$

$$x - 2 = \sqrt[3]{y}$$

$$\sqrt[3]{y} = x - 2$$

$$y = (x-2)^3$$

Your Turn

Steps	Discussion
6 Present the inverse in function notation	**Write the inverse function in function notation.**

Ex 1

$f^{-1}(x) = \sqrt{x-2}$ with $x \geq 2$

Ex 2

$g^{-1}(x) = (x-2)^3$

Your Turn

Steps	Discussion
7 Validate	**Check that $f^{-1} \circ f(x) = x$. Alternatively, graph the inverse function on the same coordinate axes as the function to see if the graphs are inverses.**

Ex 1

$$f^{-1} \circ f(x) = f^{-1}\left(f(x)\right) =$$

$$f^{-1}(x^2+2) = \sqrt{\left(x^2+2\right)-2} =$$

$$\sqrt{x^2} = |x| = x \text{ since } x \geq 0 \ \checkmark$$

Or:

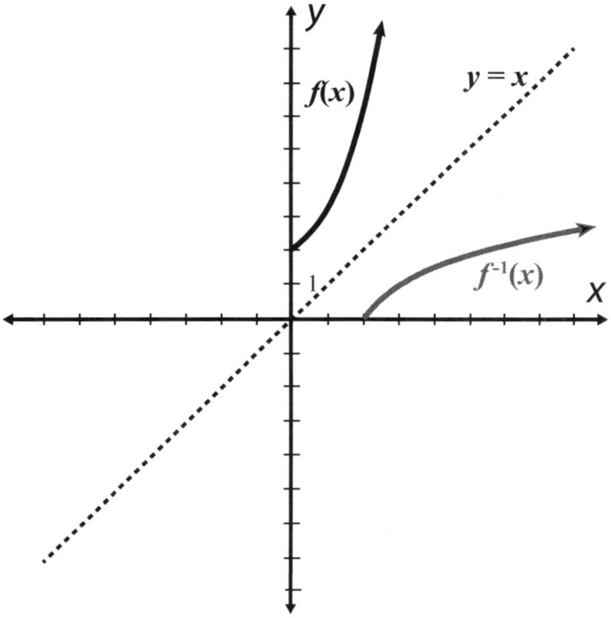

Ex 2

$$g^{-1} \circ g(x) = g^{-1}\left(g(x)\right) =$$

$$g^{-1}\left(\sqrt[3]{x}+2\right) = \left(\left(\sqrt[3]{x}+2\right)-2\right)^3 =$$

$$\left(\sqrt[3]{x}\right)^3 = x \ \checkmark$$

Or:

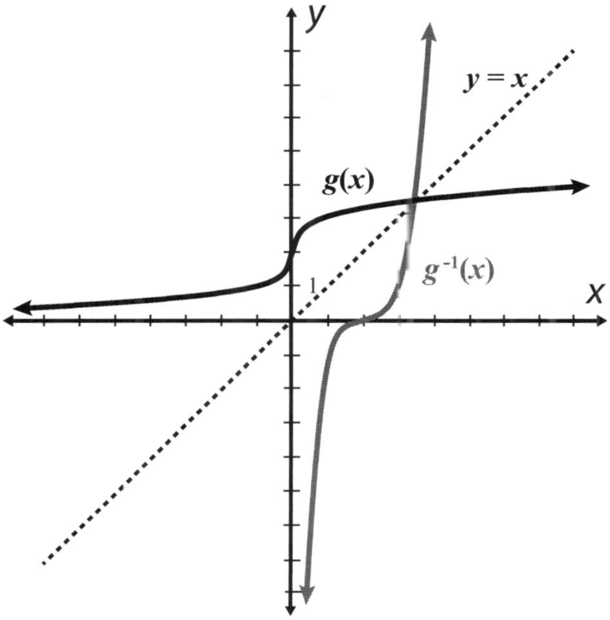

Your Turn

If you graph the inverse, use your previous graph.

Methodology 3 — 5.5

Translating a Function

	Example 1	Example 2	Your Turn
Translate the function:	$f(x) = x^2$ *to the left 3 units*	$f(x) = \sqrt{x}$ *up 2 units*	$f(x) = \lvert x \rvert$ *to the left 4 units*

Steps	Discussion
1 State the function	**Write the function in function notation.**

Ex 1

$f(x) = x^2$

Ex 2

$f(x) = \sqrt{x}$

Your Turn

Steps	Discussion
2 Graph the function	**The function graph gives you an image of what the translated graph should look like.**

Ex 1

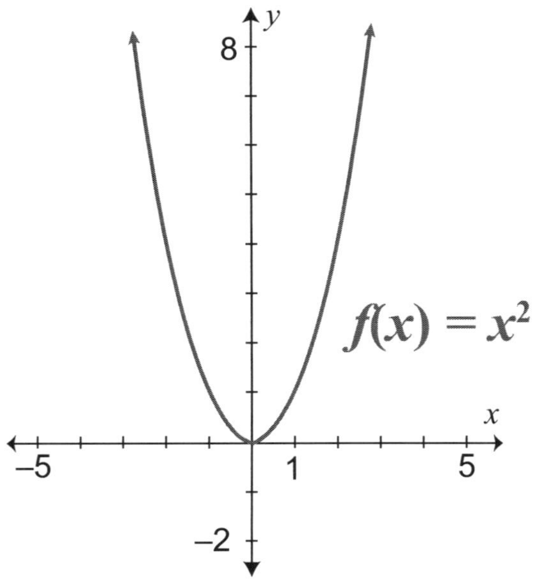

Ex 2

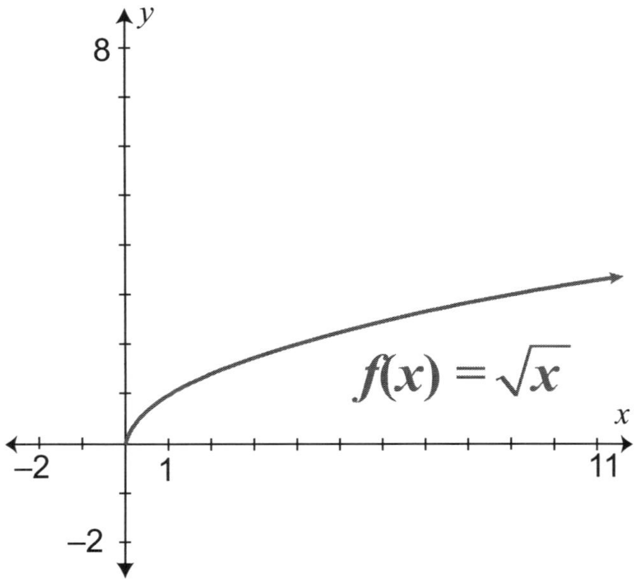

Steps	Discussion

Your Turn

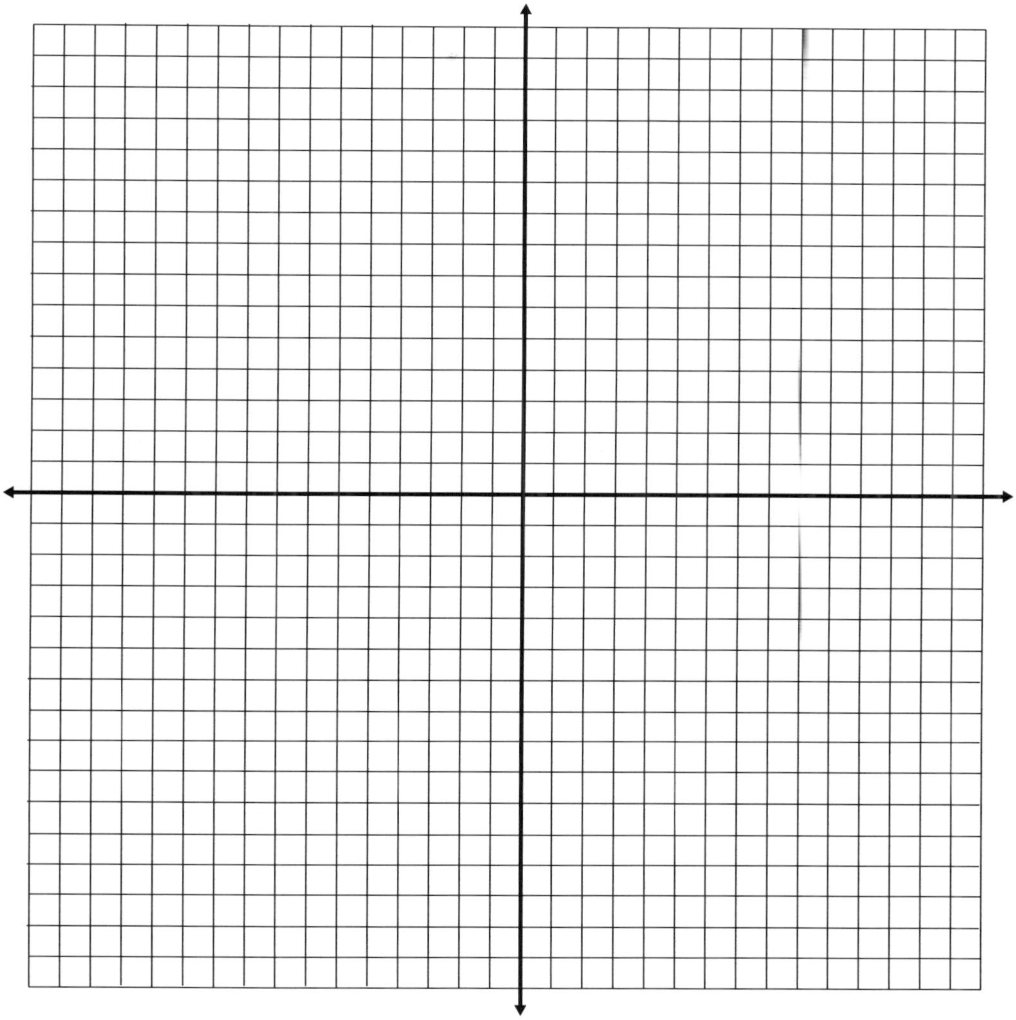

Steps	Discussion
3 Plot a point on the new function	This is a point that the graph of the translated function should pass through.

Ex 1

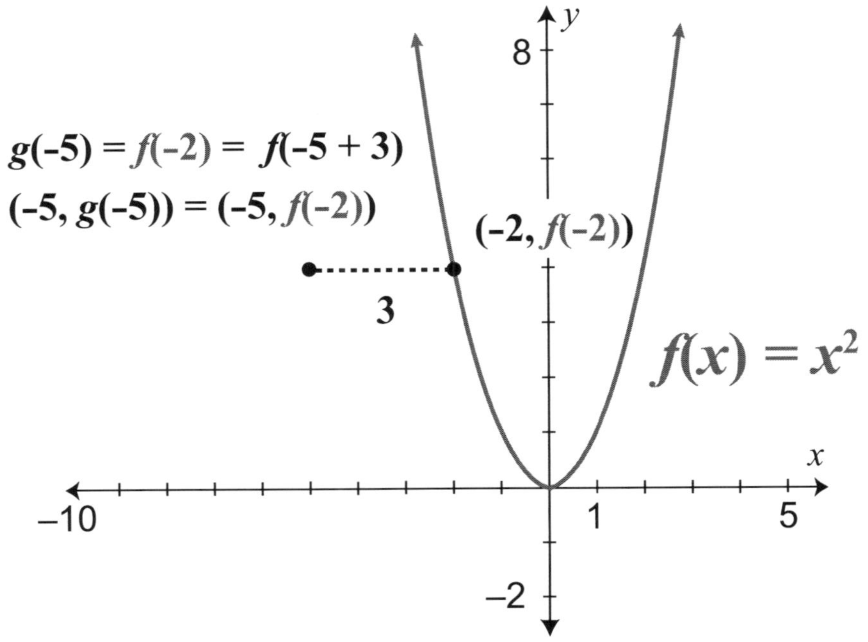

Ex 2

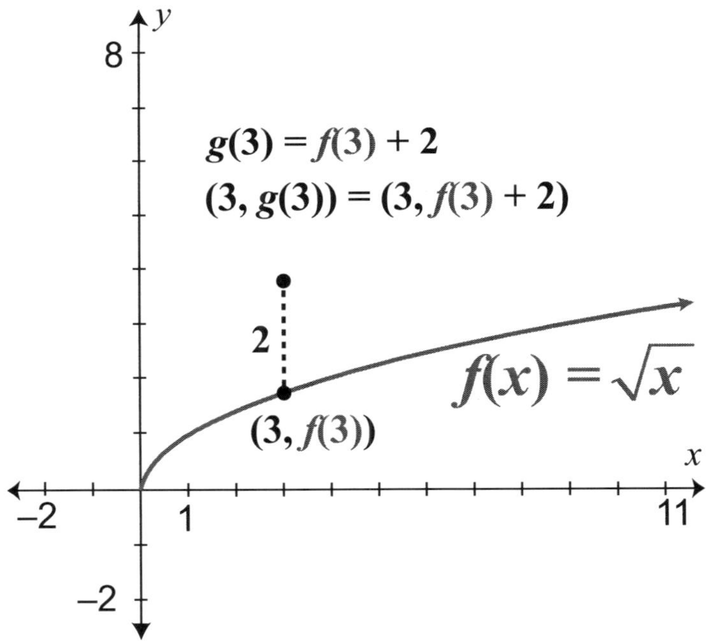

Your Turn

Use your previous graph.

Steps	Discussion
4 Determine the new function	A horizontal translation will change the argument of the function. A vertical translation will add to the value of the original function.

Ex 1

$g(x) = f(x+3)$

$g(x) = (x+3)^2$

Ex 2

$g(x) = f(x)+2$

$g(x) = \sqrt{x}+2$

Your Turn

5 Validate	Graph the new function to check if the translation is correct.

Ex 1

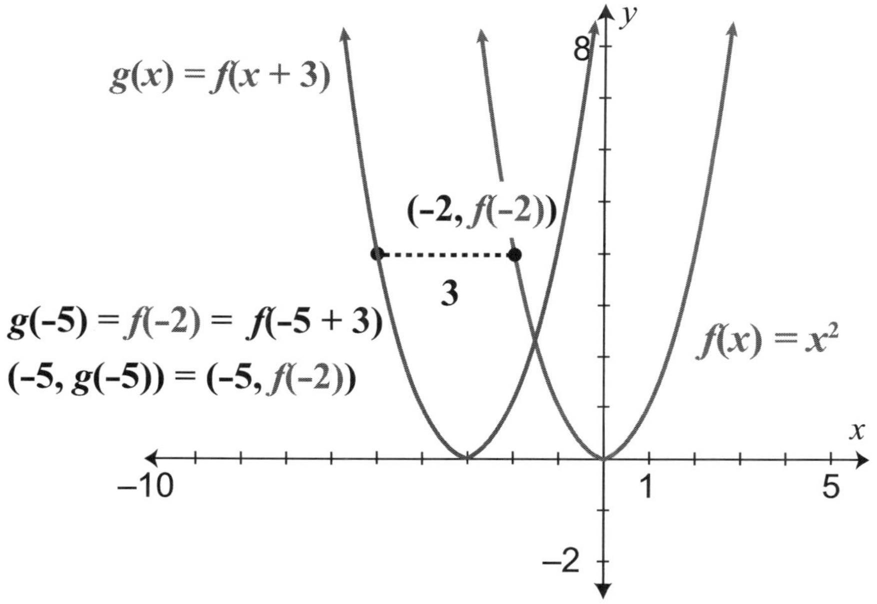

Ex 2

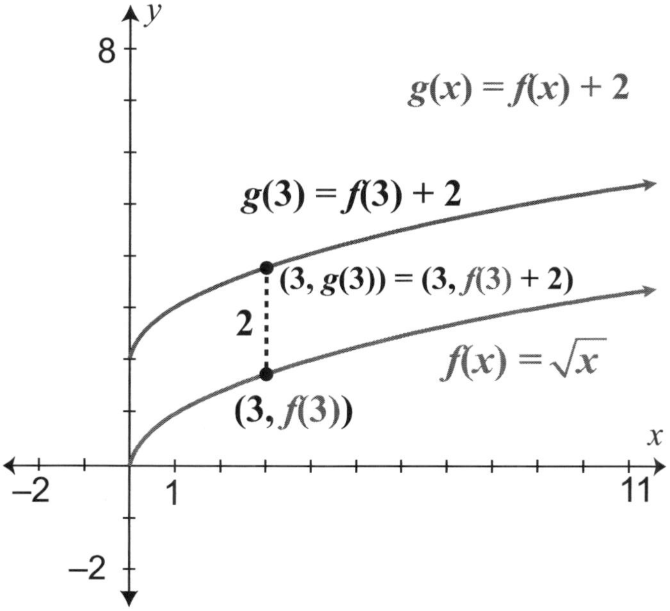

Your Turn

Use your previous graph.

Methodology 5.5

Scaling a Function

	Example 1	Example 2	Your Turn
Scale the function:	$f(x) = x^2$ *horizontally by a factor of 3*	$f(x) = \sqrt{x}$ *vertically by a factor of 3*	$f(x) = \lvert x \rvert$ *horizontally by a factor of 2*

Steps	Discussion
1 State the function	**Write the function in function notation.**

Ex 1 $f(x) = x^2$

Ex 2 $f(x) = \sqrt{x}$

Your Turn

Steps	Discussion
2 Graph the function	**The function graph gives you an image of what the scaled graph might look like.**

Ex 1

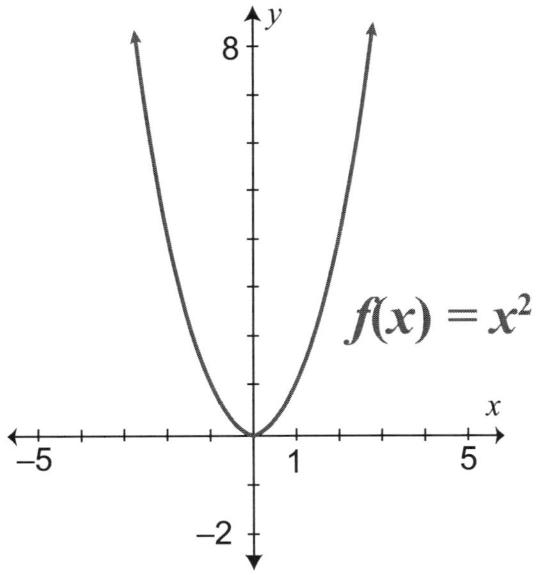

Ex 2

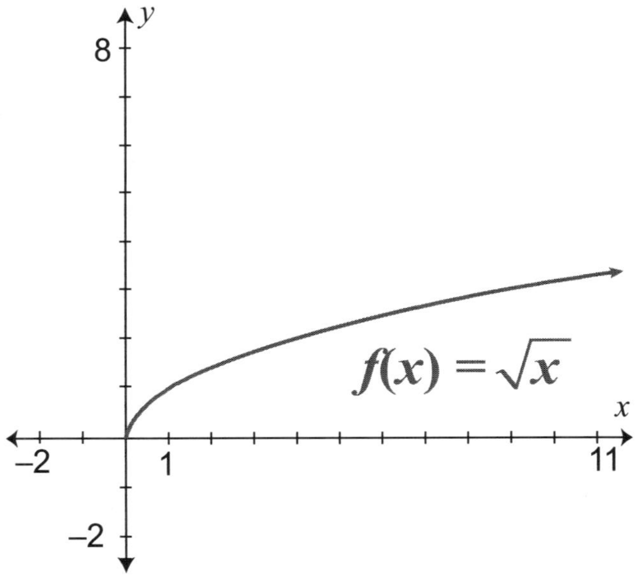

Steps	Discussion

Your Turn

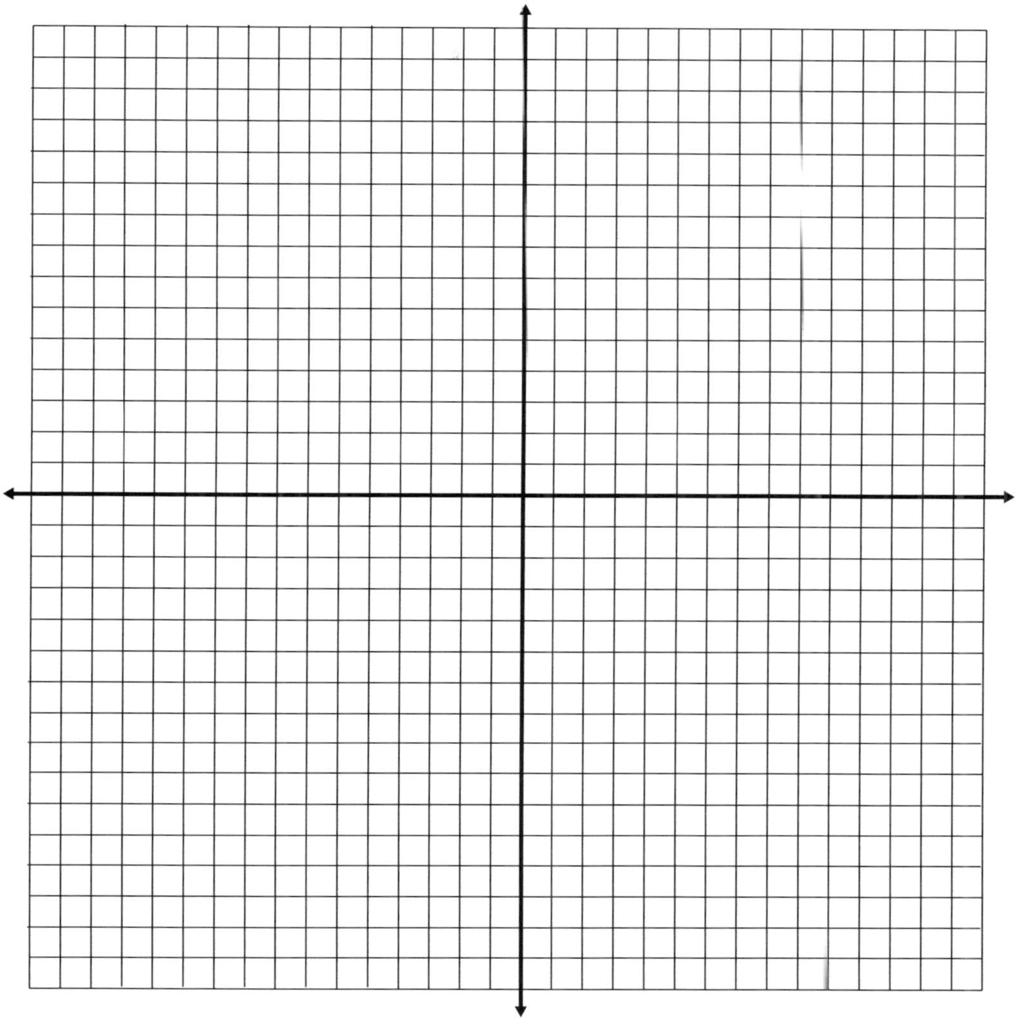

Steps	Discussion
3 Plot a point on the new function	This is a point that the graph of the scaled function should pass through.

Ex 1

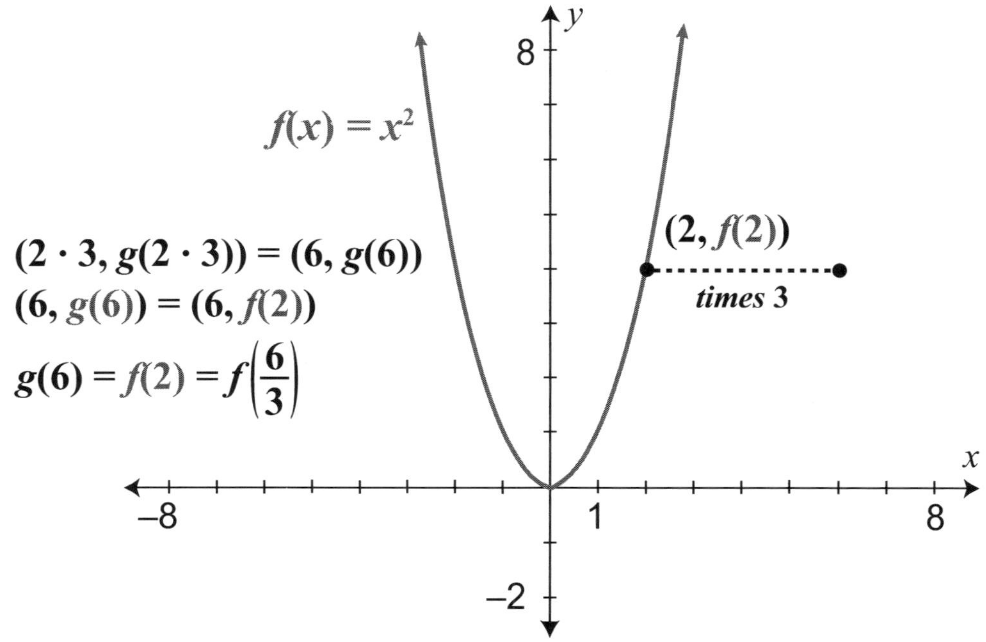

Ex 2

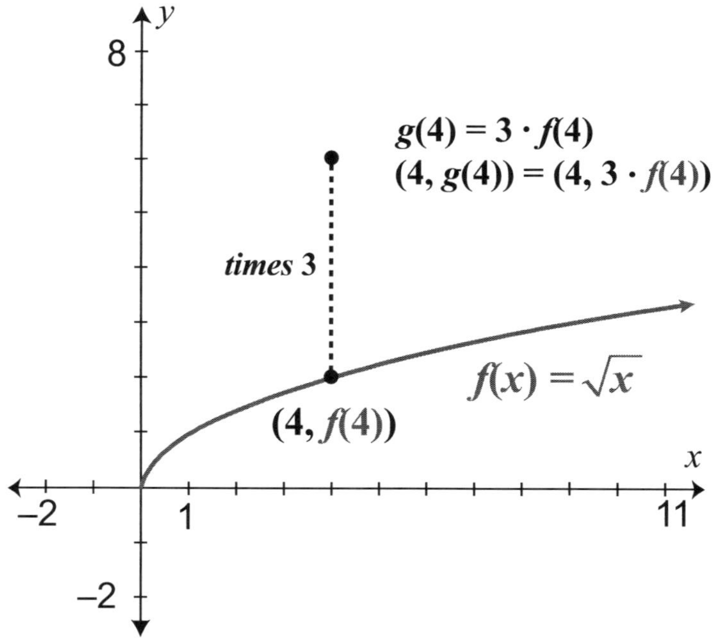

Your Turn

Use your previous graph.

Steps	Discussion
4 Determine the new function	Diving the argument of the original function by the scale factor scales the function horizontally. Multiplying the value of the original function by the scale factor scales the function vertically.

Ex 1

$$g(x) = f\left(\frac{1}{3}x\right)$$

$$g(x) = \left(\frac{x}{3}\right)^2$$

Ex 2

$$g(x) = 3 \bullet f(x)$$

$$g(x) = 3 \bullet \sqrt{x}$$

Your Turn

Steps	Discussion
5 Validate	Graph the new function to check if the scaling is correct.

Ex 1

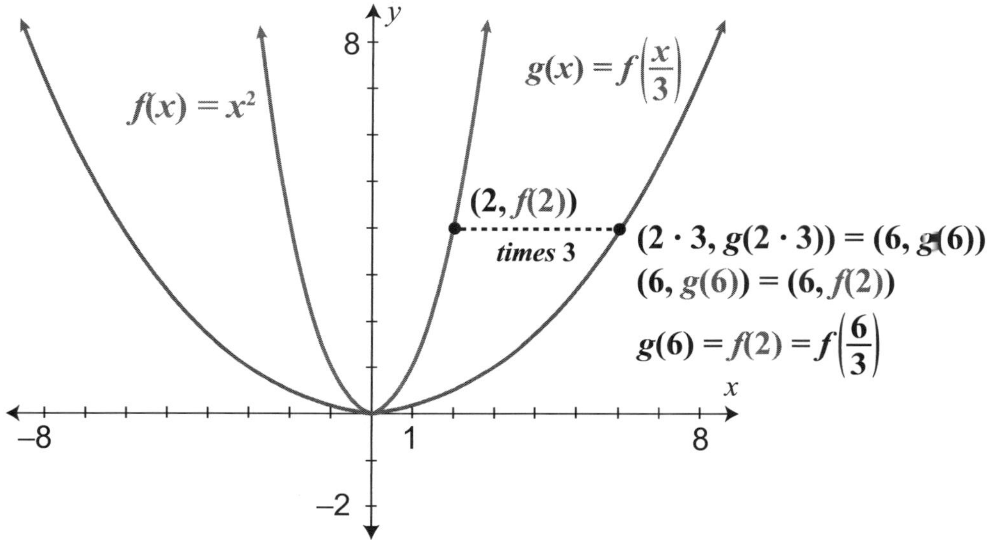

Ex 2

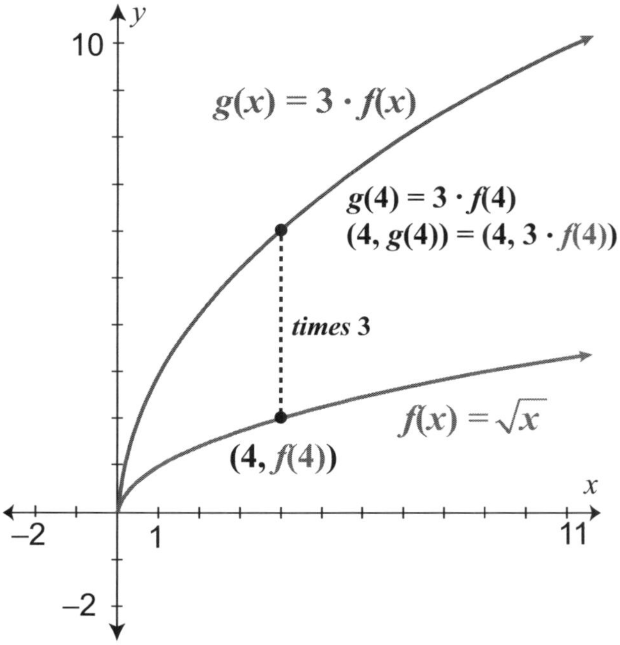

Your Turn

Use your previous graph.

Addressing Common Errors 5.5

The following table shows some of the most common errors that learners tend to make with the kinds of problems covered in this section.

ERROR: Translating a function in the wrong direction

Incorrect Process

Given the function $f(x) = \sqrt{x}$, graph the function $g(x) = f(x - 4)$.

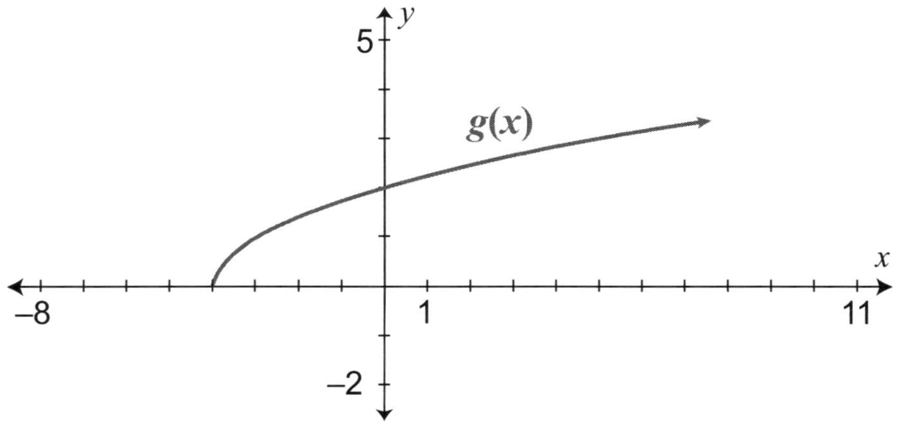

Correct Process

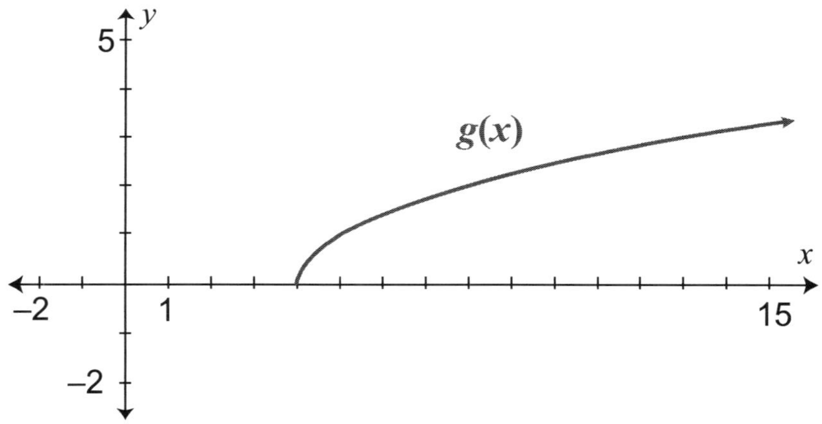

Resolution

Move the graph in the correct direction.

Validation

$g(8) = f(8-4) = f(4) = \sqrt{4} = 2$

The point (8, 2) is on the graph. ✓

ERROR: **Scaling by the wrong factor**

Incorrect Process

Scale the function $f(x)=|x|$ by a factor of three.

New function:

$g(x)=|3x|$

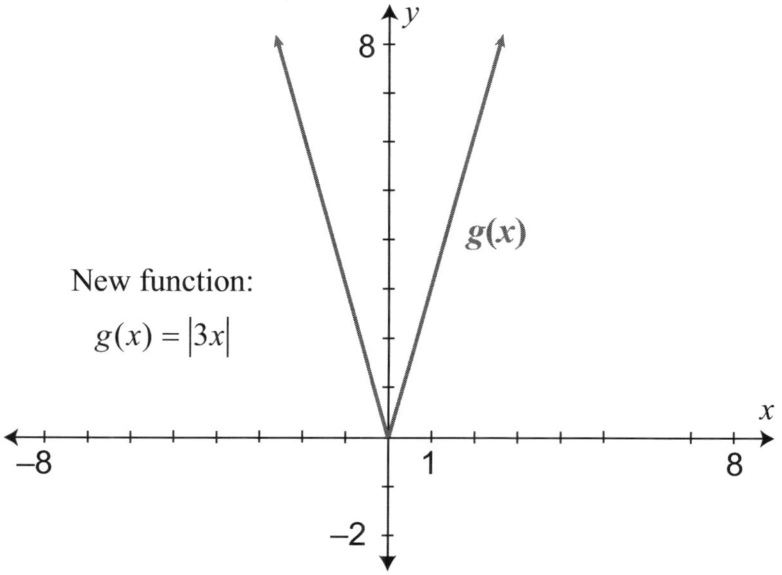

Correct Process

New function: $g(x)=f\left(\frac{x}{3}\right)$

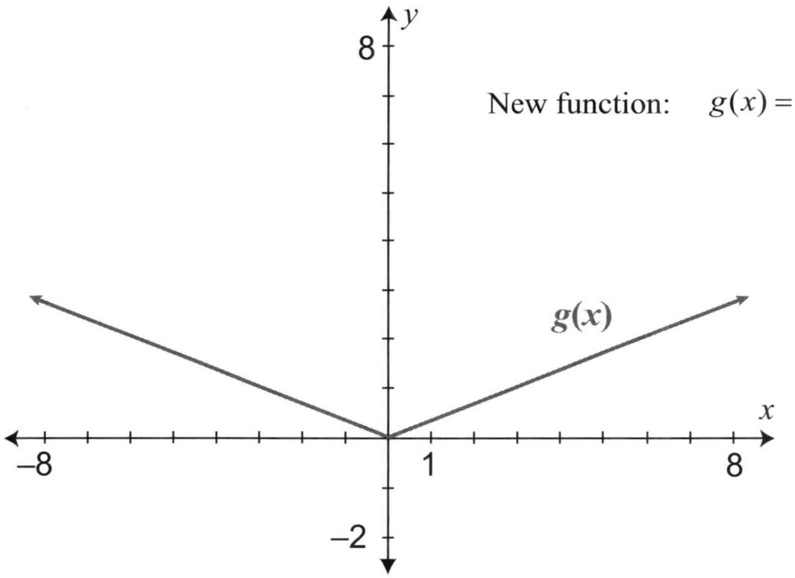

Resolution

Dividing the input of the function by 3 scales the function graph horizontally by a factor of 3.

Validation

$$g(-3)=f\left(\frac{-3}{3}\right)=f(-1)=|-1|=1$$

The point (–3, 1) is on the graph. ✓

ERROR: **Multiplying instead of composing**

Incorrect Process

Given the functions $f(x) = 2x + 3$ and $g(x) = x^2$

Find $g \circ f(x)$

$$g \circ f(x) = g(x) \bullet f(x)$$
$$= x^2 \bullet (2x+3)$$
$$= 2x^3 + 3x$$

Resolution

Use the appropriate function notation for $g \circ f(x)$.

Correct Process

$$g \circ f(x) = g(f(x))$$
$$= g(2x+3) = (2x+3)^2$$

Validation

$$g \circ f(3) = (2 \bullet 3+3)^2 = 9^2 = 81$$
$$g\left(f(3)\right) = g(9) = 9^2 = 81 \checkmark$$

Preparation Inventory 5.5

I can...	OR *Here's my question...*
☐ compose two functions.	☐
☐ compose a function with its inverse function.	☐
☐ translate a function.	☐
☐ scale a function.	☐

Critical Thinking Questions 5.5

1. How do you translate a function upwards by modifying its algebraic representation?

2. What happens to the graph of a function when you double the function values (y-values)?

3. What is the difference between adding a constant to the independent variable and multiplying the independent variable by a constant?

4. What is the difference between adding a constant to the dependent variable and multiplying the dependent variable by a constant?

5. Why is the graph of $f(x + 4)$ the graph of $f(x)$ moved to the left 4 units?

6. Why is the graph of $f\left(\frac{x}{3}\right)$ stretched horizontally rather than compressed horizontally?

Demonstrate Your Understanding 5.4

1. Find the inverse of the function $f(x) = \sqrt{3x - 2}$.

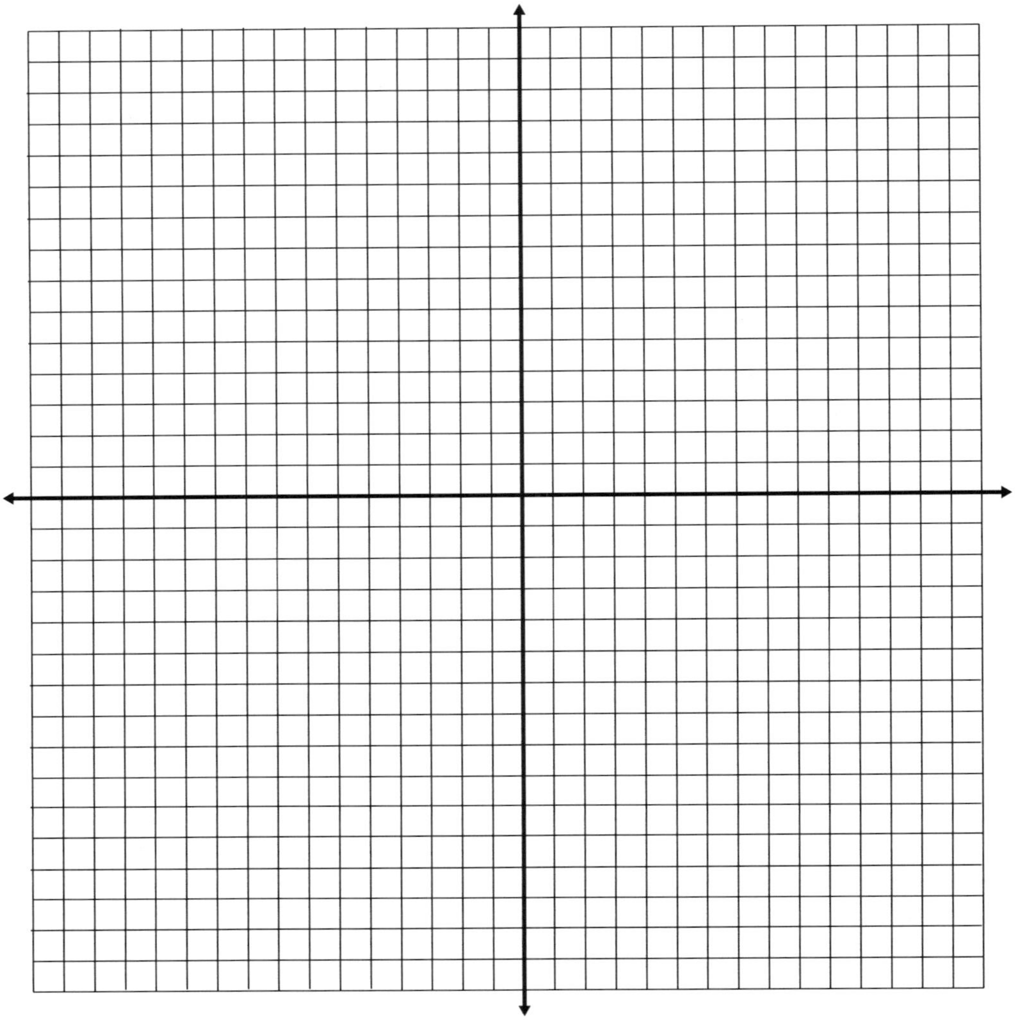

2. Validate that $g(x)=\dfrac{1}{2-x}$ is the inverse function of $f(x)=2-\dfrac{1}{x}$.

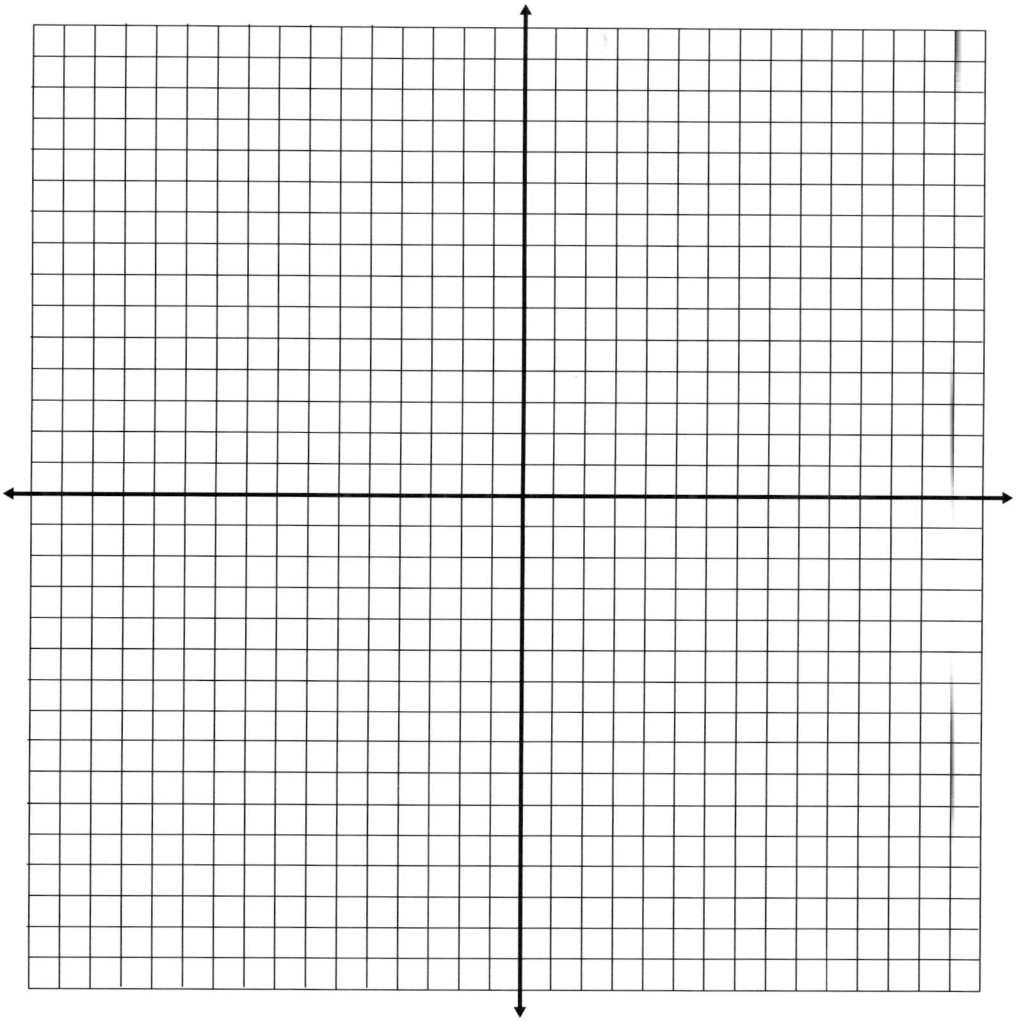

3. Find the function that is a translation of the graph of $f(x) = x^2 - 3x - 2$ to the left 2 units.

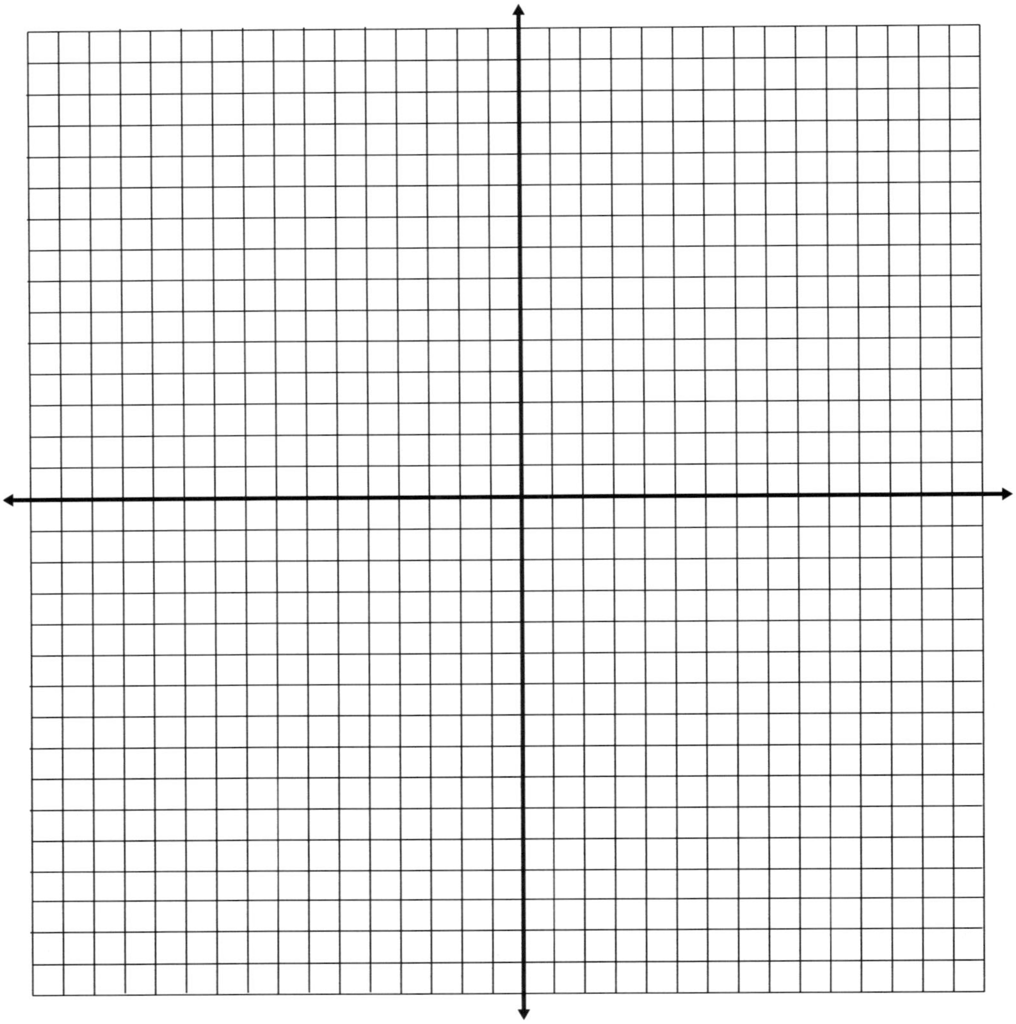

4. Find the function that is a horizontal scaling of the graph of $h(x) = \dfrac{1}{x+2}$ by a factor of $\dfrac{2}{3}$.

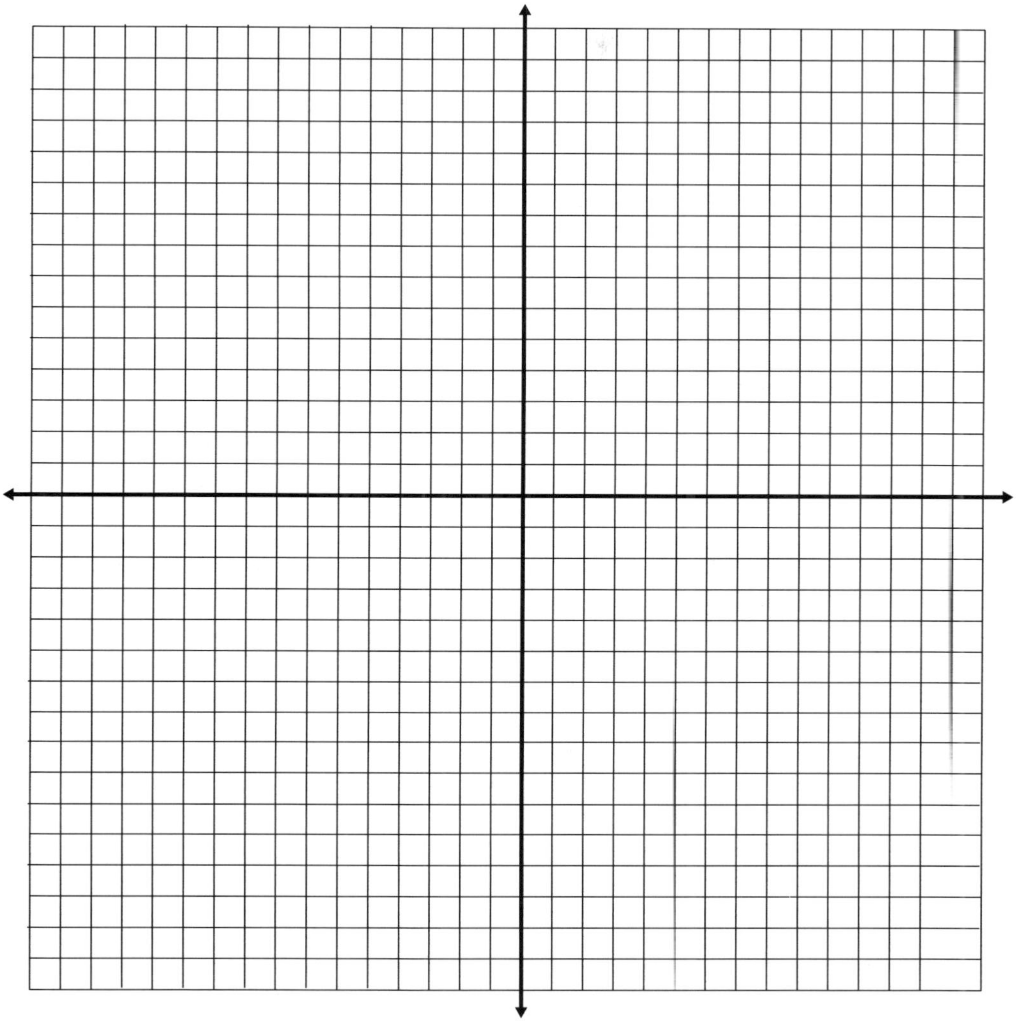

5. Find the function that is a vertical scaling of the graph of $g(x)$ by a factor of 2.

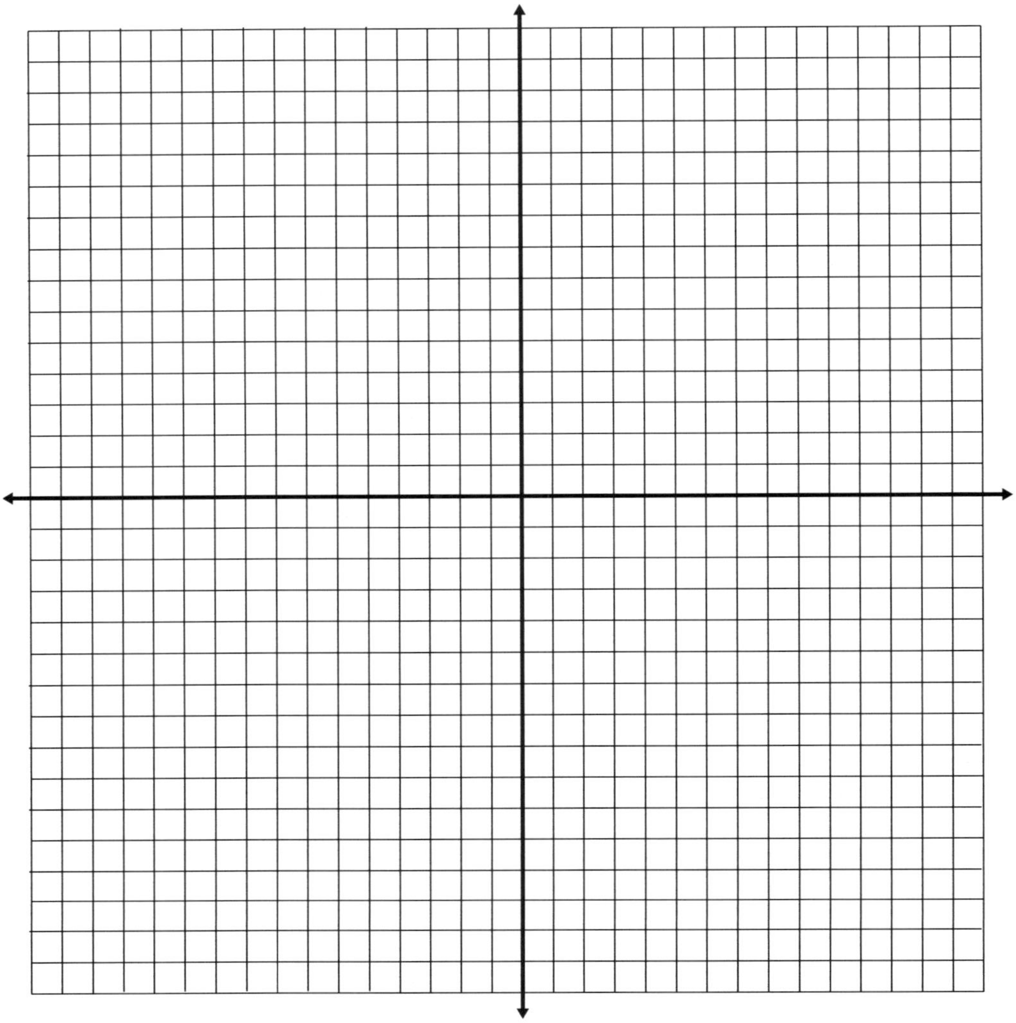

Create and Solve the Hardest Problem 5.5

Your Hardest Problem

Problem Solution

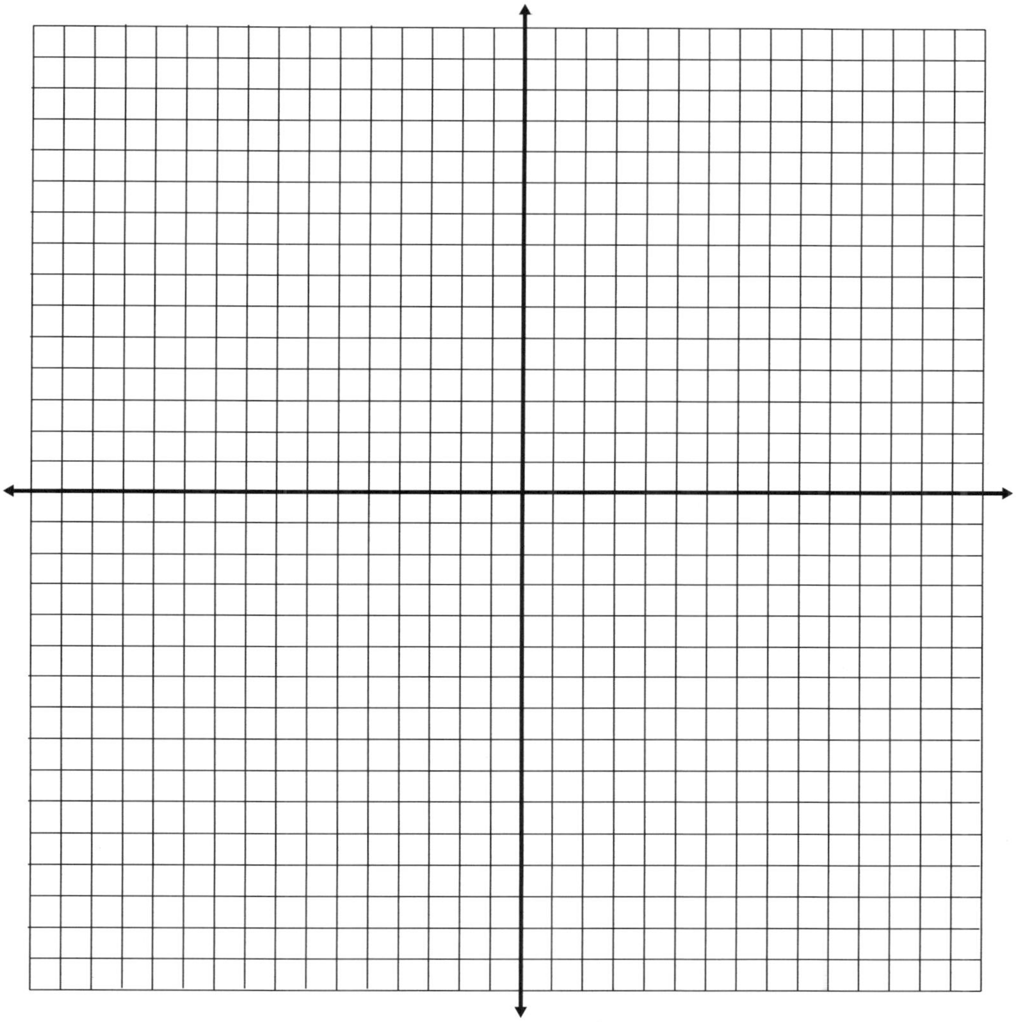

What makes a problem where you create new functions from other functions a hard problem to solve?

Identify, Describe, & Correct the Errors 5.5

Solve the problem correctly in the appropriate location. Be sure to identify and describe the error you found in the worked solution!

Problem 1

For the functions $f(x) = 5x - 2$ and $g(x) = 2x^2 - 5x + 1$, find the composition of f followed by g.

Worked Solution *(Find the error!)*

$$g \circ f(x) = g(x) \bullet f(x) =$$
$$\left(2x^2 - 5x + 1\right) \bullet (5x - 2) =$$
$$10x^3 - 4x^2 - 25x^2 + 10x + 5x - 2 =$$
$$10x^3 - 29x^2 + 15x - 2$$

Correct Process

Describe the Error

Problem 2

Find the inverse of the function $f(x) = 3 - x^2$.

Worked Solution *(Find the error!)*

$$f^{-1}(x) = \frac{1}{f(x)}$$
$$f^{-1}(x) = \frac{1}{3 - x^2}$$

Correct Process

Describe the Error

Problem 3

Define a function whose graph is the graph of $f(x) = x^2$ translated to the right (horizontally) 2 units.

Worked Solution *(Find the error!)*

$$g(x) = (x + 2)^2$$

Describe the Error

Correct Process

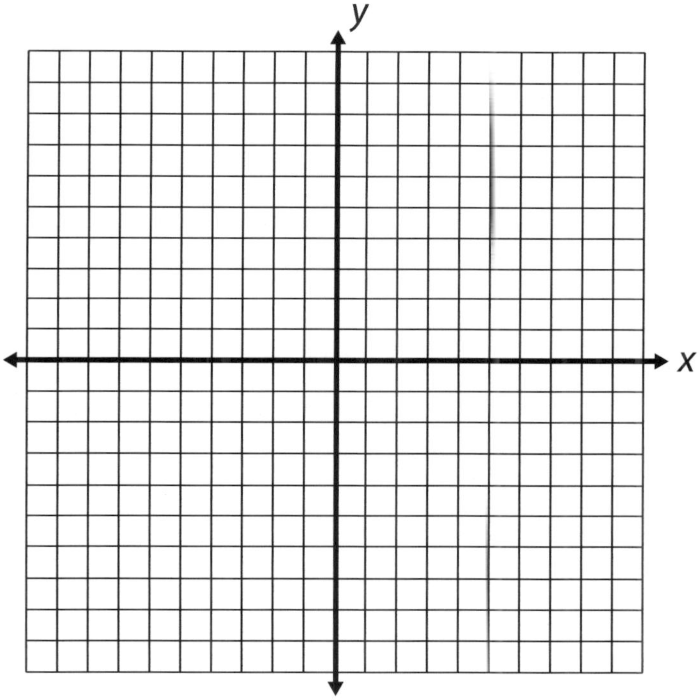

Problem 4

Define a function whose graph is the graph of $f(x) = 5^x$ translated up 2 units.

Worked Solution *(Find the error!)*

$$g(x) = 5^x - 2$$

Describe the Error

Correct Process

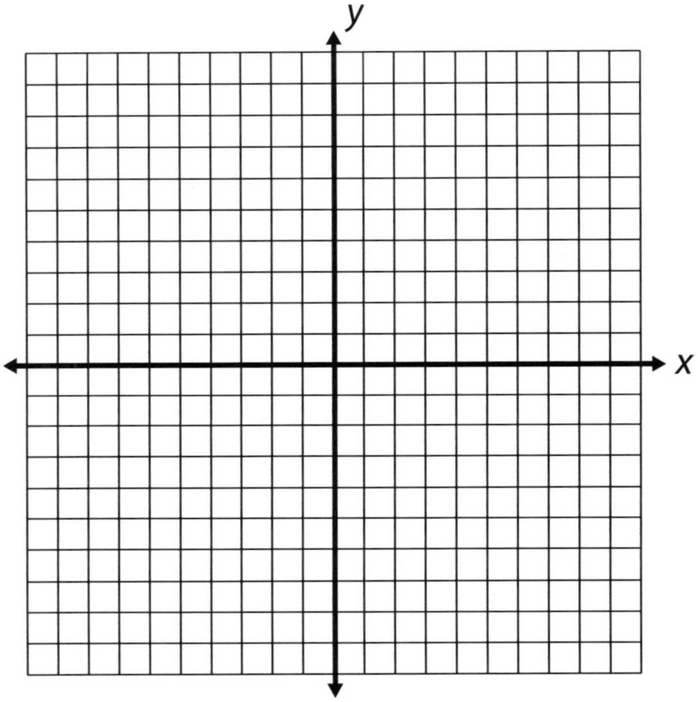

Problem 5

Define a function whose graph is the graph of $f(x)$ scaled horizontally (expanded) by a factor of 2.

Worked Solution *(Find the error!)*

$$g(x) = f(2x)$$

Describe the Error

Correct Process

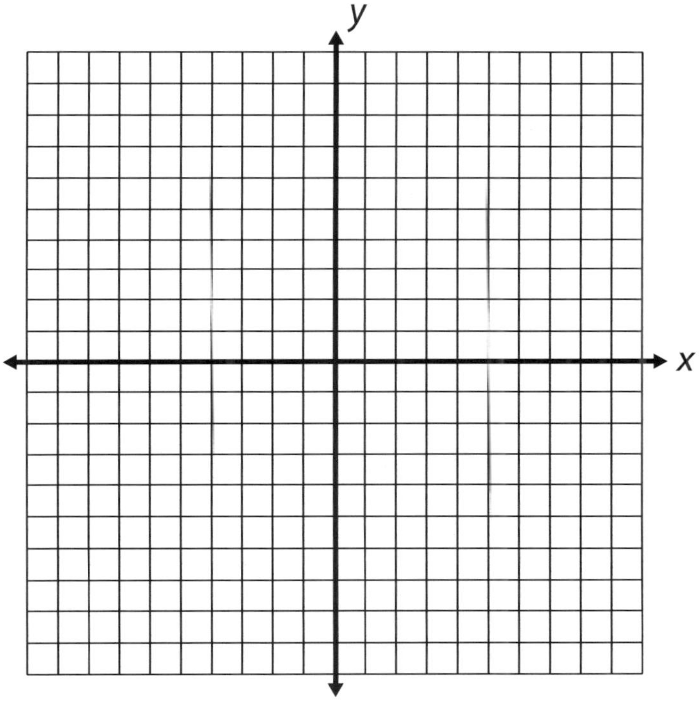

What Did You Learn About Learning Math? 5.5

1. What did you learn about graphing inverses of non-linear functions?

2. What did you learn about the similarities and differences between translating horizontally and scaling horizontally?

3. What did you learn about validating both algebraically and graphically that a function is an inverse of a given function?

4. What did you abstract from your experience in translating and scaling the graphs of functions?

Section 5.6 The Basics of Comparing Functions

Why? 5.6

Sometimes modeling a real-world situation in two different ways is possible. For example, modeling population growth can be done with exponential functions or with logistic functions. We can compare such functions to determine which one is more appropriate.

What Do You Already Know? 5.6

What do you think is the most important step in graphing a non-linear function?

What are the features of a family of functions card?

What are the intercepts of a function?

Goals 5.6

When you complete this activity, you should be able to:

1. Determine the appropriate graphing windows for comparing two functions
2. Use the points of intersection of two functions to compare the functions
3. Compare functions from various families of functions

Building Mathematical Language 5.6

New Terms: comparison window

Key Concepts 5.6

A ***comparison window*** is a graphing window (see Section 3.4) where the graphs of the functions being compared are displayed. The x-range and the y-range of the comparison window should contain the x-range and y-range of the graphing windows of both functions. Comparing and contrasting functions graphically is useful when comparing the behavior of physical phenomena.

Limitation/Caution: Not all of the behaviors of the functions may be displayed in the comparison window. Comparing two functions often requires the use of a graphing calculator or a computer with graphing software. The long-term behaviors of functions do not always appear in the graphs of functions in specific graphing windows.

Methodology 5.6

Comparing and Contrasting Functions

Comparing and contrasting two functions within the same window is useful when comparing the behavior of physical phenomena.

Limitation/Caution: Comparing two functions often requires the use of a graphing calculator or a computer with graphing software. The long-term behavior of functions does not always appear in the graphs of functions in specific graphing windows.

	Example 1	**Example 2**	**Your Turn**
Compare the given functions:	$y = 2x$ and $y = 3x$ (from the family of linear functions)	$f(x) = x^2$ and $g(x) = 2^x$ (from two different families of functions)	$f(x) = 2x$ and $g(x) = 2^x$ (from two different families of functions)

Steps	Discussion
1 Determine the graphing window	**Consider the most appropriate graphing window to use to compare the functions.**

Ex 1 Choose the window: $-10 < x < 10$ and $-7 < y < 7$

Ex 2 Choose the window: $-10 < x < 10$ and $-5 < y < 20$

Your Turn

Steps	Discussion
2 Graph functions	**Graph each function by hand or enter each function into your graphing calculator or computer software.**

Ex 1

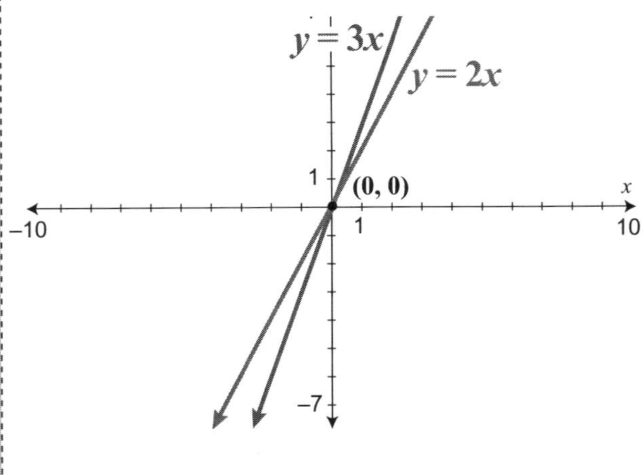

Ex 2

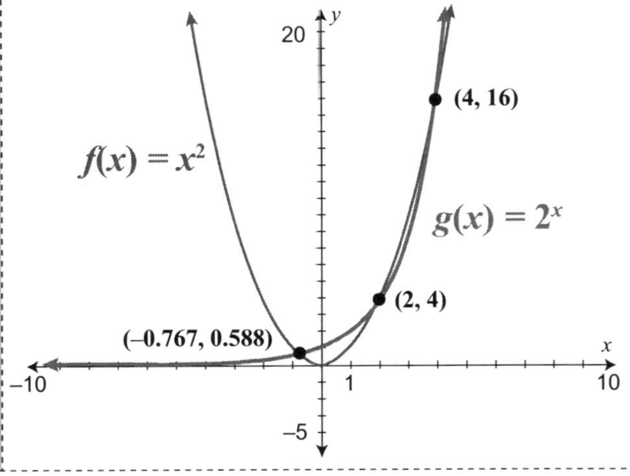

Steps	Discussion

Your Turn

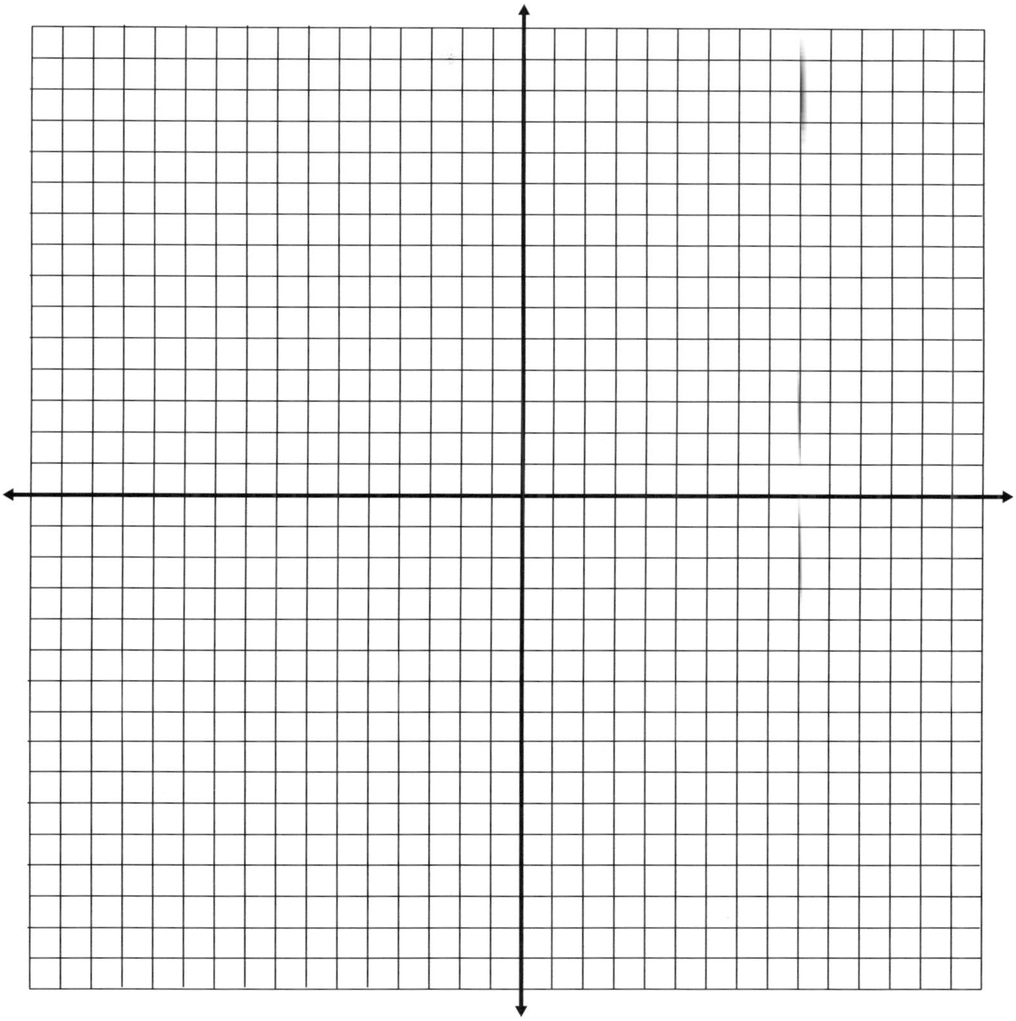

Steps	Discussion
3 Find the points where the two functions intersect	**Determine the points where the graphs of the functions intersect (if they do). This step will often require the intersection feature of a graphing calculator or computer software.**

Ex 1

The functions intersect at the origin (0, 0).

Ex 2

The functions intersect when x is 2, 4, and approximately −0.767.

Your Turn

Steps	Discussion
4 Compare the functions	**Identify the x-values where one function's y-values are greater than the other.**

Ex 1

The function $y = 3x$ is greater than the function $y = 2x$ when x is positive. The function $y = 3x$ is less than the function $y = 2x$ when x is negative.

OR

$(y = 3x) > (y = 2x)$ when $x > 0$ and $(y = 3x) < (y = 2x)$ when $x < 0$.

Ex 2

$f > g$ when x is less than approximately −0.767 and between 2 and 4.

$f < g$ when x is between approximately −0.767 and 2 and when $x > 4$.

Your Turn

Steps	Discussion
5 Compare the functions' intercepts	**Compare the functions' intercepts. Find the intercepts with a graphing calculator, if necessary.**

Ex 1

Both the functions have their x- and y-intercepts at the origin.

Ex 2

The x- and y-intercepts of f are (0, 0).

The function g has a y-intercept at (0, 1) and no x-intercept.

Your Turn

Steps	Discussion
6 Validate	**Check points near the points of intersection of the functions.**

Ex 1

Check values at $x = 1$ and 3.

$2(-1) \overset{?}{>} 3(-1)$ $\quad$ $2(3) \overset{?}{<} 3(3)$

$-2 > -3$ ✓ $\quad$ $6 < 9$ ✓

Ex 2

Check values at $x = -1,\ 1$ and 3.

$(-1)^2 \overset{?}{>} 2^{-1}$ $\quad$ $1^2 \overset{?}{<} 2^1$ $\quad$ $3^2 \overset{?}{>} 2^3$

$1 > \frac{1}{2}$ ✓ $\quad$ $1 < 2$ ✓ $\quad$ $9 > 8$ ✓

Your Turn

Note: All exponential functions with a base larger than 1 will eventually dominate (be larger than) any polynomial, no matter how high the degree of the polynomial might be.

Model 1 5.6

Compare $f(x) = \sqrt{x}$ and $g(x) = \sqrt[3]{x}$.

Step 1 Determine window The domain of f is $x \geq 0$. The domain of g is all real numbers. Choose the window: $-5 < x < 5$ and $-3 < y < 3$

Step 2 Graph functions

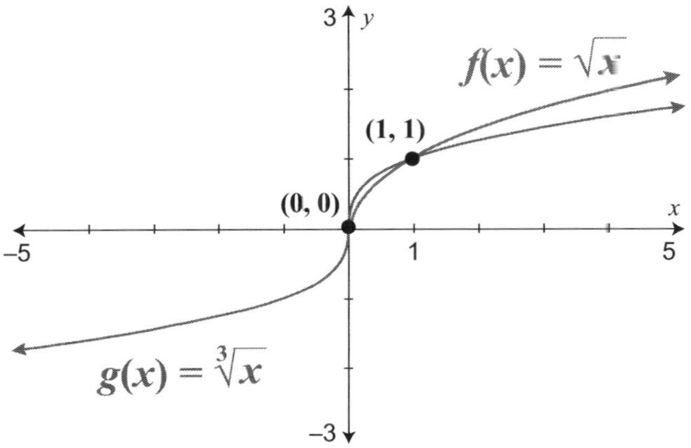

Step 3 Determine intersection(s) The functions intersect at (0, 0) and (1, 1).

Step 4 Compare the functions $g > f$ when $0 < x < 1$ and $f > g$ when $x > 1$.

Step 5 Compare intercepts Both functions have their x- and y-intercepts at (0, 0).

Step 6 Validate

Check values at $x = \frac{1}{2}$ and 2.

$\sqrt[3]{\frac{1}{2}} \overset{?}{>} \sqrt{\frac{1}{2}}$ $\quad$ $\sqrt{2} \overset{?}{>} \sqrt[3]{2}$

$0.794 > 0.707$ ✓ (approximate) $\quad$ $1.414 > 1.260$ ✓ (approximate)

Model 2 5.6

Compare $f(x) = x^3 - 20x^2 + 5x - 6$ and $g(x) = 3x^{5/2}$.

Step 1 Determine window

The domain of g is $x \geq 0$. The domain of f is all real numbers. Choose the window: $-50 < x < 50$ and $-35000 < y < 35000$

Step 2 Graph functions

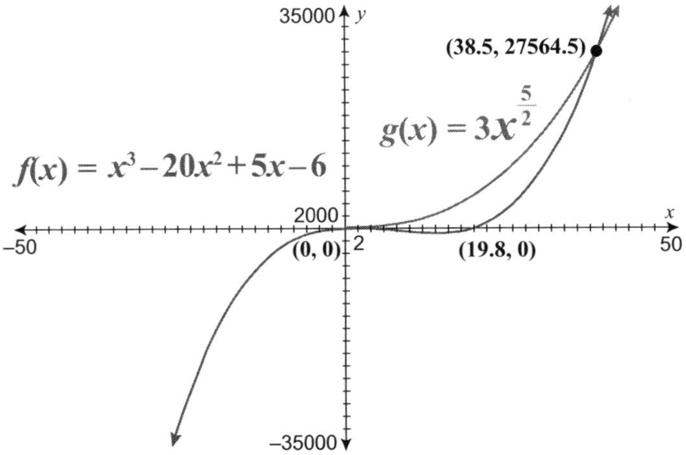

Step 3 Determine intersection(s) The functions intersect at approximately (38.5, 27564.5).

Step 4 Compare the functions $g > f$ when $0 < x < 38.5$, and $f > g$ when $x > 38.5$, approximately.

Step 5 Compare intercepts The function g has its x- and y-intercepts at (0, 0). The function f has an x-intercept at (19.8, 0), approximately and a y-intercept at (0, –6).

Step 6 Validate

g is undefined for $x < 0$.

Check values at $x = 1,\ 10,$ and 50.

$(1)^3 - 20(1)^2 + 5(1) - 6 \overset{?}{<} 3(1)^{5/2}$

$1 - 20 + 5 - 6 \overset{?}{<} 3$

$-20 < 3$ ✓

$(10)^3 - 20(10)^2 + 5(10) - 6 \overset{?}{<} 3(10)^{5/2}$

$1000 - 2000 + 50 - 6 \overset{?}{<} 53033$ (approximate)

$-954 < 53033$ (approximate) ✓

$(50)^3 - 20(50)^2 + 5(50) - 6 \overset{?}{<} 3(50)^{5/2}$

$53033 < 75244$ (approximate) ✓

Model 3 5.6

Compare $f(x) = 1 + \sqrt[3]{x}$ and $g(x) = \dfrac{3}{1 + 2^{-x}}$.

Step 1 Determine window Choose the window: $-10 < x < 10$ and $-4 < y < 4$

Step 2 Graph functions

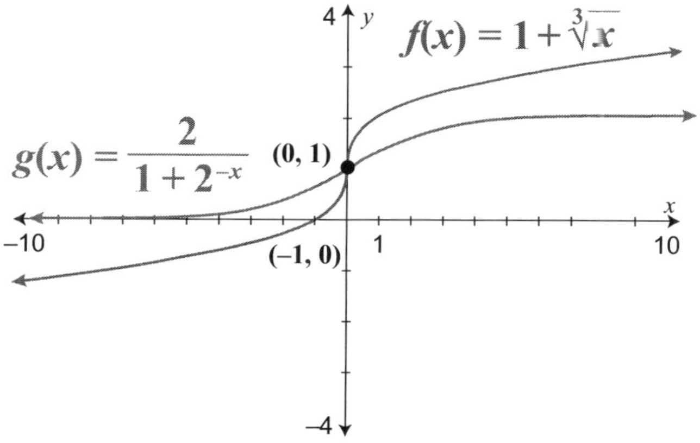

Step 3 Determine intersection(s) The functions intersect at (0, 1).

Step 4 Compare the functions $f > g$ when $x > 0$ and $f < g$ when $x < 0$.

Step 5 Compare intercepts Both functions have their y-intercepts at (0, 1). The function g has no x-intercept and the function f has an x-intercept at (–1, 0).

Step 6 Validate Check values at $x = -1$ and 1.

$$1 + \sqrt[3]{-1} \overset{?}{<} \frac{2}{1 + 2^{-(-1)}} \qquad 1 + \sqrt[3]{1} \overset{?}{>} \frac{2}{1 + 2^{-1}}$$

$$1 + (-1) \overset{?}{<} \frac{2}{3} \qquad 1 + 1 \overset{?}{>} \frac{2}{1.5}$$

$$0 < \frac{2}{3} \checkmark \qquad 2 > \frac{4}{3} \checkmark$$

Model 4 5.6

Compare $g(x) = \sqrt{x}$ and $h(x) = \log_2 x$.

Step 1 Determine window Choose the window: $-4 < x < 20$ and $-6 < y < 6$

Step 2 Graph functions

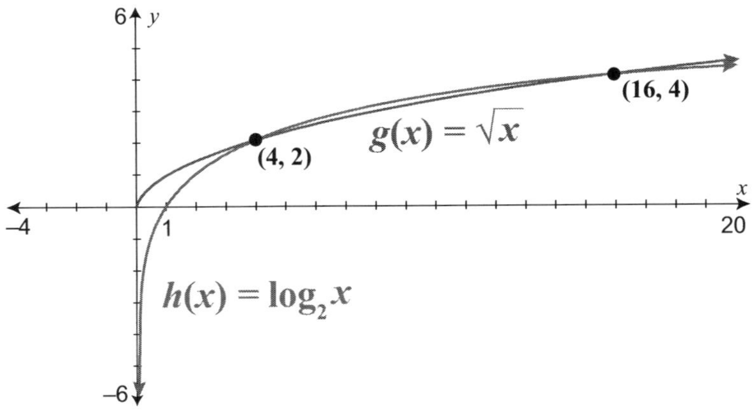

Step 3 Determine intersection(s) The functions intersect at (4, 2) and (16, 4).

Step 4 Compare the functions $h > g$ when $2 > x > 16$ and $h < g$ when $x > 16$.

Step 5 Compare intercepts The x-and y-intercepts of g are at (0, 0). The x-intercept of h is (1, 0); it has no y-intercept.

Step 6 Validate Check values at $x = 1,\ 10,\ \text{and } 20$.

$\sqrt{1} \overset{?}{>} \log_2(1)$ $\qquad$ $\sqrt{10} \overset{?}{<} \log_2(10)$

$1 > 0$ ✓ $\qquad$ $3.16228 < 3.32193$ ✓ (approximate)

$\sqrt{20} \overset{?}{>} \log_2(20)$

$4.47214 > 4.32193$ ✓ (approximate)

Model 5 5.6

Compare $f(x) = 2^{-x}$ and $r(x) = \dfrac{1}{1+x^2}$.

Step 1 Determine window

Choose the window: $-5 < x < 5$ and $-1 < y < 3$

Step 2 Graph functions

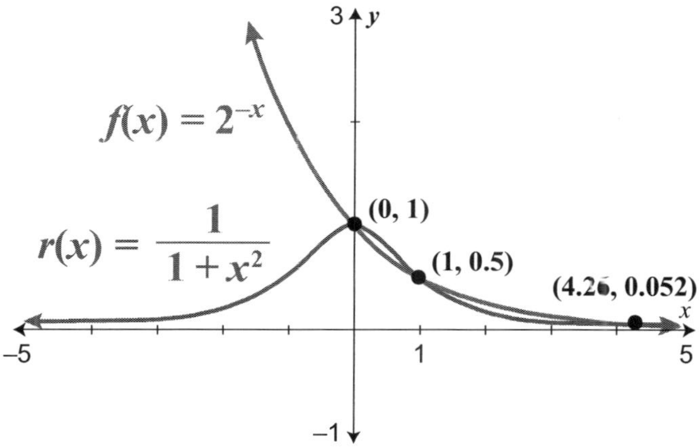

Step 3 Determine intersection(s) The functions intersect at (0, 1), (1, 0.5), and approximately (4.26, 0.052).

Step 4 Compare the functions

$f > r$ when $x < 0$ and when $1 < x < 4.26$, approximately.

$f < r$ when $0 < x < 1$ and when $x > 4.26$, approximately.

Step 5 Compare intercepts Both functions have their y-intercepts at (0, 1). Neither function has an x-intercept.

Step 6 Validate

Check values at $x = -1, \dfrac{1}{2}, 2,$ and 6.

$$2^{-(-1)} \overset{?}{>} \frac{1}{1+(-1)^2}$$

$$2 > \frac{1}{2} \checkmark$$

$$2^{-(1/2)} \overset{?}{<} \frac{1}{1+(1/2)^2}$$

$$0.707 < 0.8 \checkmark \text{(approximate)}$$

$$2^{-(2)} \overset{?}{<} \frac{1}{1+(2)^2}$$

$$\frac{1}{4} < \frac{1}{5} \checkmark$$

$$2^{-(6)} \overset{?}{<} \frac{1}{1+(6)^2}$$

$$\frac{1}{64} < \frac{1}{37} \checkmark$$

Addressing Common Errors 5.6

The following table shows some of the most common errors that learners tend to make with the kinds of problems covered in this section.

ERROR: Scaling by the wrong factor

Incorrect Process

Compare the following functions, on the interval [–10, 10]:

$$f(x) = \frac{2x^2 + 20x - 47}{x^2 + 10x - 24} \quad \text{and} \quad g(x) = \frac{1}{3}x - 3$$

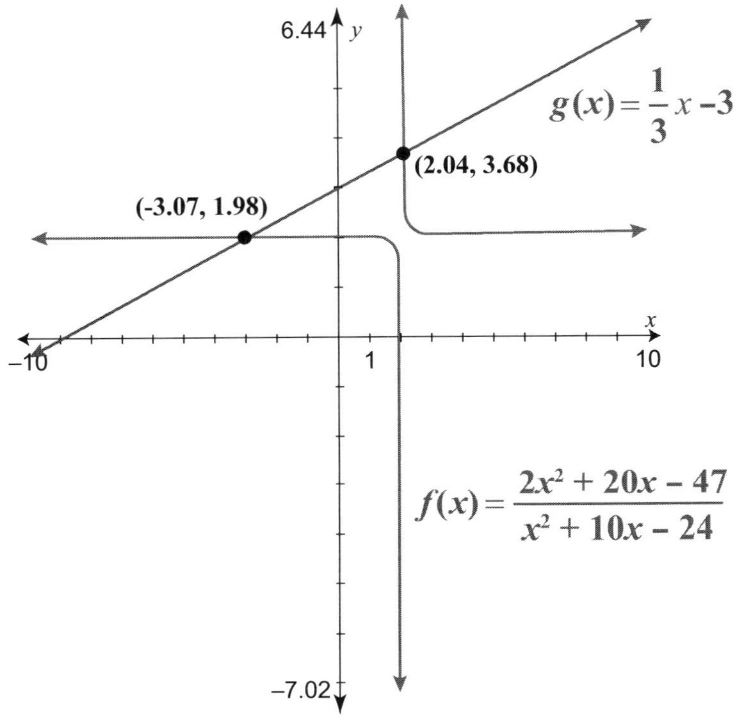

$f(x) > g(x)$ when $x < -3.07$ (approximate) and

$f(x) < g(x)$ when $x > 2.04$ (approximate)

Resolution

Understand the effect of vertical asymptotes on a function's graph.

Correct Process

$f(x) > g(x)$

when $x < -3.07$ (apprx.) or
when $2 < x < 2.04$ (apprx.)

and

$f(x) < g(x)$

when $-3.07 < x < 2$ and
$x > 2.04$ (apprx.)

Validation

Check the points –4, 0, 2.01, and 3.

$$\frac{2(-4)^2 + 20(-4) - 47}{(-4)^2 + 10(-4) - 24} \overset{?}{>} \frac{1}{3}$$

$1.97917 > 1.66667$ ✓(approximate)

$$\frac{2(0)^2 + 20(0) - 47}{(0)^2 + 10(0) - 24} \overset{?}{<} \frac{1}{3}(0) + 3$$

$\frac{47}{24} < 3$ ✓

$$\frac{2(2.01)^2 + 20(2.01) - 47}{(0)^2 + 10(2.01) - 24} \overset{?}{>} \frac{1}{3}(2.01) + 3$$

$9.13776 > 3.67$ ✓(approximate)

$$\frac{2(2)^2 + 20(2) - 47}{(2)^2 + 10(2) - 24} \overset{?}{<} \frac{1}{3}(3) + 3$$

$\frac{31}{15} < 4$ ✓

ERROR: **Reversing the comparison**

Incorrect Process

Compare the following functions:

$$f(x)=\frac{1}{2^x+1}-1 \quad \text{and} \quad g(x)=3x-4$$

$f(x) > g(x)$ when $x > 5.36$

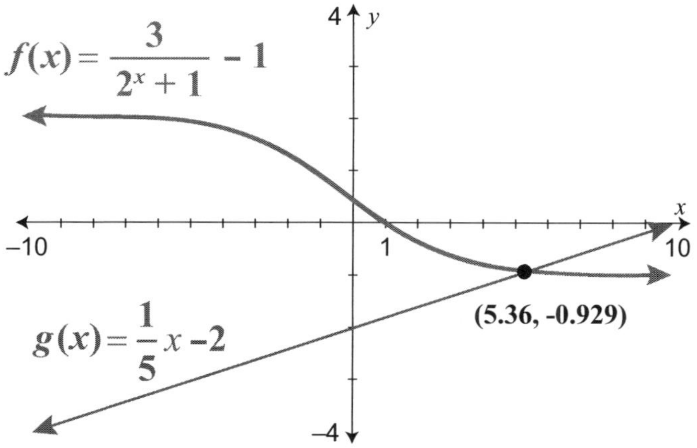

Resolution

Carefully decide where the first function is above the second.

Correct Process

$f(x) > g(x)$ when $x < 5.36$ (apprx.)

and

$f < g$ when $x > 5.36$.

Validation

Check that the intercepts are correct and that the function f is a decreasing function and g is an increasing function if the functions are analyzed left to right.

Preparation Inventory 5.6

I can...	OR ***Here's my question...***
☐ identify an appropriate comparison window for two functions.	☐
☐ find the intercepts of any graph.	☐
☐ determine all the points of intersection of the two functions.	☐
☐ determine the x-values (intervals) for which one function exceeds the another.	☐

Critical Thinking Questions 5.6

1. Why would you compare two functions?

2. What issues should you consider when you're deciding on a comparison window?

3. Why are intersections points of function graphs important when comparing functions?

4. Why are vertical asymptotes important when comparing two functions?

5. When comparing power and exponential functions, what key factor should you keep in mind?

6. When two functions are compared and one function is increasing and the other decreasing, can they have more than one point of intersection? Explain.

Demonstrate Your Understanding 5.6

1. Compare $f(x) = x$ and $g(x) = \log_2 x$.

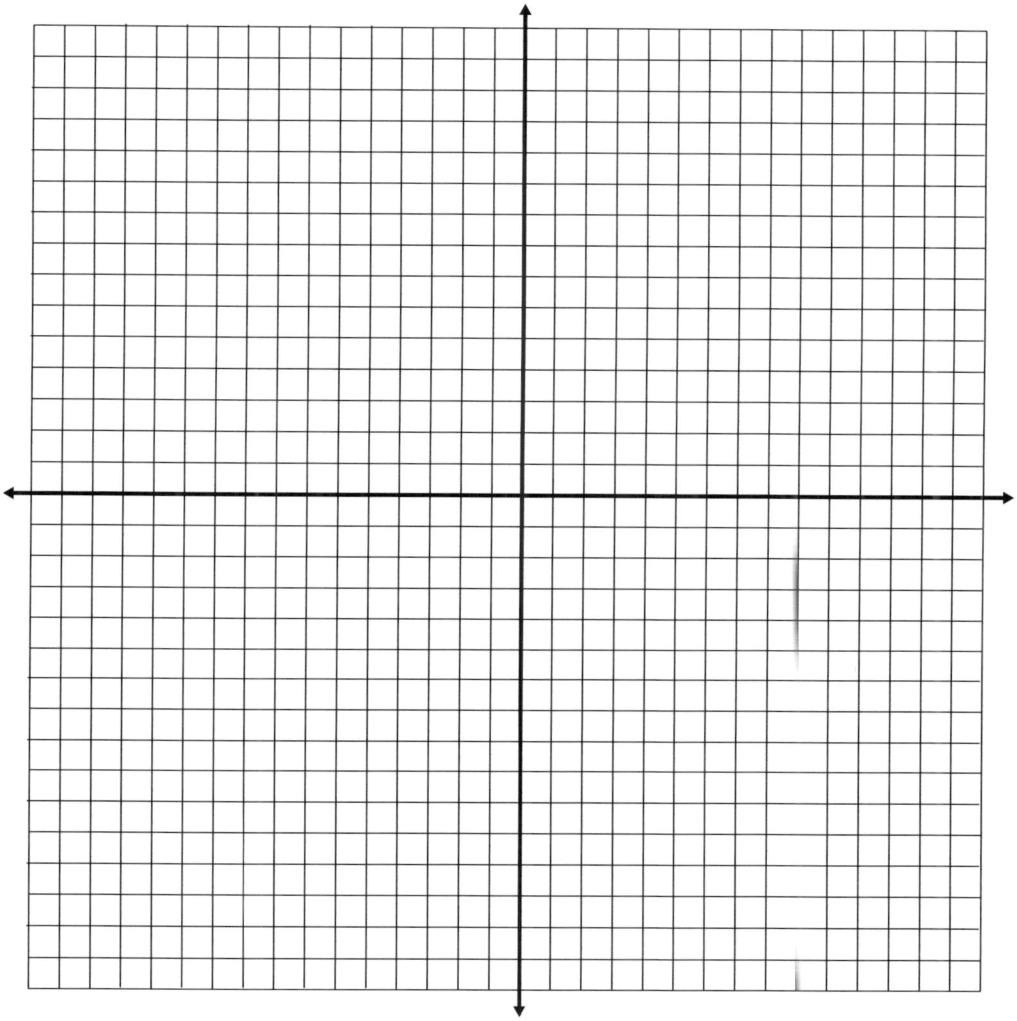

2. Compare $f(x) = x^2$ and $g(x) = x^3$.

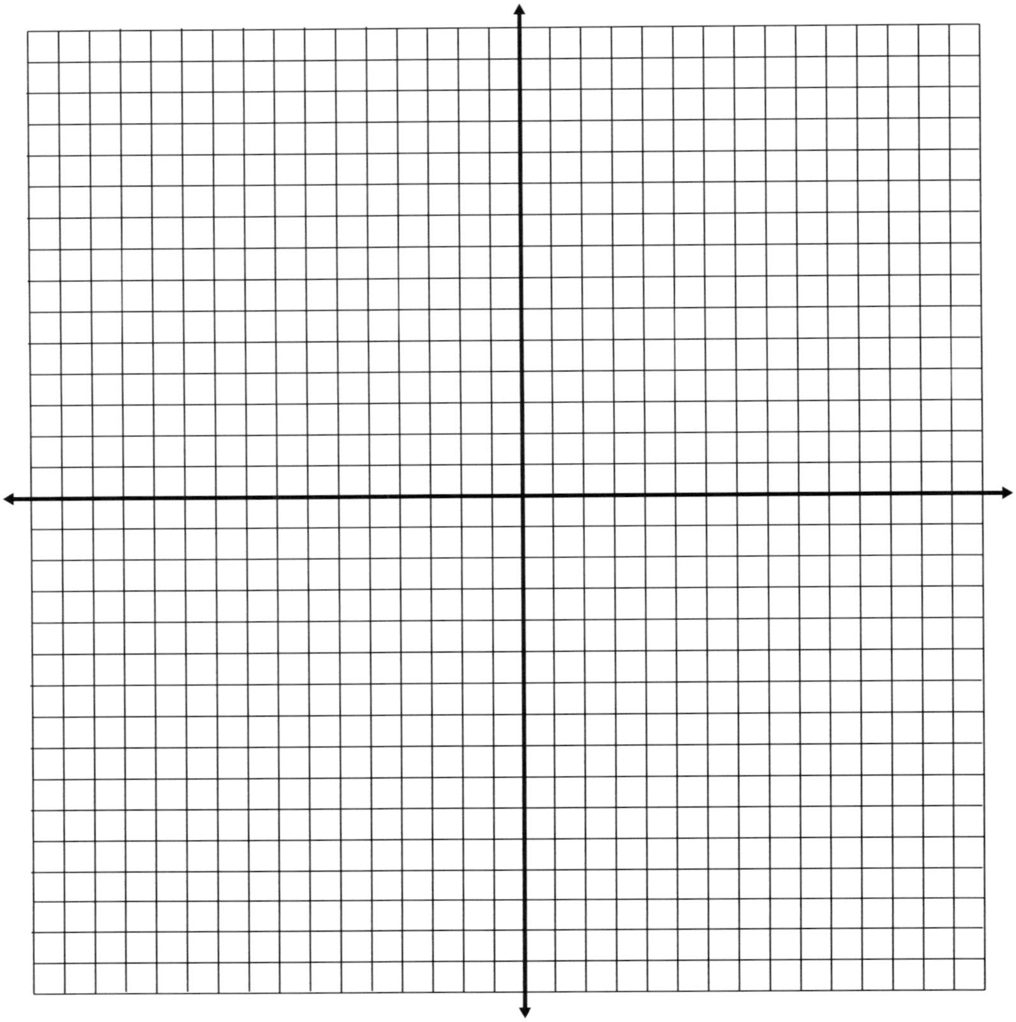

3. Compare $y = \dfrac{1}{x}$ and $y = x$.

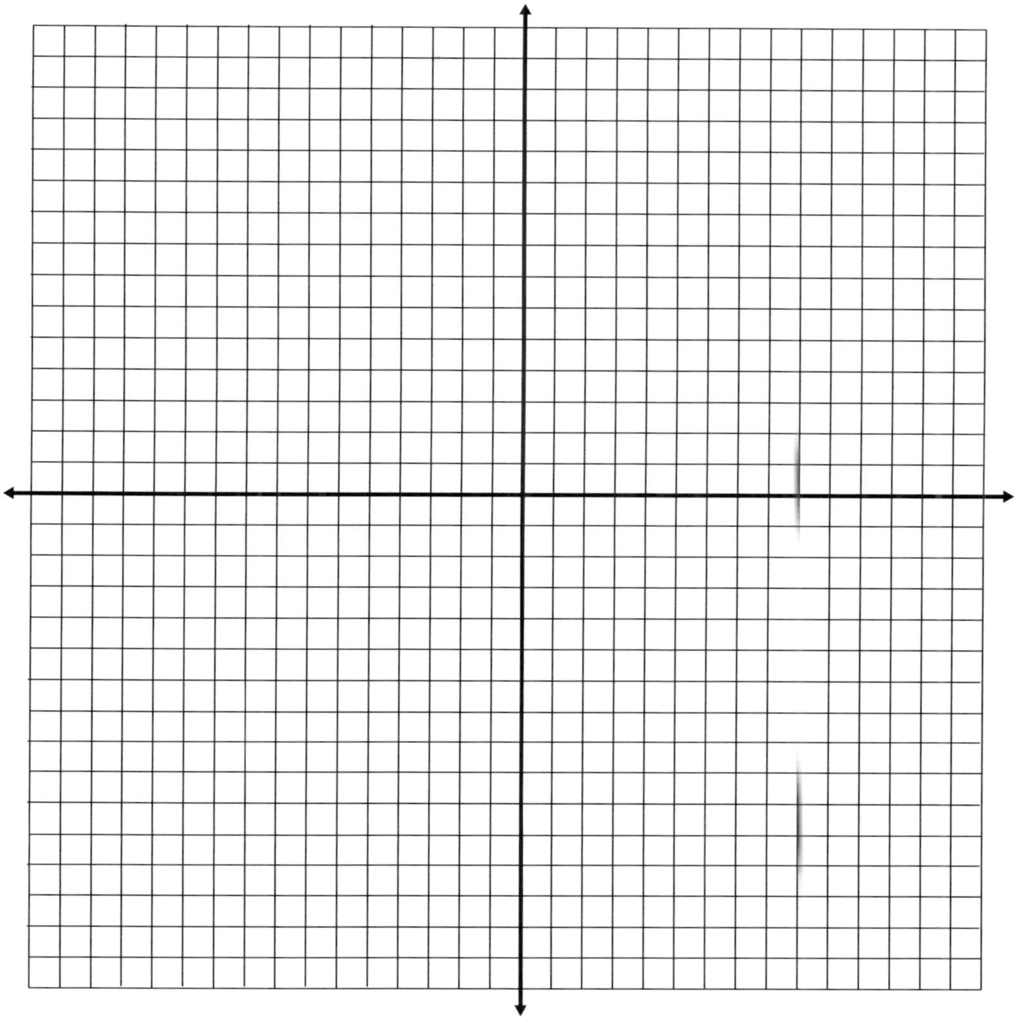

4. Compare $f(x) = 2^x$ and $g(x) = \dfrac{1}{1+x}$ with $x \geq 0$.

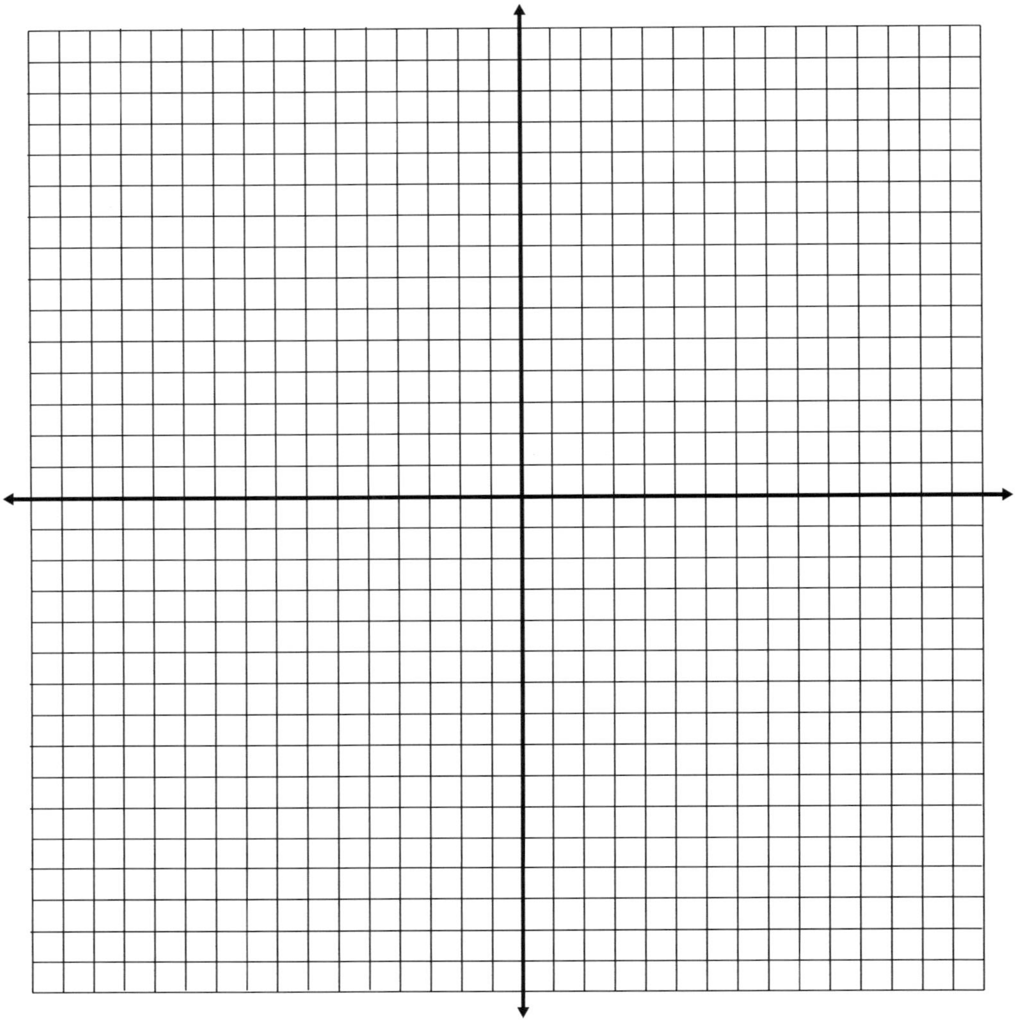

Create and Solve the Hardest Problem 5.6

Your Hardest Problem

Problem Solution

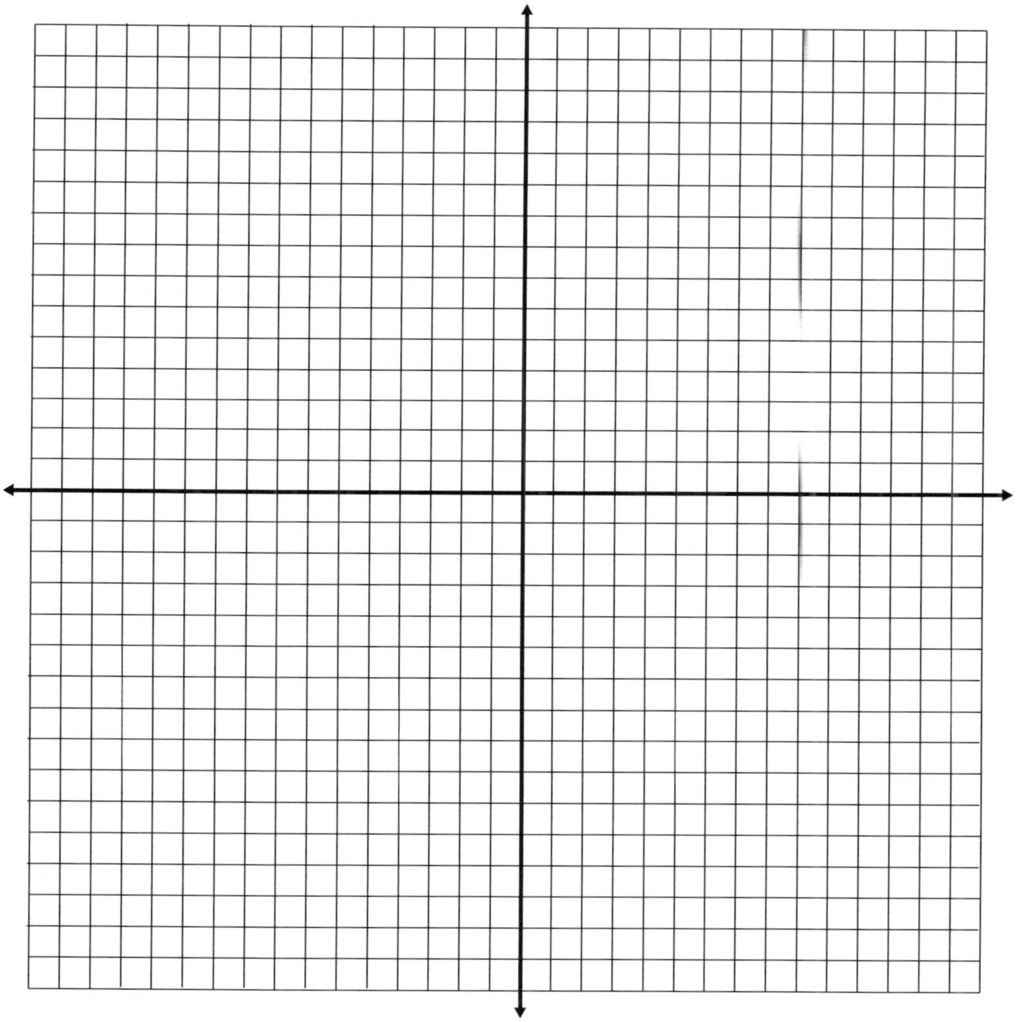

What makes a problem where you must compare functions a hard problem to solve?

Identify, Describe, & Correct the Errors 5.6

Solve the problem correctly in the appropriate location. Be sure to identify and describe the error you found in the worked solution!

Problem 1

Compare:

$f(x) = x$ and $g(x) = \frac{1}{10}x^2$

Worked Solution *(Find the error!)*

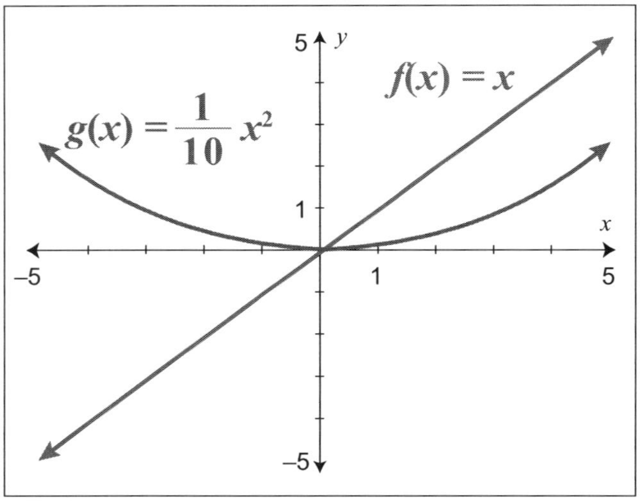

Window: $-5 < x < 5, -5 < y < 5$

Intersection: (0, 0)

Domains: $g > f$ when $x > 0$

Intercepts: x- and y-intercepts at (0, 0)

Describe the Error

Correct Process

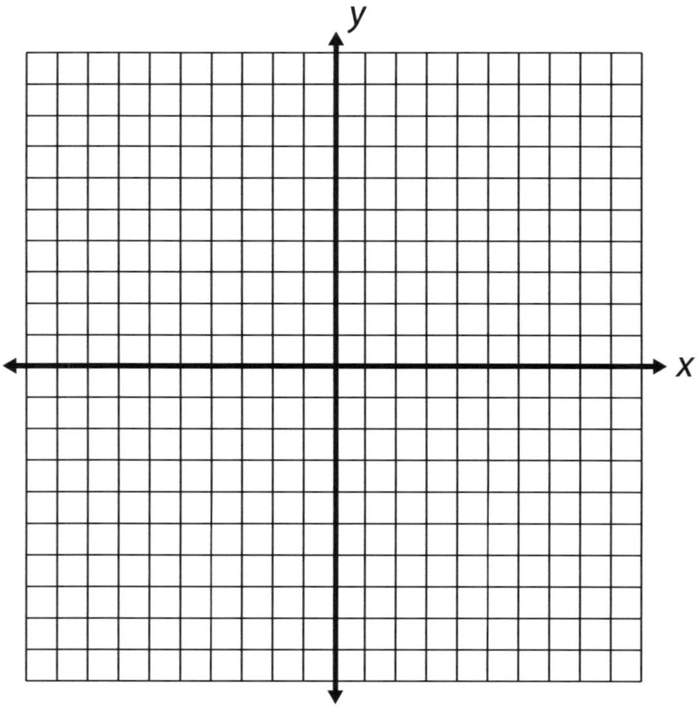

Problem 2

Find an appropriate graphing window for comparing two functions:

$$f(x) = \frac{x^5}{16} - \frac{7x^4}{4} + 18x^3 - 88x^2 + 208x - 188$$

$$\text{and } g(x) = \frac{10}{x-15}$$

Describe the Error

Worked Solution *(Find the error!)*

An appropriate graphing window is [–10, 10] for x and [–10, 10] for y.

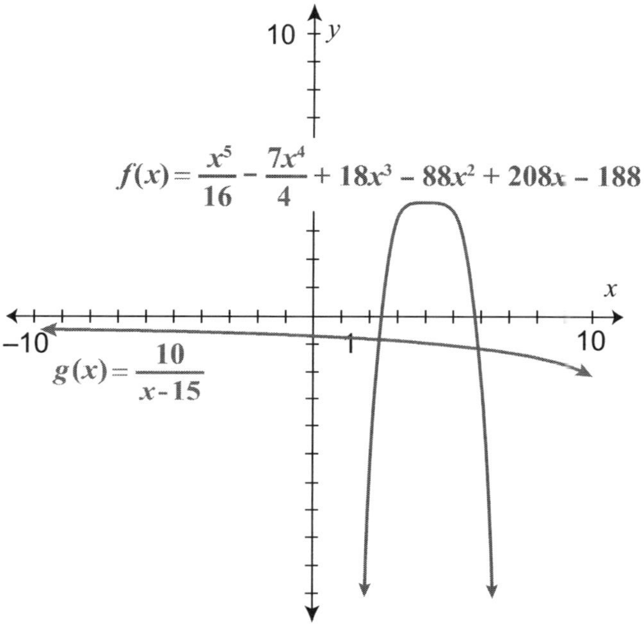

Correct Process

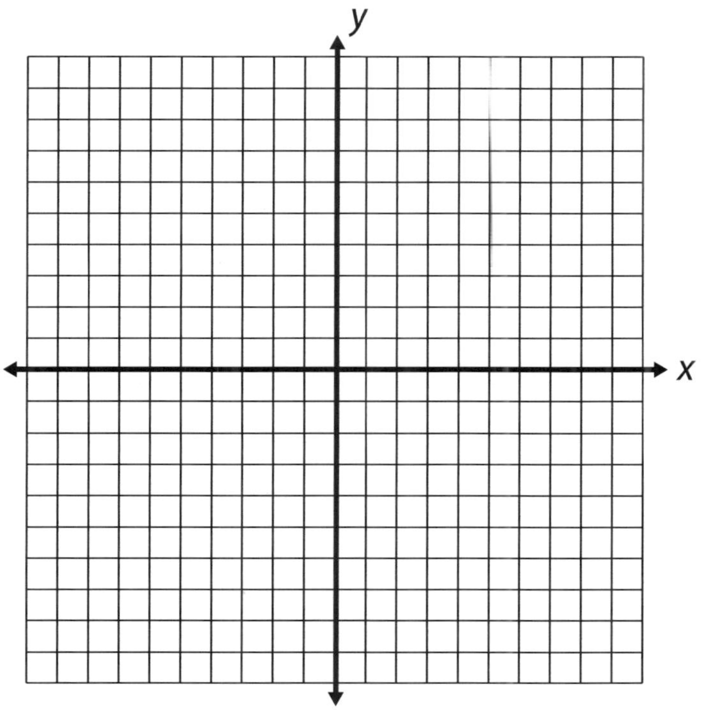

Problem 3

Compare the functions $d_1(t)$ and $d_2(t)$ on the interval $0 \le t \le 5$.

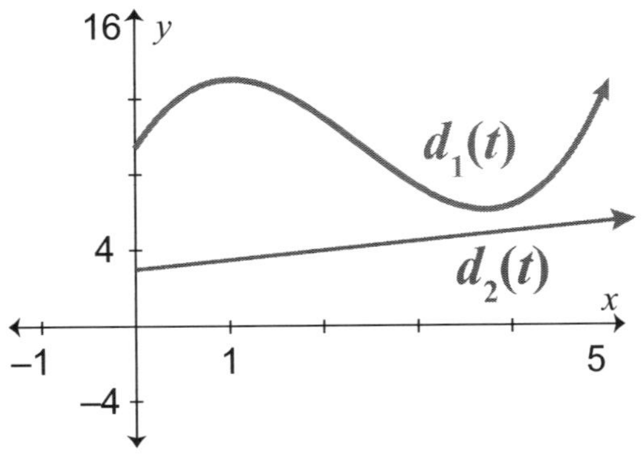

Worked Solution *(Find the error!)*

The function $d_2(t)$ is greater than the function $d_1(t)$ on the interval $0 \le t \le 5$.

Correct Process

Describe the Error

What Did You Learn About Learning Math? 5.6

1. What did you find difficult about determining when one function was larger than another?

2. How do you ensure that you have all the important features of both functions in your graphing window?

3. When comparing two functions, how are points of intersection helpful and how are intercepts helpful?

Concept Map — Chapter 5

Terminology from the chapter
check the terms used in your map

__absolute value
__asymptotes
__behavior of a function
__comparison window
__domain
__end behavior of a function
__exponential
__families of functions
__function
__function behavior
__function composition
__function notation
__inverse function
__inverse relation
__linear
__logarithmic
__logistic
__one-to-one function
__polynomial
__quadratic
__radical
__range
__rational
__reflection
__relation
__scaling
__translation

Your instructor will specify how many terms from the chapter you should use in your concept map.

CHAPTER 6

In this chapter you will continue to develop your ability to solve a variety of types of equations using the Solving Advanced Equations Methodology. This methodology uses ideas that you have mastered throughout this course including transforming an equation to an equivalent form, inverse functions, graphing, and validating. In addition, you will learn graphical means of approximating solutions to equations that cannot be solved symbolically.

6.1 Solving Absolute Value and Rational Equations

6.2 Transforming Equations

6.3 Solving an Equation

6.4 Additional Tools for Solving Equations

Section 6.1 Solving Absolute Value and Rational Equations

Why? 6.1

Phenomena in the real world can be modeled with absolute value equations and rational equations. For example, it is known that the intensity of light is inversely proportional to the square of the distance from its light source. So, the intensity divided by the reciprocal of the square of the distance is always a constant.

In other words, $\dfrac{\text{intensity}}{\left(\dfrac{1}{\text{distance}^2}\right)}$ is always the same.

What Do You Already Know? 6.1

How do you define the absolute value of x?

How do you add two rational expressions?

How do you validate the solution to an equation?

Goals 6.1

When you complete this activity, you should be able to:

1. Solve absolute value equations
2. Solve rational equations
3. Explain what extraneous solutions to rational equations are

Building Mathematical Language 6.1

New Terms:

- absolute value equation
- eliminating the denominator
- extraneous solutions
- reduced solution set
- rational equation

Key Concepts 6.1

When solving an ***absolute value equation*** or a ***rational equation*** symbolically, you are trying to transform an equation into a (usually linear or quadratic) equation in one variable that can be solved. Equations involving absolute value are transformed into equations without absolute value. ***Denominators are eliminated*** from rational equations.

Limitation/Caution: The process of solving an equation can lead to solutions that do not work within the context of the original equation. Such solutions are called ***extraneous solutions***. The solution to the original equation with the extraneous solutions removed is called the ***reduced solution set***. In addition, there are many equations that cannot be solved using symbolic manipulations. Often such equations can be solved using graphing software on handheld devices or computers.

Methodology 1 — 6.1

Solving Absolute Value Equations

	Example 1	Example 2	Your Turn
Solve:	$\lvert 2x-3\rvert = 5$	$\lvert 4x-7\rvert = -3$	$\lvert 2-3x\rvert = 5$

Steps	Discussion
1 Remove the absolute value functions	Use the definition of absolute value: $\lvert x\rvert = \begin{cases} x, \text{if } x \ge 0 \\ -x, \text{if } x < 0 \end{cases}$ An absolute value can never be a negative number.

Ex 1

$2x-3=5$ or $2x-3=-5$

Ex 2

Since absolute values cannot be negative, there is no solution to this equation.

Your Turn

2 Solve for the variable	Use the Methodology for Solving Equations.

Ex 1

$2x=8$ or $2x=-2$

$x=4$ or $x=-1$

Ex 2

No solution.

Your Turn

3 Validate	Show that the solution set satisfies the absolute value equation.

Ex 1

$\lvert 2x-3\rvert = 5$

$\lvert 2(4)-3\rvert = 5$ or $\lvert 2(-1)-3\rvert = 5$

$\lvert 8-3\rvert = 5$ or $\lvert -2-3\rvert = 5$

$\lvert 5\rvert = 5$ or $\lvert -5\rvert = 5$

$5=5$ or $5=5$ ✓

Ex 2

In the original equation, observe that the absolute value of the left side could never be equal to a negative number.

Your Turn

Methodology 2 — 6.1

Eliminating Denominators in a Rational Equation

	Example 1	Example 2	Your Turn
Eliminate the denominator:	$\frac{2}{x}+\frac{3}{2x-1}=2$	$\frac{3}{x+2}-\frac{x+1}{x^2-4x-12}=5$	$\frac{3}{x^2-5x-6}+\frac{2}{x-1}=-1$

Steps	Discussion
1 Determine the LCM of the denominators	Factor the denominators and create the LCM.

Ex 1

$$\frac{2}{x}+\frac{3}{2x-1}=2$$

Already factored.

$\text{LCM}=x(2x-1)$

Ex 2

$$\frac{3}{x+2}-\frac{x+1}{x^2-4x-12}=5$$

$$\frac{3}{x+2}-\frac{x+1}{(x-6)(x+2)}=5$$

$\text{LCM}=(x-6)(x+2)$

Your Turn

Steps	Discussion
2 Multiply by the LCM	Multiply both sides of the equation by the LCM to eliminate the denominators.

Ex 1

$$x(2x-1)\bullet\left(\frac{2}{x}+\frac{3}{2x-1}=2\right)$$

$$2(2x-1)+3x=2x(2x-1)$$

$$4x-2+3x=4x^2-2x$$

$$7x-2=4x^2-2x$$

Ex 2

$$(x-6)(x+2)\bullet\left(\frac{3}{x+2}-\frac{x+1}{x^2-4x-12}=5\right)$$

$$(x-6)(x+2)\bullet\frac{3}{x+2}-(x-6)(x+2)\bullet\frac{x+1}{(x-6)(x+2)}=5\bullet(x-6)(x+2)$$

$$3(x-6)-(x+1)=5\left(x^2-4x-12\right)$$

$$3x-18-x-1=5x^2-20x-60$$

$$2x-19=5x^2-20x-60$$

Steps	Discussion
2 Multiply by the LCM (con't)	Multiply both sides of the equation by the LCM to eliminate the denominators.

Your Turn

Steps	Discussion
3 Determine the excluded values	Determine the value(s) of the variable that make the denominator(s) zero.

Ex 1

The denominators are x and $(2x-1)$.

The values that make these denominators zero and should be excluded are:

$x=0$ and $x=\dfrac{1}{2}$.

Ex 2

The denominators are $(x+2)$ and $(x-6)(x+2)$.

The values that make these denominators zero and should be excluded are:

$x=6$ and $x=-2$.

Your Turn

Steps	Discussion
4 Validate that the denominators were correctly eliminated	**Transform the original equation and the simplified equation into equations with a zero on the right side. Substitute a non-excluded value in for the variable. The original equation result multiplied by the LCM should be the simplified result.**

Ex 1

$$\frac{2}{x}+\frac{3}{2x-1}-2=0$$

Let $x=5$.

$$\frac{2}{(5)}+\frac{3}{2(5)-1}-2$$

$$\frac{2}{5}+\frac{3}{9}-2$$

$$\frac{18}{45}+\frac{15}{45}-\frac{90}{45}$$

$$\frac{33-90}{45}$$

$$\frac{-57}{45}\ \checkmark$$

$$7x-2-4x^2+2x=0$$

$$7(5)-2-4(5)^2+2(5)$$

$$35-2-100+10$$

$$33-90$$

$$-57\ \checkmark$$

Ex 2

$$\frac{3}{x+2}-\frac{x+1}{x^2-4x-12}-5=0$$

Let $x=3$.

$$\frac{3}{(3+2)}-\frac{3+1}{3^2-(4\bullet 3)-12}-5$$

$$\frac{3}{5}-\frac{4}{-15}-5$$

$$\frac{9}{15}+\frac{15}{15}-\frac{75}{15}$$

$$\frac{62}{-15}\ \checkmark$$

$$2x-19-5x^2+20x+60$$

$$2(3)-19-5(3)^2+20(3)+60$$

$$6-19-45+60+60$$

$$62\ \checkmark$$

Your Turn

Methodology 3 — 6.1

Solving Rational Equations

	Example 1	Example 2	Your Turn
Solve:	$\frac{1}{x-3}+1=\frac{6}{x^2-9}$	$\frac{x}{x+3}-\frac{x}{x-2}=\frac{10}{x^2+x-6}$	$\frac{3}{x-2}+\frac{2x}{x^2-3x+2}=6$

Steps	Discussion
1 Eliminate the denominators	Use the Methodology for Eliminating Denominators.

Ex 1

$$\frac{1}{x-3}+1=\frac{6}{x^2-9}$$

$$\left(\frac{1}{x-3}\right)(x+3)(x-3)+1(x+3)(x-3)=\left(\frac{6}{(x+3)(x-3)}\right)(x+3)(x-3)$$

$$\left(\frac{1}{\cancel{x-3}}\right)(x+3)\cancel{(x-3)}+1(x+3)(x-3)=\left(\frac{6}{\cancel{(x+3)(x-3)}}\right)\cancel{(x+3)(x-3)}$$

$$(x-3)+(x+3)(x-3)=6$$

Ex 2

$$\frac{x}{x+3}-\frac{x}{x-2}=\frac{10}{x^2+x-6}$$

$$(x+3)(x-2)\left(\frac{x}{x+3}-\frac{x}{x-2}=\frac{10}{(x+3)(x-2)}\right)$$

$$x(x-2)-x(x+3)=10$$

Your Turn

Steps	Discussion
2 Solve the equation	Use the Methodology for Solving Equations.

Ex 1

$x^2 + x - 6 = 6$

$x^2 + x - 12 = 0$

$(x+4) \bullet (x-3) = 0$

$x + 4 = 0$ or $x - 3 = 0$

$x = -4$ or $x = 3$

Both of these numbers are solutions of
$x + 3 + x^2 - 9 = 6$

$-4 + 3 + (-4)^2 - 9 \stackrel{?}{=} 6$

$-1 + 16 - 9 \stackrel{?}{=} 6$

$6 = 6$

$3 + 3 + (3)^2 - 9 \stackrel{?}{=} 6$

$3 + 3 + 9 - 9 \stackrel{?}{=} 6$

$6 = 6$ ✓

Ex 2

$x^2 - 2x - x^2 - 3x = 10$

$-5x = 10$

$x = -2$

$(-2)^2 - 2(-2) - (-2)^2 - 3(-2) = 4 + 6 = 10$,
so -2 is a solution of the simplified equation.

Your Turn

Steps	Discussion
3 Eliminate the extraneous solution(s), if any	**Determine which solutions satisfy the original equation. Graphing software can assist you with this task.**

Ex 1

$$x+3+x^2-9=6$$
$$x^2+x-6=6$$
$$x^2+x-12=0$$

Let $x=-4$

$$\frac{1}{x-3}+1\overset{?}{=}\frac{6}{x^2-9}$$
$$\frac{1}{(-4)-3}+1\overset{?}{=}\frac{6}{(-4)^2-9}$$
$$\frac{1}{-7}+\frac{7}{7}\overset{?}{=}\frac{6}{16-9}$$
$$\frac{6}{7}=\frac{6}{7}\ \checkmark$$

Let $x=3$

$$\frac{1}{(3)-3}+1=\frac{6}{(3)^2-9}$$
$$\frac{1}{0}=\frac{6}{0}\quad \text{undefined}$$

Therefore, $x=3$ is an extraneous solution.

Ex 2

Let $x=-2$

$$\frac{-2}{-2+3}-\frac{-2}{-2-2}\overset{?}{=}\frac{10}{(-2)^2+(-2)-6}$$
$$-2-\frac{1}{2}\overset{?}{=}\frac{10}{-4}$$
$$-\frac{5}{2}=-\frac{5}{2}\ \checkmark$$

Your Turn

Steps	Discussion
4 Present the solution	**Present the validated reduced solution set by eliminating the extraneous solutions.**

Ex 1 The solution is $x=-4$.

Ex 2 The solution is $x=-2$.

Your Turn

Addressing Common Errors 6.1

The following table shows some of the most common errors that learners tend to make with the kinds of problems covered in this section.

ERROR: Failing to check for extraneous solutions

Incorrect Process

Solve:

$$\frac{6}{x^2+x-2}-\frac{2}{x-1}=1$$

$$\frac{6}{(x+2)(x-1)}-\frac{2}{x-1}=1$$

$$\text{LCM}=(x+2)(x-1)$$

$$(x+2)(x-1)\bullet\left(\frac{6}{(x+2)(x-1)}-\frac{2}{x-1}=1\right)$$

$$(x+2)(x-1)\bullet\frac{6}{(x+2)(x-1)}-(x+2)(x-1)\bullet\frac{2}{x-1}=1(x+2)(x-1)$$

$$6-2(x+2)=(x+2)(x-1)$$

$$6-2x-4=x^2+x-2$$

$$2-2x=x^2+x-2$$

$$0=x^2+3x-4$$

$$0=(x+4)(x-1)$$

$$x+4=0 \text{ or } x-1=0$$

$$x=-4 \text{ or } x=1$$

Correct Process

$x=4$ or -1. Check each in the original equation:

Let $x=-4$

$$\frac{6}{(-4)^2+(-4)-2}-\frac{2}{(-4)-1}\stackrel{?}{=}1$$

$$\frac{6}{16-4-2}-\frac{2}{-5}\stackrel{?}{=}1$$

$$\frac{6}{10}+\frac{2}{5}\stackrel{?}{=}1$$

$$\frac{3}{5}+\frac{2}{5}\stackrel{?}{=}1;\quad 1=1\ ✓$$

Let $x=1$

$$\frac{6}{(1)^2+(1)-2}-\frac{2}{(1)-1}\stackrel{?}{=}1$$

$$\frac{6}{0}-\frac{2}{0}\stackrel{?}{=}1$$

$x=1$ is not a solution.

Resolution

Check for extraneous solutions.

Validation

The correct process here includes the validation.

ERROR: **Incorrectly removing the absolute value signs**

Incorrect Process

Solve: $|2x-5|=7$

$|2x-5|=7$

$2x-5=7$

$2x=12$

$x=6$

Resolution

Correctly use the definition of absolute value:

$$|x|=\begin{cases} x, \text{ if } x \geq 0 \\ -x, \text{ if } x < 0 \end{cases}$$

Correct Process

$|2x-5|=7$

$2x-5=7$ or $2x-5=-7$

$2x=12$ or $2x=-2$

$x=6$ or $x=-1$

Validation

Let $x=6$ or $x=-1$

$|2(6)-5| \stackrel{?}{=} 7$ or $|2(-1)-5| \stackrel{?}{=} 7$

$|12-5| \stackrel{?}{=} 7$ or $|-2-5| \stackrel{?}{=} 7$

$|7| \stackrel{?}{=} 7$ or $|-7| \stackrel{?}{=} 7$

$7=7$ ✓ or $7=7$ ✓

ERROR: **Eliminating the denominators of a rational equation by multiplying only one side of the equation by the LCM**

Incorrect Process

Eliminate the denominators of

$$\frac{2}{x-3}+5x=3$$

$\frac{2}{x-3}+5x=3$

LCM $=x-3$

$(x-3)\left(\frac{2}{x-3}+5x=3\right)$

$2+5x^2-15x=3$

$5x^2-15x+2=3$

Correct Process

$\frac{2}{x-3}+5x=3$

LCM $=x-3$

$(x-3)\left(\frac{2}{x-3}+5x=3\right)$

$2+5x^2-15x=3 \cdot (x-3)$

$5x^2-15x+2=3x-9$

Resolution

Multiply both sides of the equation by the LCM.

ERROR: Eliminating the denominators of a rational equation by multiplying only one side of the equation by the LCM (con't)

Validation

Using the quadratic formula, you can solve the simplified equation. Substitute solutions back into the original equation to validate.

$5x^2 - 18x + 11 = 0;$

a = 5, b = –18, c = 11

Use a calculator to solve.

Solutions: $x = 2.82, x = 0.78$

$$\frac{2}{2.82-3}+5(2.82) \stackrel{?}{=} 3$$
$$-\frac{2}{0.18}+14.1 \stackrel{?}{=} 3$$
$$-11.11+14.1 \stackrel{?}{=} 3$$
$$2.99 \approx 3 \checkmark$$

$$\frac{2}{0.78-3}+5(0.78) \stackrel{?}{=} 3$$
$$-\frac{2}{2.22}+3.9 \stackrel{?}{=} 3$$
$$-.901+3.9 \stackrel{?}{=} 3$$
$$2.999 \approx 3 \checkmark$$

Preparation Inventory 6.1

I can...	OR *Here's my question...*
☐ solve an absolute value equation.	☐
☐ eliminate denominators.	☐
☐ check for extraneous solutions.	☐
☐ solve a rational equation.	☐

Critical Thinking Questions 6.1

1. How does solving an absolute value inequality (Section 3.2) differ from solving an absolute value equation?

2. Why is the conjunction “or” rather than “and” used when solving an absolute value inequality?

3. How do you clear denominators?

4. How do you determine the expression to use when clearing denominators?

5. How do you determine the values to exclude from the solution set of a rational equation?

6. How do you validate a solution set?

7. Why do you clear denominators?

8. Why do you exclude values from a solution set?

9. What is the nature of a rational equation that allows you to "cross multiply"?

10. How would you create a rational equation that has no solutions?

11. How might you begin to solve an absolute value inequality with two absolute value functions?

Demonstrate Your Understanding 6.1

1. Determine the excluded values: $\dfrac{2}{y-5} - \dfrac{y}{y^2 - 8y - 20} = \dfrac{y-3}{2}$

2. Solve: $|3 - 2x| = 4$

3. Solve the following equations for the designated variable:

   a) $s + \dfrac{2}{3-s} = 2$ (for s)

b) $\frac{x+17}{5} - \frac{x}{2} = 2x$ (for x)

c) $\frac{4x}{x+2} - \frac{3x}{x-1} = \frac{x^2 - 8x - 4}{x^2 + x - 2}$ (for x)

Create and Solve the Hardest Problem 6.1

Your Hardest Problem

Problem Solution

What makes a problem where you solve absolute value or rational equations hard?

Identify, Describe, & Correct the Errors 6.1

Solve the problem correctly in the appropriate location. Be sure to identify and describe the error you found in the worked solution!

Problem 1

Solve for x: $|x - 5| = 3$

Worked Solution *(Find the error!)*

$$|x - 5| = 3$$
$$x - 5 = 3$$
$$x = 8$$

Correct Process

Describe the Error

Problem 2

Solve for x: $|-2x + 5| = 7$

Worked Solution *(Find the error!)*

$$|-2x + 5| = 7$$
$$-(-2x + 5) = 7 \text{ and } -2x + 5 = 7$$
$$2x - 5 = 7 \text{ and } -2x = 2$$
$$2x = 12 \text{ and } x = -1$$
$$x = 6 \text{ and } x = -1$$

But a number cannot be both positive and negative.

There is no solution.

Correct Process

Describe the Error

Problem 3

Solve for y:

$$\frac{4y}{y-4} = 2 + \frac{16}{y-4}$$

Worked Solution *(Find the error!)*

$$\frac{4y}{y-4} = 2 + \frac{16}{y-4}$$

$$\text{LCM} = y-4$$

$$(y-4)\frac{4y}{y-4} = (y-4)(2) + (y-4)\frac{16}{y-4}$$

$$4y = 2y - 8 + 16$$

$$2y = 8$$

$$y = 4$$

Describe the Error

Correct Process

Problem 4

Eliminate the denominators:

$$\frac{2}{3(x-4)} + \frac{x}{x-2} = 7$$

Worked Solution *(Find the error!)*

$$\frac{2}{3(x-4)} + \frac{x}{x-2} = 7$$

$$\text{LCD} = (x-4)(x-2)$$

$$(x-4)(x-2)\frac{2}{3(x-4)} + (x-4)(x-2)\frac{x}{x-2} = 7(x-4)(x-2)$$

$$2(x-2) + (x-4)x = 7(x-4)(x-2)$$

Describe the Error

Correct Process

Problem 4 con't

Eliminate the denominators: $\frac{2}{3(x-4)} + \frac{x}{x-2} = 7$

Problem 5

Eliminate the denominators:

$$\frac{x}{2x-3} - \frac{4}{x+2} = 7$$

Describe the Error

Worked Solution *(Find the error!)*

$$\frac{x}{2x-3} - \frac{4}{x+2} = 7$$

$$\text{LCM} = (2x-3)(x+2)$$

$$(2x-3)(x+2)\frac{x}{2x-3} - (2x-3)(x+2)\frac{4}{x+2} = 7$$

$$(x+2)x - (2x-3)(4) = 7$$

Correct Process

Problem 5 con't

Eliminate the denominators: $\frac{x}{2x-3} - \frac{4}{x+2} = 7$

What Did You Learn About Learning Math? 6.1

1. What did you learn about deciding how to approach solving an equation?

2. What did you learn about using prior knowledge when solving equations?

3. How is validating the solution set of a rational equation the same as validating the solution set of an absolute value problem?

Section 6.2 Transforming Equations

Why? 6.2

In the real world, not all equations are "pretty," that is, easy to solve by previously learned methods. Some equations, in order to be solved, should first be transformed into an equivalent equation, or at least an equation that includes the solution set of the original equation. Section 6.3 will deal with solving such equations after they are transformed.

What Do You Already Know? 6.2

What are equivalent equations?

What is an inverse function?

What are extraneous solutions?

What is a one-to-one function?

Goals 6.2

When you complete this activity, you should be able to:

1. Use an inverse function to remove an embedded function
2. Transform equations so that a chosen function no longer appears

Building Mathematical Language 6.2

New Terms: embedded functions transformed equations

Key Concepts 6.2

Equivalent equations can be created in several ways. We have seen how adding, subtracting, multiplying, or dividing both sides of an equation by a non-zero value results in an equivalent equation. Applying a one-to-one function to both sides of an equation also results in an equivalent equation.

Limitation/Caution: Applying a function that is not one-to-one to both sides of an equation does not always result in an equivalent equation that has the same solution set as the original equation. However, the new equation will contain all of the solutions of the original equation.

Example: The equation $x - 1 = 3$ has only the number 4 in its solution set (the simplified equation is $x = 4$). If you square the original equation $(x - 1 = 3)$ by applying the function $f(x) = x^2$ to both sides, the result is:

$$(x-1)^2 = 3^2$$

$$x^2 - 2x + 1 = 9$$

$$x^2 - 2x - 8 = 0$$

If we solve this equation: $x^2 - 2x - 8 = 0$

$$(x+2)(x-4) = 0$$

$$x = -2 \text{ and } x = 4$$

We see that both 4 **and** −2 are members of the solution set. Thus the solution set for the new equation contains all members of the solution set for the original equation, plus an additional member.

A ***transformed equation*** is the result of applying one function (such as the inverse) to another function that is in an equation.

An ***embedded function*** in an equation is a non-polynomial function in the equation. For example, the function $f(x) = \sqrt{x}$ is embedded in the equation $x + \sqrt{2x+5} = 5$.

Methodology 6.2

Transforming an Equation with Embedded Functions

Equations with embedded functions can be transformed so that the embedded functions no longer appear. Note that in equations, functions are denoted by just their defining expressions.

Limitation/Caution: If there are two different non-polynomial functions in an equation, it may not be possible to remove them both.

	Example 1	**Example 2**	**Your Turn**
Transform the given equation so that the selected function does not appear:	Equation $4x + 2\sqrt{x} = 5\sqrt{x}$ Function $f(x) = \sqrt{x}$	Equation $\log_2(x+1) = 3 - \log_2(x-1)$ Function $f(x) = \log_2(x)$	Equation $3e^x - 2 = 5x^2 - e^x$ Function $f(x) = e^x$

Steps	**Discussion**
1 Identify the function(s) in the equation	**Identify the family of functions that the embedded function(s) belong to.**

Ex 1: The function(s) are from the radical family of functions.

Ex 2: The function(s) are from the logarithmic family of functions.

Your Turn:

Steps	Discussion
2 Graph the function	Pull the function out of the equation and graph it.

Ex 1

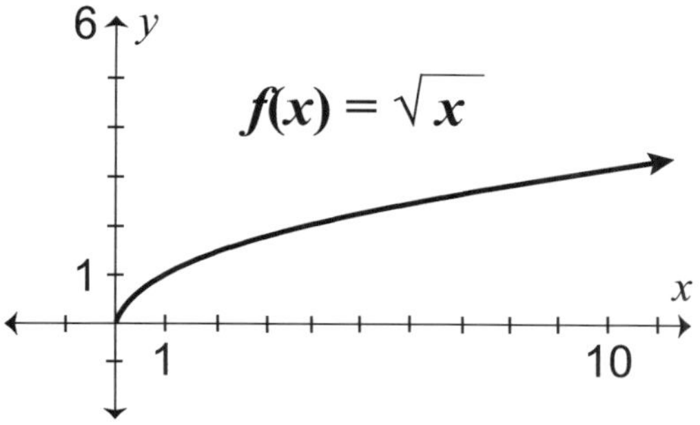

Ex 2

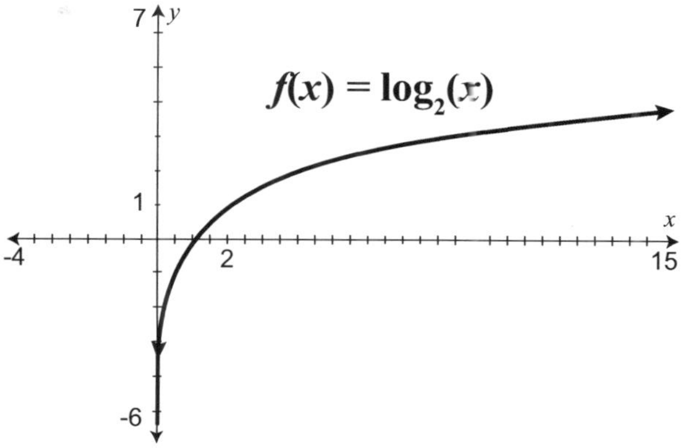

Your Turn

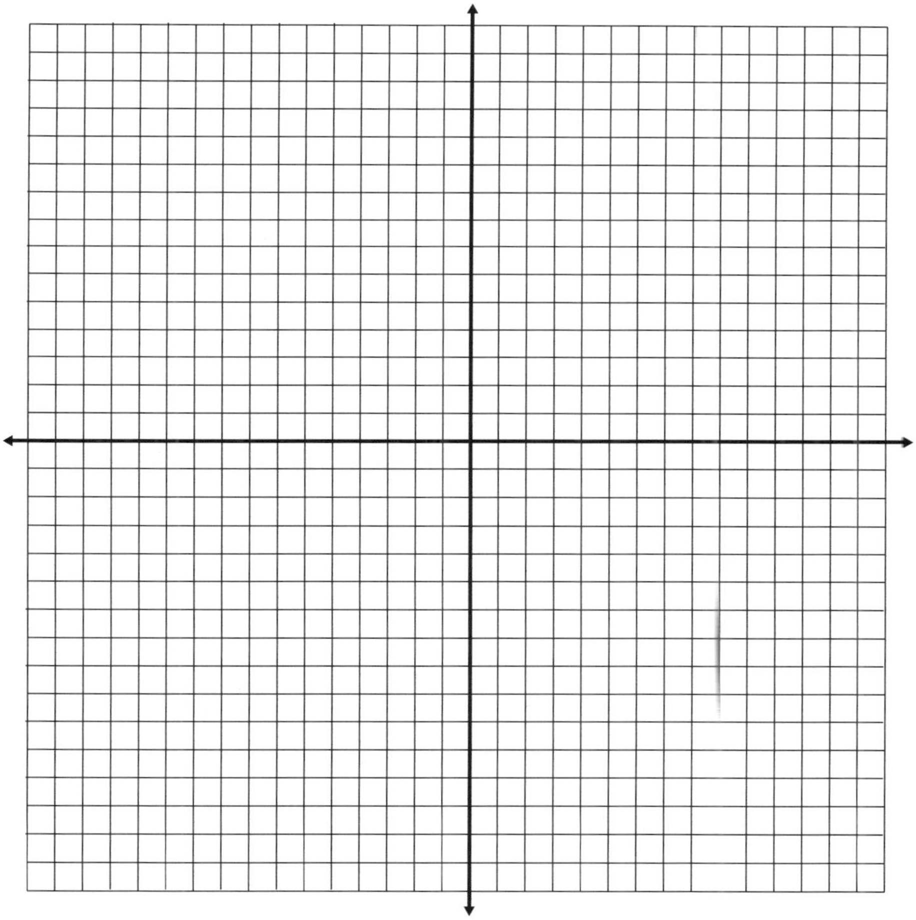

3 Identify the domain	Determine the domain of the embedded function graphically or symbolically.

Ex 1 $x \geq 0$

Ex 2 $x \geq 0$

Your Turn

Steps	Discussion
4 Decide if the function is one-to-one	**If the function is one-to-one, then its inverse will be a one-to-one function.**

Ex 1

Yes, it is one-to-one.

Ex 2

Yes, it is one-to-one.

Your Turn

Steps	Discussion
5 Isolate the function	**Bring all occurrences of the same function to one side.**

Ex 1

$$4x + 2\sqrt{x} = 5\sqrt{x}$$
$$4x = 5\sqrt{x} - 2\sqrt{x}$$

Ex 2

$$\log_2(x+1) = 3 - \log_2(x-1)$$
$$\log_2(x+1) + \log_2(x-1) = 3$$

Your Turn

Steps	Discussion
6 Simplify the function expression	**Factor out the function using the Distributive Property or use properties of the function to reduce the occurrences of the embedded function.**

Ex 1

$$4x = 5\sqrt{x} - 2\sqrt{x}$$
$$4x = (5-2)\sqrt{x}$$
$$4x = 3\sqrt{x}$$
$$\frac{4}{3}x = \sqrt{x}$$

Ex 2

$$\log_2\left((x+1)(x-1)\right) = 3$$
$$\log_2\left(x^2 - 1\right) = 3$$

Your Turn

Steps	Discussion
7 Take the inverse of each side of the equation	Apply the inverse of the identified function from Step 1.

Ex 1

$$\left(\frac{4}{3}x\right)^2 = \left(\sqrt{x}\right)^2$$

The function inverse is x^2 with $x \geq 0$.

Ex 2

$$2^{\log_2\left(x^2-1\right)} = 2^3$$

The inverse of the logarithmic function base 2 ($\log_2(x)$) is the exponential function base 2 (2^x).

Your Turn

Steps	Discussion
8 Simplify the equation by applying the rule that $f^{-1} \circ f(x) = x$	Simplifying the equation in this way removes the function. You may have to apply the inverse function several times.

Ex 1

$$\frac{16}{9}x^2 = x$$

Ex 2

$$x^2 - 1 = 8$$
$$x^2 = 9$$

Your Turn

Steps	Discussion
9 Write down the domain of possible values	The equation with the function removed may have false solutions.

Ex 1

$$x \geq 0$$

Ex 2

$x > 1$ since both $(x + 1)$ and $(x - 1)$ must be positive as inputs to $f(x) = \log_2(x)$.

Your Turn

Steps	Discussion
10 Repeat Step 2 through Step 8 if necessary	
Ex 1: Not necessary.	Ex 2: Not necessary.
Your Turn	
11 Validate your work	Check that all occurrences of the function have been removed.
Ex 1: All occurrences of the function have been removed. ✓	Ex 2: All occurrences of the function have been removed. ✓
Your Turn	

Note: A shortcut for step 7 involves applying the definition of logarithm: $\log_2(a) = b$ if and only if $2^b = a$, which is equivalent to applying exponentiation to the logarithm function. Step 7 would then read $2^3 = x^2 - 1$.

Model 1 6.2

Transform the equation $\sqrt{x-2}+\sqrt{2x+3}=4$ so that the radical function no longer appears.

Step 1 **Identify family of functions** — The radical family of functions

Step 2 **Graph function**

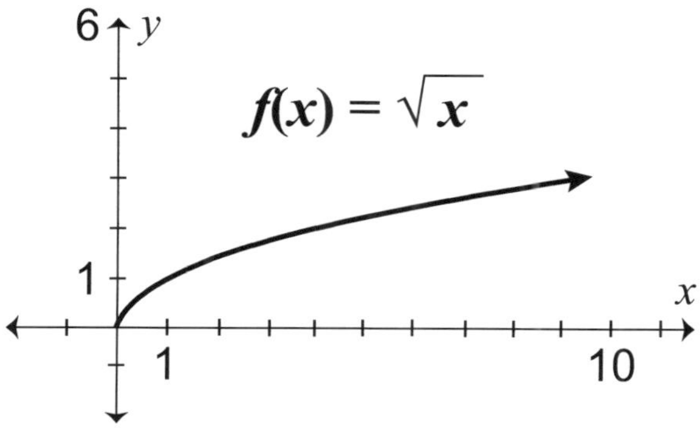

Step 3 **Identify domain** — $x \geq 2$

Step 4 **Is the function one-to-one?** — Yes.

Step 5 **Isolate the function**

$$\sqrt{x-2}+\sqrt{2x+3}=4$$
$$\sqrt{x-2}=4-\sqrt{2x+3}$$

Step 6 **Simplify function expression** — It is simplified.

Step 7 **Take inverse of both sides**

$$\left(\sqrt{x-2}\right)^2=\left(4-\sqrt{2x+3}\right)^2$$

Step 8 **Simplify by applying rule $f^{-1} \circ f(x) = x$**

$$x-2=16-8\sqrt{2x+3}+2x+3$$

Step 9 **Write down domain of possible values** — $x \geq 2$

Step 10 **Repeat Steps 2 – 8 as needed** — Not all radicals are removed; **repeat steps 2 – 8.**

Step 2 **Graph function**

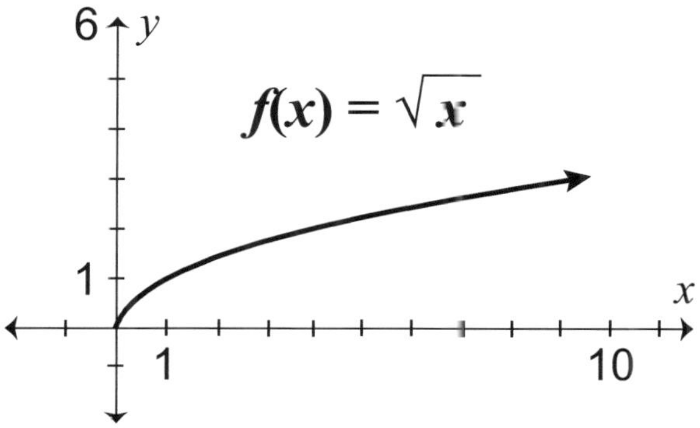

Step 3 **Identify domain**

$$2x+3 \geq 0$$
$$2x \geq -3$$
$$x \geq \frac{-3}{2}$$

Step 4 **Is the function one-to-one?** Yes.

Step 5 **Isolate the function**

$$x - 2 = 16 - 8\sqrt{2x+3} + 2x + 3$$
$$x - 21 - 2x = -8\sqrt{2x+3}$$

Step 6 **Simplify function expression**

$$-x - 21 = -8\sqrt{2x+3}$$
$$x + 21 = 8\sqrt{2x+3}$$

Step 7 **Take inverse of both sides**

$$(x+21)^2 = \left(8\sqrt{2x+3}\right)^2$$

Step 8 **Simplify by applying rule $f^{-1} \circ f(x) = x$**

$$x^2 + 42x + 441 = (64)(2x+3)$$

Step 9 **Write down domain of possible values**

$$x \geq 2 \text{ and } x \geq -\frac{3}{2} \text{ or just } x \geq 2$$

Step 10 **Repeat Steps 2 – 8 as needed** Not necessary.

Step 11 **Validate** All occurrences of the function have been removed. ✓

Note: In solving complicated equations with multiple square root functions, you may miss some constraints. For example, if $\sqrt{x-2} = 4 - \sqrt{2x+3}$, then $4 - \sqrt{2x+3} \geq 0$ since the square root symbol $\left(\sqrt{\ }\right)$ yields a positive number, which leads to $x \leq \frac{13}{2}$.

Therefore, you should check all the solutions you obtain in situations where you apply functions that are not one-to-one.

Addressing Common Errors 6.2

The following table shows some of the most common errors that learners tend to make with the kinds of problems covered in this section.

ERROR: Not removing all occurrences of the embedded function

Incorrect Process	Resolution
Transform the equation $\log_3(x) = 2 + \log_3(x-2)$ $3^{\log_3(x)} = 3^{2+\log_3(x-2)}$ $x = 3^{2+\log_3(x-2)}$	*Isolate the function first.*

ERROR: Not removing all occurrences of the embedded function (con't)

Correct Process

$$\log_3(x) = 2 + \log_3(x-2)$$
$$\log_3(x) - \log_3(x-2) = 2$$
$$\log_3\left(\frac{x}{x-2}\right) = 2$$
$$3^2 = \frac{x}{x-2} \quad \text{or} \quad 9 = \frac{x}{x-2}$$

Validation

Apply log to both sides of the transformed equation:

$$\log_3 9 = \log_3 \frac{x}{x-2}$$
$$2 = \log_3(x) - \log_3(x-2)$$
$$2 + \log_3(x-2) = \log_3(x)$$
$$\log_3(x) = 2 + \log_3(x-2),$$ which is the original equation.

ERROR: Not correctly identifying the domain of the embedded function

Incorrect Process

Transform the equation $\sqrt{x} + \sqrt{x-2} = 4$ to remove the radical function.

$\sqrt{x} = 4 - \sqrt{x-2}$

$f(x) = \sqrt{x}$ Domain is $x \geq 0$

$\sqrt{x-2} = 4 - \sqrt{x}$

$g(x) = \sqrt{x-2}$ Domain is $x \geq 2$

Domain is $x \geq 0$ or $x \geq 2$.

Resolution

The variable domain must fit all embedded functions.

Correct Process

In order for both square roots to be defined, x must be ≥ 0 AND x must be ≥ 2. The values that meet both criteria would be $x \geq 2$.

Validation

Let $x = 1$.

Note that $\sqrt{1-2} = \sqrt{-1}$ which is not a real number and therefore not in the domain of the equation.

Let $x = 4$.

Note that both $\sqrt{4}$ and $\sqrt{2}$ are real numbers and hence in the domain.

Preparation Inventory 6.2

I can...	OR *Here's my question...*
☐ identify an embedded function.	☐
☐ find the inverse of an embedded function.	☐
☐ remove an embedded function.	☐

Critical Thinking Questions 6.2

1. What are three functions that have domains consisting of ***not*** all real numbers?

2. Solving an equation that has been transformed from an original equation may result in solutions that are not solutions to the original equation. How do you identify these "false solutions"?

3. How do you transform an equation so that a chosen function no longer appears?

4. How do you validate that you have correctly transformed an equation so that a chosen function no longer appears?

5. What are some examples of equations that may need to be transformed so that embedded functions no longer appear? Provide at least three examples.

Demonstrate Your Understanding 6.2

Transform the given equation so that the chosen function no longer appears:

1. Equation: $\sqrt{2x} = 2 + \sqrt{x-4}$ Function: $f(x) = \sqrt{x}$

2. Equation: $\sqrt{2a-8} - \sqrt{a-3} - 1 = 0$ Function: $f(a) = \sqrt{a}$

3. Equation: $\log_2(y) = 5 + \log_2(y + 4)$ Function: $f(y) = \log_2(y)$

4. Equation: $3^t - 4 = 6$ Function: $f(t) = 3^t$

5. Equation: $\sqrt[3]{2x - 5} = 3$ Function: $f(x) = \sqrt[3]{x}$

Create and Solve the Hardest Problem 6.2

Your Hardest Problem

Problem Solution

What makes a problem where you must transform equations a hard problem to solve?

Identify, Describe, & Correct the Errors 6.2

Solve the problem correctly in the appropriate location. Be sure to identify and describe the error you found in the worked solution!

Problem 1

Transform the equation so that the function no longer appears:

Equation	*Function*
$\sqrt{x-5}=4$	$f(x)=\sqrt{x}$

Worked Solution *(Find the error!)*

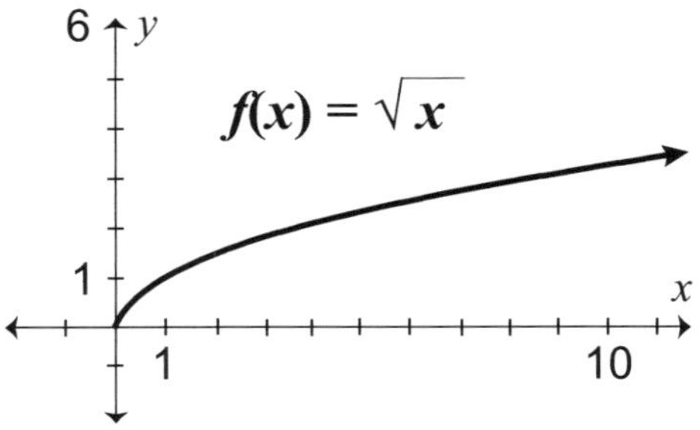

$\sqrt{x-5}=4$

$x-5=4$

Correct Process

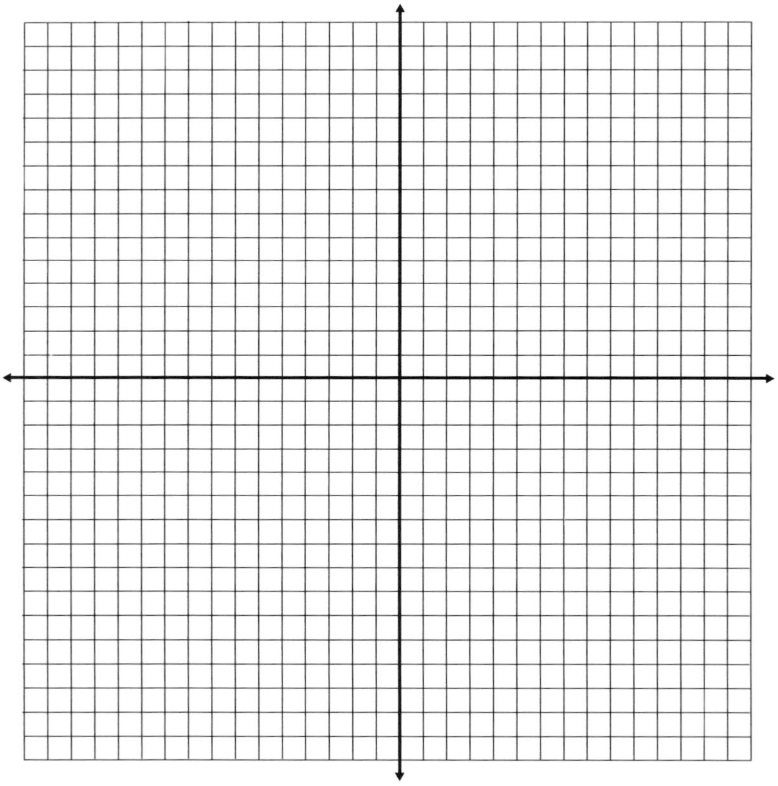

Describe the Error

Problem 2

Transform the equation so that the function no longer appears:

Equation	*Function*
$10^{2x} = 60$	$f(x) = 10^x$

Worked Solution *(Find the error!)*

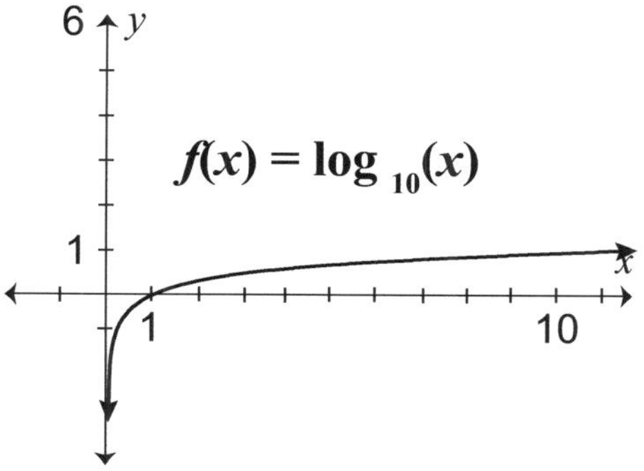

$10^{2x} = 60$

$\log_{10}\left(10^{2x}\right) = 60$

$2x = 60$

Correct Process

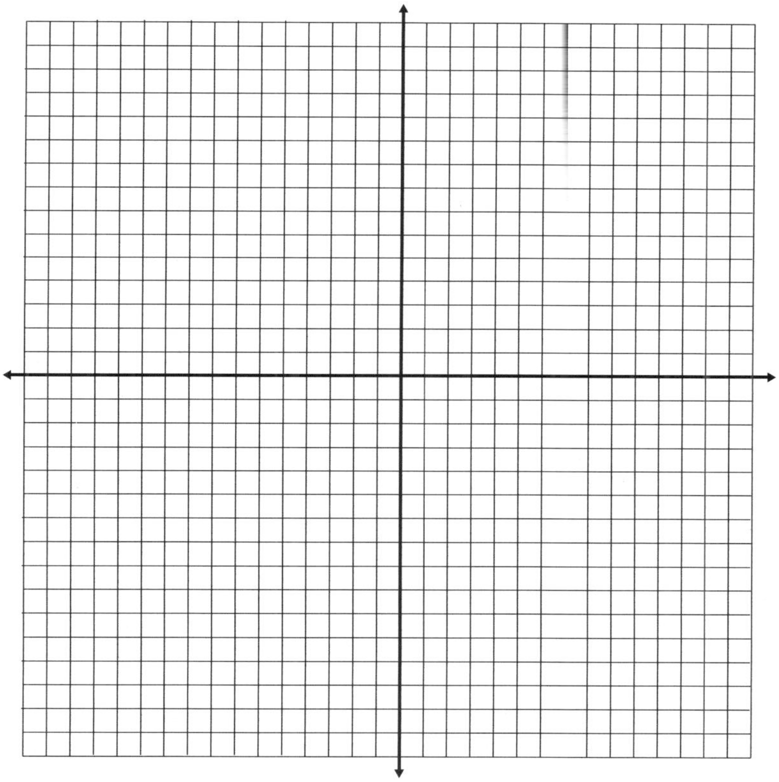

Describe the Error

Problem 3

Transform the equation so that the function no longer appears:

Equation	*Function*
$x^{3/5} - 2 = 4$	$f(x) = x^{3/5}$

Worked Solution *(Find the error!)*

$f(x) = \frac{5}{3}x$

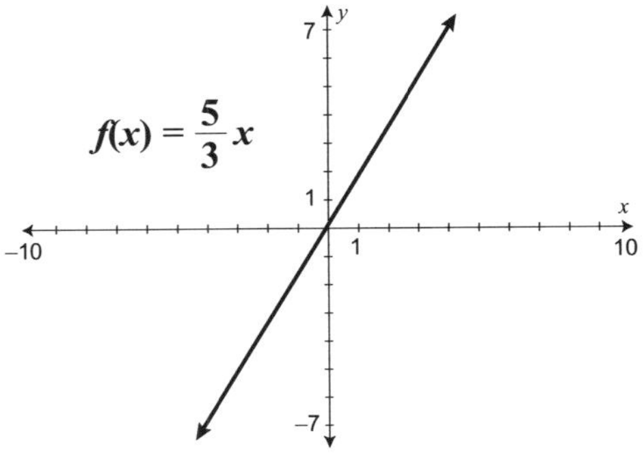

$$x^{3/5} - 2 = 4$$

$$x^{3/5} = 6$$

$$x = 6\left(\frac{5}{3}\right)$$

$$x = 10$$

Correct Process

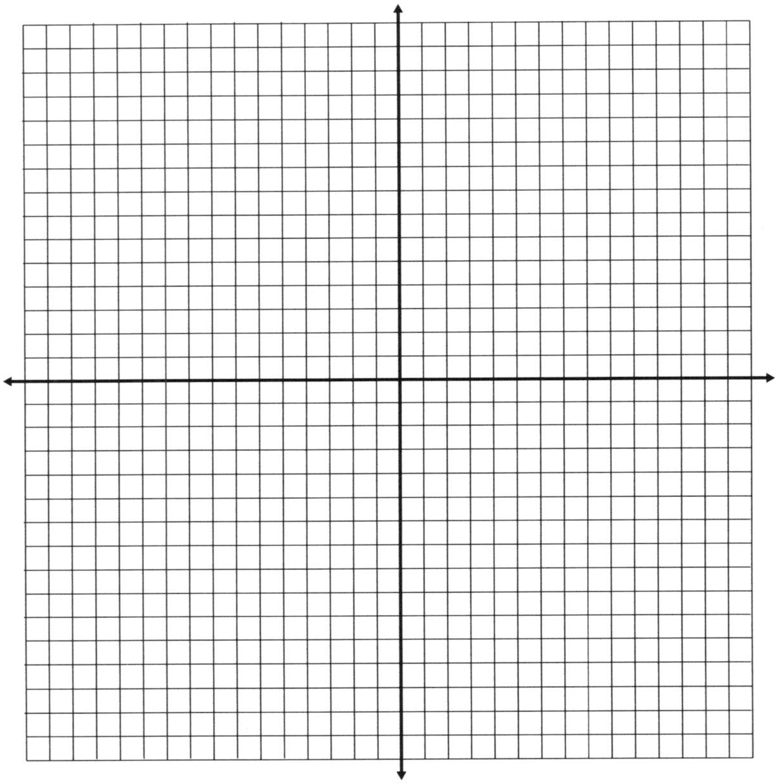

Describe the Error

Problem 4

Transform the equation so that the function no longer appears:

Equation

$$\log_2(x+1) + \log_2(x-2) = 2$$

Function

$$f(x) = \log_2(x)$$

Worked Solution *(Find the error!)*

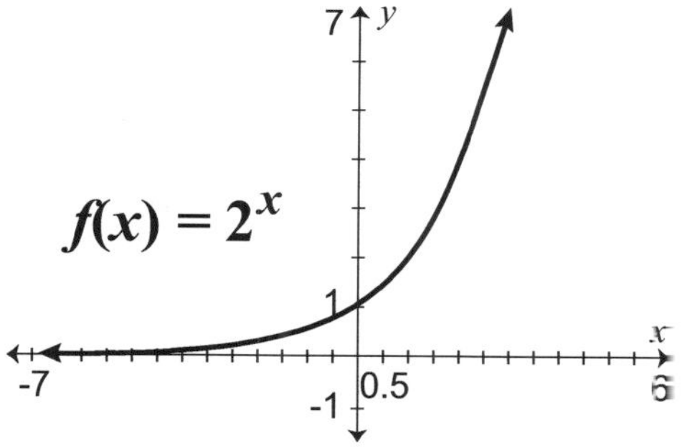

$$\log_2(x+1) + \log_2(x-2) = 2$$

$$\log_2(x+1)(x-2) = 2$$

$$2^{\log_2(x+1)(x-2)} = 2$$

$$(x+1)(x-2) = 2$$

$$x^2 - x - 2 = 2$$

$$x^2 - x - 6 = -4$$

Describe the Error

Correct Process

What Did You Learn About Learning Math? 6.2

1. How do you know that you have the domain of a function correct?

2. How do you visualize the inverse of a function?

3. How do you identify an embedded function?

Section 6.3 — Solving an Equation

Why? 6.3

Equations occur in many forms. Some of them can be solved by symbolic manipulation. For example, you can estimate when the population of your city will double using exponential equations.

What Do You Already Know? 6.3

How do you solve linear equations?

Describe two ways you can solve a quadratic equation.

How do you eliminate a function from an equation?

How do you solve rational equations?

Goals 6.3

When you complete this activity, you should be able to:

1. Apply prior knowledge in solving equations
2. Use transformations in solving equations
3. Use a general methodology for solving equations
4. Be persistent in solving equations

Methodology 6.3

Solving Advanced Equations

Equations containing more than one function can often be solved by removing one of the functions using that function's inverse function, transforming the equation by substitution, or by removing denominators.

Limitation/Caution: Not all equations can be solved in these ways. In particular, this methodology may not work on all equations containing more than one type of function. Consider, for example, $x^2 - 3 = \log_2(x)$, which contains both a polynomial and a logarithmic function. Such functions can be solved (approximately) using graphical techniques (see Section 6.4).

Methodology 6.3

Solving Advanced Equations (con't)

	Example 1	Example 2	Your Turn
Solve the given equation:	$\frac{\sqrt{x+2}}{3} = 2(x-3)$	$\sqrt{2-x} = 5$	$2^x + 5 = 11$

Steps	Discussion
1 Determine the variable	**Decide which variable in the equation we are solving for.**

Ex 1: Solve for x.

Ex 2: Solve for x.

Your Turn:

Steps	Discussion
2 Remove parentheses (grouping symbols)	**Parentheses that are used as grouping symbols should be removed. Do not remove the parentheses that are part of function notation.**

Ex 1:

$$\frac{\sqrt{x+2}}{3} = 2(x-3)$$

$$\frac{\sqrt{x+2}}{3} = 2x-6$$

Ex 2: Already done.

Your Turn:

Steps	Discussion
3 Clear the fractions	**Use the methodology for clearing the fraction components.**

Ex 1: $\sqrt{x+2} = 6x-18$

Ex 2: Already done.

Your Turn:

Steps	Discussion
4 Transform the equation	**Simplify the equation using the Methodology for Transforming an Equation with Embedded Functions.**

Ex 1:

Since the inverse of the square root function is the function

$$g(x) = x^2,\ x \geq 0$$

with the following conditions:

$x+2 \geq 0$	and	$6x-18 \geq 0$
$x \geq -2$		$6x \geq 18$
		$x \geq 3$

x must be greater than 3.

Squaring both sides: $x+2 = (6x-18)^2$

Ex 2:

Since the inverse of the square root function is the function

$$g(x) = x^2,\ x \geq 0$$

with the following condition:

$$2-x \geq 0 \text{ or } 2 \geq x$$

x must be less than or equal to 2.

Squaring both sides: $2-x = 25$

Steps	Discussion

Your Turn

5 Solve the polynomial equation	Use the Methodologies for Solving Linear and Polynomial Equations.

Ex 1

$$x + 2 = 36x^2 - 216x + 324$$

$$0 = 36x^2 - 217x + 322$$

$$x = \frac{-(-217) \pm \sqrt{(-217)^2 - 4(36)(322)}}{2(36)}$$

$$x \approx 2.64095 \; or \; x \approx 3.38683$$

Ex 2

$$-x = 23$$

$$x = -23$$

Your Turn

Steps	Discussion
6 Validate the solution set	Be sure that all the solutions presented are part of the solution set by checking each solution.

Ex 1

Check both results:

$$\frac{\sqrt{x+2}}{3} = 2(x-3)$$

$x \approx 2.64095$

$$\frac{\sqrt{2.64095+2}}{3} \stackrel{?}{=} 2(2.64095-3)$$

$$\frac{\sqrt{4.64095}}{3} \stackrel{?}{=} 2(-0.35905)$$

$0.71810 \neq -0.71810$

$x \approx 0.71810$ is **not** a solution (is extraneous). ✓

$x \approx 3.38683$

$$\frac{\sqrt{3.38683+2}}{3} \stackrel{?}{=} 2(3.38683-3)$$

$$\frac{\sqrt{5.38683}}{3} \stackrel{?}{=} 2(0.38683)$$

$0.77365 \approx 0.77366$

$x \approx 3.38683$ is an approximate solution. ✓

Ex 2

Check the result:

$\sqrt{2-x} = 5$

$x = -23$

$\sqrt{2-(-23)} \stackrel{?}{=} 5$

$\sqrt{2+23} \stackrel{?}{=} 5$

$\sqrt{25} \stackrel{?}{=} 5$

$5 = 5$

$x = 5$ is a solution. ✓

Your Turn

Model 1 6.3

Solve for x: $\dfrac{\log_3(x+2)}{3} = \dfrac{2}{3}$

Step 1 **Determine the variable**

Solve for x.

Step 2 **Remove parentheses**

No grouping symbols or parentheses present, except those that are part of function notation.

Step 3 **Clear the fractions**

$$\frac{\log_3(x+2)}{3} = \frac{2}{3}$$
$$\log_3(x+2) = 2$$

Step 4 **Transform the equation**

The inverse of $f(x) = \log_3(x)$ is $f^{-1}(x) = 3^x$.

$x + 2 > 0 \rightarrow x > -2$ because the domain of the logarithm function is positive real numbers.

We apply the exponential function with base 3 to both sides:

$$3^{\log_3(x+2)} = 3^2$$
$$x + 2 = 9$$

We could also have used the equivalent form of the logarithmic equation to yield the same transformed equation.

Step 5 **Solve the equation**

$x = 7$ is the solution. Note that $x > -2$, as required.

Step 6 **Validate the solution set**

$$\frac{\log_3(x+2)}{3} = \frac{2}{3}$$
$$x = 7$$
$$\frac{\log_3(7+2)}{3} \stackrel{?}{=} \frac{2}{3}$$
$$\frac{\log_3(9)}{3} \stackrel{?}{=} \frac{2}{3}$$
$$\frac{2}{3} = \frac{2}{3} \quad \checkmark$$

Information for Solving Some Types of Equations

Equation Type	Inverse	Reduced Solution Set	Properties or Concepts	Most Common Error	Second Most Common Error
Exponential	Logarithmic	No reduction	Properties of exponents and logarithms	Taking the inverse of one side only	Misusing the properties of exponents and logarithms
Logarithmic	Exponential	Argument(s) must be greater than 0	Properties of exponents and logarithms	Taking the inverse of one side only	Misusing the properties of exponents and logarithms
Radical	Power	If the index is even, then the argument(s) under the radical must be greater than 0.	$(a + b)^2 = a^2 + 2ab + b^2$	Taking the inverse of one side only	Taking the wrong power as the inverse
Power	Radical	If the denominator of the power is even, then the argument(s) of the power must be greater than 0.	Properties of exponents	Forgetting that in some cases the argument has to be non-negative	Taking the wrong radical (power) as the inverse
Absolute Value	No inverse	No reduction	Compound equation concept	Removing the absolute value signs without regard to domain	Incorrectly presenting the solution set with incorrect inequalities
Rational	No inverse	Reduce the solution set by eliminating the numbers that make the denominators of the rational equation 0.	Addition and multiplication of rational expressions	Failing to multiply both sides of the equation by the LCM	Computing an incorrect LCM

Solving Special Equation Types

Equations involving any of the types of functions in the table can sometimes be solved by transforming equations or eliminating denominators.

Limitation/Caution: Such equations often cannot be solved by symbolic manipulations, though they may be solved (approximately) by graphing techniques (see Section 6.4).

The techniques for solving any of these types of equations can involve substitutions, using properties of a function, transforming an equation, eliminating denominators. In all cases, you are using the Methodology for Solving Advanced Equations. The following models focus on the transformation of an equation into a simpler form that can be solved.

Model 2: Solving an Exponential Equation 6.3

Solve for x: $2^x - 1 = 3 \cdot 2^x - 17$

Use the methodology

When simplifying this equation, it may be helpful to substitute a variable (z) in for 2^x and then substitute 2^x back in for z.

$$2^x - 1 = 3 \cdot 2^x - 17$$

$$z - 1 = 3z - 17$$

$$-2z = -16$$

$$z = 8$$

$$2^x = 8$$

$$\log_2(2^x) = \log_2(8)$$

$$x = 3$$

Validate

$$2^3 - 1 \stackrel{?}{=} 3 \cdot 2^3 - 17$$

$$8 - 1 \stackrel{?}{=} 24 - 17$$

$$7 = 7 \quad \checkmark$$

Model 3: Solving an Logarithmic Equation 6.3

Solve for x: $2\log_5(x + 1) = -1$

Use the methodology

$$2\log_5(x+1) = -1$$

$$\log_5(x+1)^2 = -1$$

$(x+1)^2 = 5^{-1}$ (Using the equivalent form of the equation)

$$(x+1)^2 = \frac{1}{5}$$

$$x+1 = \pm\sqrt{\frac{1}{5}}$$

$$x = -1 \pm \sqrt{\frac{1}{5}}$$

Validate:

$$2\log_5(x+1) = -1$$

$$2\log_5\left(-1 \pm \sqrt{\frac{1}{5}} + 1\right) \stackrel{?}{=} -1$$

$2\log_5\left(\sqrt{\frac{1}{5}}\right) \stackrel{?}{=} -1$ since the argument of log must be greater than 0.

$$\log_5\left(\sqrt{\frac{1}{5}}\right)^2 \stackrel{?}{=} -1$$

$$\log_5\left(\frac{1}{5}\right) \stackrel{?}{=} -1$$

$$5^{-1} \stackrel{?}{=} \frac{1}{5}$$

$$\frac{1}{5} = \frac{1}{5} \checkmark$$

The solution is $x = \sqrt{\frac{1}{5}}$.

Model 4: Solving a Logarithmic Equation 6.3

Solve for x: $\dfrac{x^2 + \log(x^2)}{2} = 18 + \log(x)$

Use the methodology

$x^2 + \left(\log(x^2)\right) = 36 + 2\left(\log(x)\right)$

$x^2 + 2\log(x) = 36 + 2\log(x)$

$x^2 = 36$

$x = \pm 6$ but $x > 0$, so $x = 6$.

Validate:

$\dfrac{x^2 + \log(x^2)}{2} = 18 + \log(x)$

$\dfrac{6^2 + \log(6^2)}{2} \stackrel{?}{=} 18 + \log(6)$

$\dfrac{36 + 1.5564}{2} \stackrel{?}{\approx} 18 + .7782$

$18.7782 \approx 18.7782$ ✓

Model 5: Solving a Radical Equation 6.3

Solve for a: $\sqrt[3]{2a-5} + 4 = 7$

Use the methodology

$\sqrt[3]{2a-5} + 4 = 7$

$\sqrt[3]{2a-5} = 3$

$\left(\sqrt[3]{2a-5}\right)^3 = 3^3$

$2a - 5 = 27$

$2a = 32$

$a = 16$

Validate:

$\sqrt[3]{2a-5} + 4 = 7$

$\sqrt[3]{2 \bullet 16 - 5} + 4 \stackrel{?}{=} 7$

$\sqrt[3]{32 - 5} + 4 \stackrel{?}{=} 7$

$\sqrt[3]{27} + 4 \stackrel{?}{=} 7$

$3 + 4 \stackrel{?}{=} 7$

$7 = 7$ ✓

Model 6: Solving a Power Function Equation 6.3

Solve for x: $5 - x^{2/3} = 2$

Use the methodology

$$5 - x^{2/3} = 2$$
$$-x^{2/3} = -3$$
$$x^{2/3} = 3$$
$$\left(x^{2/3}\right)^{3/2} = 3^{3/2}$$
$$x = 3^{3/2} = 3\sqrt{3} \approx 5.196$$

Validate:

$$5 - x^{2/3} = 2$$
$$5 - 5.196^{2/3} \overset{?}{\approx} 2$$
$$5 - 2.99994 \overset{?}{\approx} 2$$
$$2.00006 \approx 2 \quad \checkmark$$

Model 7: Solving an Absolute Value Equation 6.3

Solve for s: $\dfrac{4\left(|s-3|+2\right)}{5} = 8$

Use the methodology

$$\frac{4\left(|s-3|+2\right)}{5} = 8$$
$$\frac{4|s-3|+8}{5} = 8$$
$$4|s-3| + 8 = 40$$
$$4|s-3| = 40 - 8$$
$$4|s-3| = 32$$
$$|s-3| = \frac{32}{4}$$
$$s - 3 = -8 \text{ or } s - 3 = 8$$
$$s = 3 - 8 \text{ or } s = 3 + 8$$
$$s = -5 \text{ or } s = 11$$

Validate:

$$\frac{4\left(|s-3|+2\right)}{5} = 8$$

$s = -5$

$$\frac{4\left(|-5-3|+2\right)}{5} \overset{?}{=} 8$$
$$\frac{4(8+2)}{5} \overset{?}{=} 8$$
$$8 = 8 \quad \checkmark$$

or

$s = 11$

$$\frac{4\left(|11-3|+2\right)}{5} \overset{?}{=} 8$$
$$\frac{4(8+2)}{5} \overset{?}{=} 8$$
$$8 = 8 \quad \checkmark$$

Model 8: Solving a Rational Equation 6.3

Solve for x: $\dfrac{x}{2x-1}+\dfrac{4}{x+1}=2$

Use the methodology

$$\frac{x}{2x-1}+\frac{4}{x+1}=2$$

$$\text{LCM} = (2x-1)(x+1)$$

$$(2x-1)(x+1)\left(\frac{x}{2x-1}+\frac{4}{x+1}\right)=(2x-1)(x+1)(2)$$

$$x(x+1)+4(2x-1)=2(2x^2+x-1)$$

$$x^2+x+8x-4=4x^2+2x-2$$

$$x^2+9x-4=4x^2+2x-2$$

$$0=3x^2-7x+2$$

$$0=(3x-1)(x-2)$$

$$3x-1=0 \text{ or } x-2=0$$

$$3x=1 \text{ or } x=2$$

$$x=\frac{1}{3} \text{ or } x=2$$

Validate:

$$\frac{x}{2x-1}+\frac{4}{x+1}=2$$

$$x=\frac{1}{3}$$

$$\frac{\left(\frac{1}{3}\right)}{2\left(\frac{1}{3}\right)-1}+\frac{4}{\left(\frac{1}{3}\right)+1}\stackrel{?}{=}2$$

$$\frac{\left(\frac{1}{3}\right)}{\frac{2}{3}-1}+\frac{4}{\frac{4}{3}}\stackrel{?}{=}2$$

$$\frac{\left(\frac{1}{3}\right)}{-\frac{1}{3}}+4\bullet\frac{3}{4}\stackrel{?}{=}2$$

$$-1+3=2$$

$$2=2 \checkmark$$

or

$$\frac{x}{2x-1}+\frac{4}{x+1}=2$$

$$x=2$$

$$\frac{2}{2\bullet2-1}+\frac{4}{2+1}\stackrel{?}{=}2$$

$$\frac{2}{4-1}+\frac{4}{3}\stackrel{?}{=}2$$

$$\frac{2}{4-1}+\frac{4}{3}\stackrel{?}{=}2$$

$$\frac{2}{3}+\frac{4}{3}\stackrel{?}{=}2$$

$$\frac{6}{3}\stackrel{?}{=}2$$

$$2=2 \checkmark$$

Addressing Common Errors 6.3

The following table shows some of the most common errors that learners tend to make with the kinds of problems covered in this section.

ERROR: Transforming only one side of an equation

Incorrect Process

Solve for x

$\sqrt{x-5}=4$

$\sqrt{x-5}=4$

$x-5=4$

$x=9$

Resolution

Square both sides of the equation.

Correct Process

$\sqrt{x-5}=4$

$x-5=16$

$x=21$

Validation

$\sqrt{x-5}=4$

$\sqrt{(21)-5}\stackrel{?}{=}4$

$\sqrt{16}\stackrel{?}{=}4$

$4=4$ ✓

ERROR: Forgetting to check for extraneous solutions

Incorrect Process

Solve for x

$\sqrt{2x+3}+7=4$

$\sqrt{2x+3}+7=4$

$\sqrt{2x+3}=-3$

$2x+3=9$

$2x=6$

$x=3$

Correct Process

$\sqrt{2x+3}+7=4$

$\sqrt{2x+3}=-3$

$2x+3=9$

$2x=6$

$x=3$

Validate: $\sqrt{2(3)+3}+7\stackrel{?}{=}4$

$\sqrt{6+3}+7\stackrel{?}{=}4$

$\sqrt{9}+7\stackrel{?}{=}4$

$3+7\stackrel{?}{=}4$ and $-3+7\stackrel{?}{=}4$

$10\neq 4$ (no) $\quad 4=4$ (yes)

This only validates for a value of –3, not 3. Thus there is no solution to the original equation.

Resolution

Check for extraneous solutions.

Validation

Validation is part of the correct process of checking for extraneous solutions.

ERROR: **Misusing the law of exponents**

Incorrect Process

Solve for t

$3^{4t} = 36$

$3^{4t} = 36$

$3^{t} = 9$

$3^{t} = 3^{2}$

$t = 2$

Correct Process

$3^{4t} = 36$

$\log_3\left(3^{4t}\right) = \log_3(36)$

$4t \bullet \log_3(3) = \log_3(36)$

$4t \bullet 1 = \log_3(36)$

$t = \dfrac{\log_3(36)}{4}$

Resolution

Use logarithms and laws of exponents correctly.

Validation

$3^{4t} = 36$

$3^{4\frac{\log_3(36)}{4}} \stackrel{?}{=} 36$

$3^{\log_3(36)} \stackrel{?}{=} 36$

$36 = 36$ ✓

Preparation Inventory 6.3

I can...	OR *Here's my question...*
☐ determine the inverse function I need to use to transform a function.	☐
☐ correctly transform an equation into a linear or quadratic equation when possible.	☐
☐ determine extraneous solutions to an equation.	☐
☐ describe why extraneous solutions arise.	☐

Critical Thinking Questions 6.3

1. What are at least three key differences between $f(x) = x^n$ and $f(x) = n^x$?

2. What are the inverse functions for the functions given by the expressions e^x, 10^x, $\ln(x)$ and $\log(x)$?

3. What are the steps in calculating specific values for logarithms and exponents, whether in base ten or base e on your calculator?

4. How would you illustrate that the function $f(x) = \log(x)$ is the inverse function of $g(x) = 10^x$ graphically, numerically, and symbolically?

5. Which function grows faster: $y = x^n$ or $y = n^x$ and why?

6. How do you transform an equation containing an exponential or logarithmic function into an equation that you know how to solve?

7. How do you validate the solutions to exponential and logarithmic equations using your calculator?

Demonstrate Your Understanding 6.3

Solve each of the following equations for x.

1. $2 + \log_5(x) = 4$

2. $\log_7(x) + 3 = 5$

3. $2^x = \frac{1}{8}$

4. $5^x = 3$

5. $\frac{1}{x} + \frac{1}{x+2} = \frac{3}{4}$

6. $\sqrt{x-4}-5=7$

Create and Solve the Hardest Problem 6.3

Your Hardest Problem

Problem Solution

What makes a problem where you must solve an advanced equation a hard problem to solve?

Identify, Describe, & Correct the Errors 6.3

Solve the problem correctly in the appropriate location. Be sure to identify and describe the error you found in the worked solution!

Problem 1

Solve for x: $\log_3(x+2)=7$

Worked Solution *(Find the error!)*

$$\log_3(x+2)=7$$

$$\log_3 x=5$$

$$3^5=x$$

$$x=243$$

Correct Process

Describe the Error

Problem 2

Solve for x: $2^{x+3}=11$

Worked Solution *(Find the error!)*

$$2^{x+3}=11$$

$$2^x=11-3$$

$$2^x=8$$

$$\log_2 2^x=\log_2 8$$

$$x=3$$

Correct Process

Describe the Error

Problem 3

Solve for y: $\log_5 (3y) = 6$

Worked Solution *(Find the error!)*

$$\log_5 (3y) = 6$$
$$\log_5 y = 2$$
$$5^2 = y$$
$$y = 25$$

Correct Process

Describe the Error

Problem 4

Solve for x: $\ln(x^3) - \ln(3) = 5$

Worked Solution *(Find the error!)*

$$\ln (x^3) = 8$$
$$x^3 = 8$$
$$x = 2$$

Correct Process

Describe the Error

Problem 5

Solve for y: $\sqrt{y+3} = 5$

Worked Solution *(Find the error!)*

$$\sqrt{y+3} = 5$$
$$y + 3 = 5$$
$$y = 2$$

Correct Process

Describe the Error

What Did You Learn About Learning Math? 6.3

1. How are the various ways the model equations were solved similar?

2. How are the various ways the model equations were solved different?

3. What did you learn about different equation solving contexts that will be useful in the future?

Section 6.4 Additional Tools for Solving Equations

Why? 6.4

Some equations can be transformed into standard polynomial forms that we already know how to solve. Many other equations cannot be solved using symbolic manipulations and inverse function transformations. This section will give you additional tools for solving such types of equations.

What Do You Already Know? 6.4

Why do you solve a quadratic equation by transforming the equation into an equivalent equation with a zero on one side?

How do you determine the graphing window for the function $f(x) = \dfrac{2}{x-25}$?

What can you say about the x-value of the intersection point of the graphs of two functions?

What can you say about the second coordinate of a point where a function crosses the x-axis? What can you say about the first coordinate?

Goals 6.4

When you complete this activity, you should be able to:

1. Symbolically manipulate an equation so that it can be solved more easily by graphing
2. Solve equations of one variable using a graphing utility
3. Validate approximate solutions obtained using a graphing utility

Building Mathematical Language 6.4

New Terms: graphing utility

Key Concepts 6.4

A ***graphing utility*** can be a graphing calculator, a smartphone or tablet app, or a mathematical computer software package you use to investigate the graphs of functions.

Limitation/Caution: Such electronic graphing utilities usually use decimal approximations when graphing functions. Thus, solutions to equations obtained using such graphing utilities usually are approximations of the actual solutions.

Methodology 1 — 6.4

Solving Advanced Equations Graphically

Most equations in one variable, contrary to popular belief, cannot be solved by symbolic methods. Most polynomial equations of degree higher than four cannot be solved by such methods. Most equations that have more than one type of function cannot be solved by such methods either. However, most equations in one variable that have solutions can be solved using a graphing utility.

Limitation/Caution: The solutions obtained using a graphing utility will be approximate (although sometimes they *look* exact).

	Example 1	**Example 2**	**Your Turn**
Solve for x:	$3x = x^2 - 5$	$x^6 - 3x^2 = 5x + 3$	$x^2 - 5\log_2 x = 7$

Steps	**Discussion**
1 Enter the equation	**Enter the left side as one function and the right side as a second function.**

Ex 1

$3x = x^2 - 5$

$f(x) = 3x$

$g(x) = x^2 - 5$

Ex 2

$x^6 - 3x^2 = 5x + 3$

$f(x) = x^6 - 3x^2$

$g(x) = 5x + 3$

Your Turn

Steps	**Discussion**
2 Graph the functions	**Decide on an appropriate graphing window and graph the functions.**

Ex 1

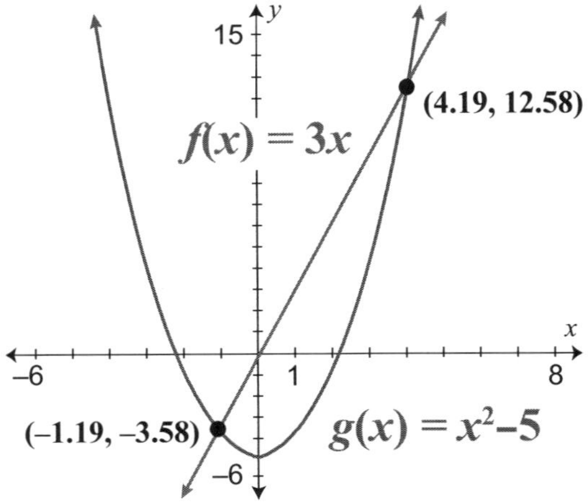

Ex 2

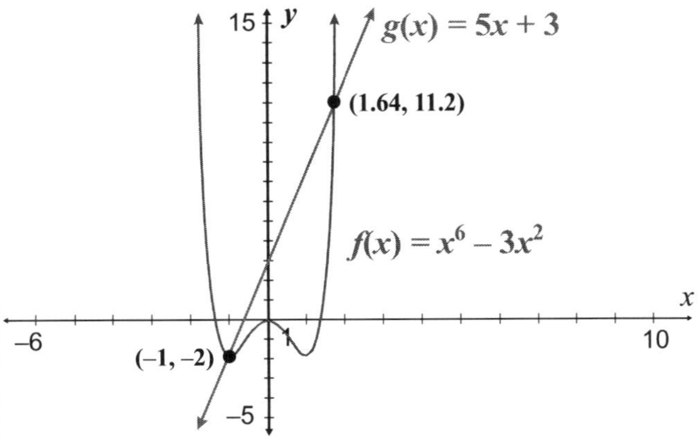

Your Turn

Use your graphing utility.

Steps	Discussion
3 Determine the solutions	**The coordinates of the point(s) where the graphs of the functions intersect are the solutions to the equation.**

Ex 1	Ex 2	Your Turn
The points of intersection are (–1.19, –3.58) and (4.19, 12.58).	The points of intersection are (–1, –2) and (1.64, 11.2).	

Steps	Discussion
4 Present the solution set	**Write down the x-coordinates of the points of intersection as the approximate solutions to the equation.**

Ex 1	Ex 2	Your Turn
$x \approx -1.19$ or $x \approx 4.19$	$x = -1$ or $x \approx 1.64$	

Steps	Discussion
5 Validate the solution set	**Check by substituting the solution values in the original equation.**

Ex 1

$$3x = x^2 - 5$$

$$3(-1.19) \overset{?}{\approx} (-1.19)^2 - 5$$

$$-3.57 \approx -3.58 \checkmark$$

$$3(4.19) \overset{?}{\approx} (4.19)^2 - 5$$

$$12.57 \approx 12.56 \checkmark$$

Ex 2

$$x^6 - 3x^2 = 5x + 3$$

$$(-1)^6 - 3(-1)^2 \overset{?}{=} 5(-1) + 3$$

$$1 - 3 \overset{?}{=} -5 + 3$$

$$-2 = -2 \checkmark$$

$$(1.64)^6 - 3(1.64)^2 \overset{?}{\approx} 5(1.64) + 3$$

$$11.4 \approx 11.2 \checkmark$$

Your Turn

Model 1 6.4

Solve for x: $3 \bullet 2^x - 4x^2 = 5$

Step 1 **Enter the equation**

$$f(x) = 3 \bullet 2^x - 4x^2$$
$$g(x) = 5$$

Step 2 **Graph the functions**

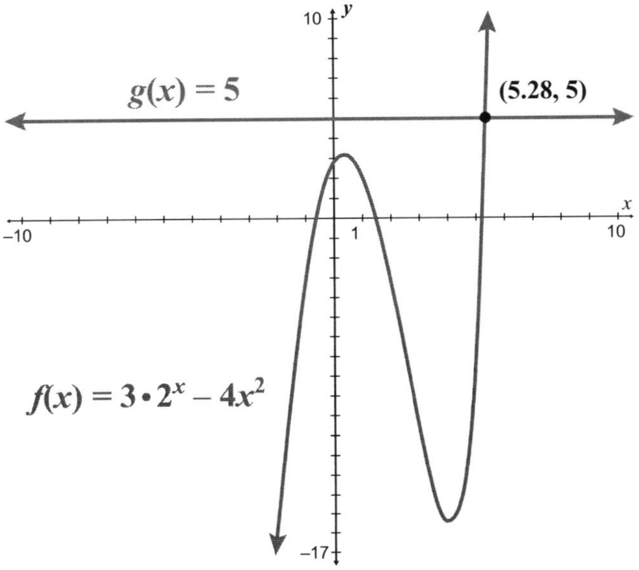

Step 3 **Determine solutions**

The point of intersection is (5.28, 5)

Step 4 **Present solution set**

$$x \approx 5.28$$

Step 5 **Validate solution set**

$$f(5.28) = 3 \bullet 2^x - 4x^2$$
$$f(5.28) = 3 \bullet 2^{5.28} - 4 \bullet 5.28^2 \approx 5.05$$
$$g(5.28) = 5$$
$$5 \approx 5.05 \ \checkmark$$

The two sides of the equation are approximately equal when $x \approx 5.28$.

Model 2 6.4

Solve for x: $3^x - 5x^2 - 7 = 0$

Step 1 **Enter the left side of the equation as the first function and the right side as the second function**

$f(x) = 3^x - 5x^2 - 7$

$g(x) = 0$

Step 2 **Graph the functions**

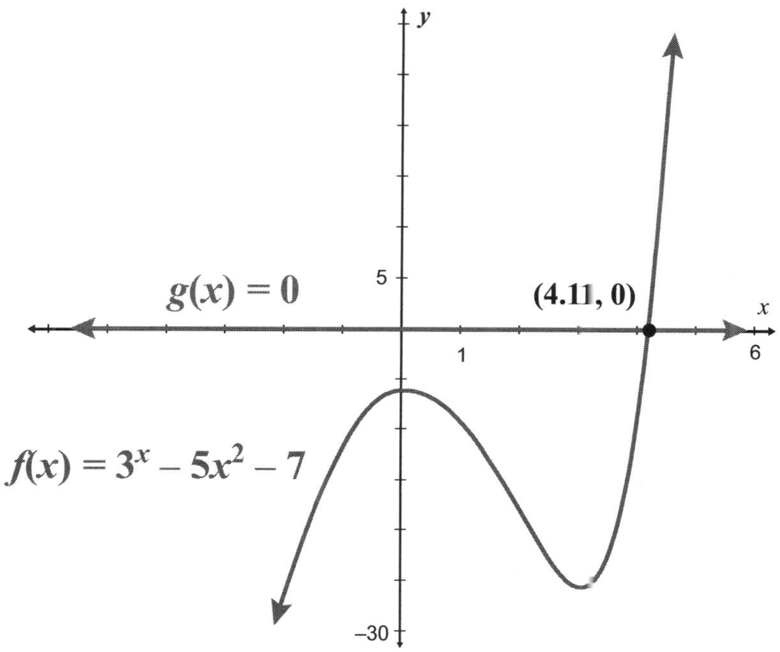

Step 3 **Determine solutions**

The point of intersection is (4.11, 0)

Step 4 **Present solution set**

$x \approx 4.11$

Step 5 **Validate solution set**

$$f(4.11) = 3^{4.11} - 5 \cdot (4.11)^2 - 7$$
$$= 91.4047 - 84.4605 - 7$$
$$= -0.0558 \approx 0 \checkmark$$
$$g(4.11) = 0 \checkmark$$

The two sides of the equation are approximately equal when $x \approx 4.11$.

Methodology 2 — 6.4

Solving an Equation in Standard Form

Some equations may look like they are impossible to solve exactly, but if you observe that they can be converted into a linear or quadratic equation by a clever substitution or transformation, they become solvable.

	Example 1	**Example 2**	**Your Turn**
Solve:	$x^4 - 4x^2 - 12 = 0$	$x^6 + 3x^3 - 4 = 0$	$x^4 = 5x^2 - 6$

Steps	**Discussion**
1 Determine standard form	**Determine whether the equation can be converted into linear or quadratic form.**

Ex 1: It is in quadratic form, $\left(x^2\right)^2 - 4x^2 - 12 = 0$

Ex 2: It is in quadratic form, $\left(x^3\right)^2 + 3\left(x^3\right) - 4 = 0$

Your Turn:

2 Determine the substitution	**Determine the substitution that will result in the linear or quadratic standard form.**

Ex 1: $z = x^2$

Ex 2: $z = x^3$

Your Turn:

3 Substitute	**The coordinates of the point(s) where the graphs of the functions intersect are the solutions to the equation.**

Ex 1: $z^2 - 4z - 12 = 0$

Ex 2: $z^2 + 3z - 4 = 0$

Your Turn:

4 Solve	**Use the Methodology for Solving Equations to solve the transformed equation.**

Ex 1:

$(z-6)(z+2) = 0$

$z - 6 = 0$ or $z + 2 = 0$

$z = 6$ or $z = -2$

Ex 2:

$(z+4)(z-1) = 0$

$z + 4 = 0$ or $z - 1 = 0$

$z = -4$ or $z = 1$

Your Turn:

Steps	Discussion
5 Substitute back the original variable	**Substitute the expression you removed in Step 2 back in for the variable.**

Ex 1

$x^2 = 6 \quad \text{or} \quad x^2 = -2$

Ex 2

$x^3 = -4 \text{ or } x^3 = 1$

Your Turn

Steps	Discussion
6 Solve	**Solve the new equations.**

Ex 1

$x = \pm\sqrt{6} \quad \text{or} \quad \text{No real solutions.}$

Ex 2

$x = \sqrt[3]{-4} \text{ or } x = 1$

Your Turn

Steps	Discussion
7 Validate	**Validate the solution(s) by substituting back into the original equation.**

Ex 1

$$x^4 - 4x^2 - 12 = 0$$
$$x = \pm\sqrt{6} \quad \text{so} \quad x^2 = 6$$
$$6^2 - 4(6) - 12 \stackrel{?}{=} 0$$
$$36 - 24 - 12 \stackrel{?}{=} 0$$
$$36 - 36 \stackrel{?}{=} 0$$
$$0 = 0 \quad \checkmark$$

Ex 2

$$x^6 + 3x^3 - 4 = 0$$
$$x = \sqrt[3]{-4}$$
$$\left(\sqrt[3]{-4}\right)^6 + 3\left(\sqrt[3]{-4}\right)^3 - 4 \stackrel{?}{=} 0$$
$$(-4)^2 + 3(-4) - 4 \stackrel{?}{=} 0$$
$$16 - 12 - 4 \stackrel{?}{=} 0$$
$$0 = 0 \quad \checkmark$$

$$x = 1$$
$$1^6 + 3(1)^3 - 4 \stackrel{?}{=} 0$$
$$1 + 3 - 4 \stackrel{?}{=} 0$$
$$0 = 0 \quad \checkmark$$

Steps	Discussion
7 Validate (con't)	**Validate the solution(s) by substituting back into the original equation.**

Your Turn

Model 3 6.4

Solve: $x + \sqrt{x} = 2$

Step 1 **Determine standard form** Quadratic form

Step 2 **Determine substitution** $z^2 = x$ or $z = \sqrt{x}$

Step 3 **Substitute**

$$z^2 + z = 2$$
$$z^2 + z - 2 = 0$$

Step 4 **Solve**

$$(z + 2)(z - 1) = 0$$
$$z + 2 = 0 \text{ or } z - 1 = 0$$
$$z = -2 \text{ or } z = 1$$

Step 5 **Return to original variable** $\sqrt{x} = -2$ or $\sqrt{x} = 1$

Step 6 **Solve** The only real solution is $x = 1$.

Step 7 **Validate**

$$x + \sqrt{x} = 2$$
$$1 + \sqrt{1} \stackrel{?}{=} 2$$
$$2 = 2 \ \checkmark$$

Addressing Common Errors 6.4

The following table shows some of the most common errors that learners tend to make with the kinds of problems covered in this section.

ERROR: **Using the second coordinates of points of intersection of the graphs of the two functions**

Incorrect Process

Solve for x $\quad x^2 + 3x = \frac{1}{2}x + 1$

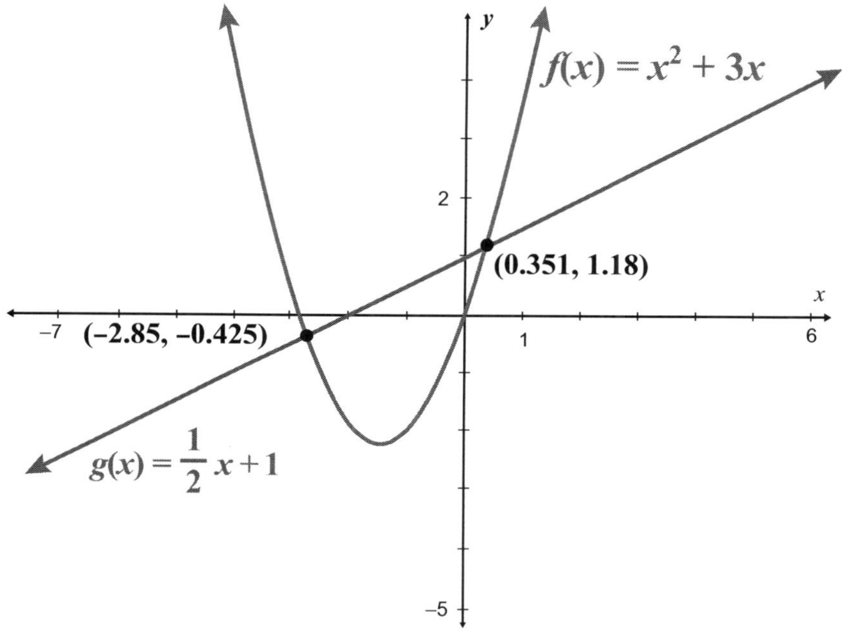

The solution set is $x \approx -0.425$ or $x \approx 1.18$

Correct Process

With intersection coordinates of $(\mathbf{-2.85}, -0.425)$ and $(\mathbf{0.351}, 1.18)$, the solution set is $x \approx -2.85$ or $x \approx 0.351$

Validation

Let $x = -2.85$ or $x = 0.351$

$$(-2.85)^2 + 3(-2.85) \stackrel{?}{=} \frac{1}{2}(-2.85) + 1$$

$-0.427 \approx -0.425$ ✓

or

$$(0.351)^2 + 3(0.351) \stackrel{?}{=} \frac{1}{2}(0.351) + 1$$

$1.176 \approx 1.175$ ✓

Resolution

Use the first coordinate of points of intersection of the graphs of the two functions.

ERROR: Using the wrong graphing window

Incorrect Process

Solve for x

$$\frac{1}{x^2-17x+30}+1=0$$

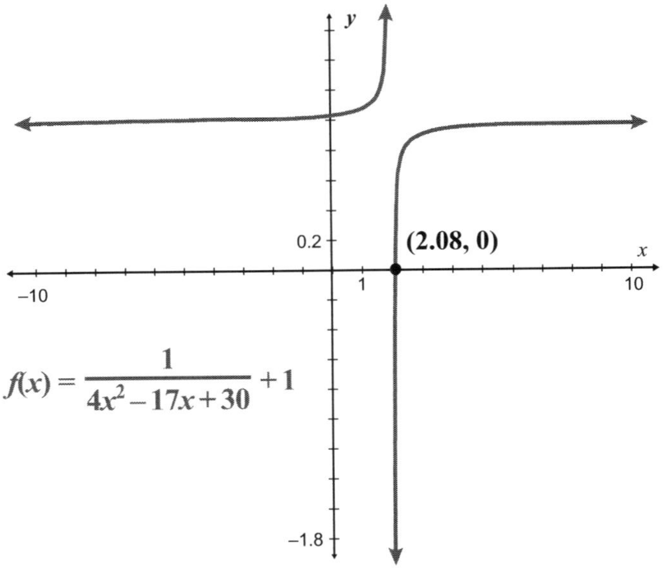

The solution is

$x \approx 2.08$

Correct Process

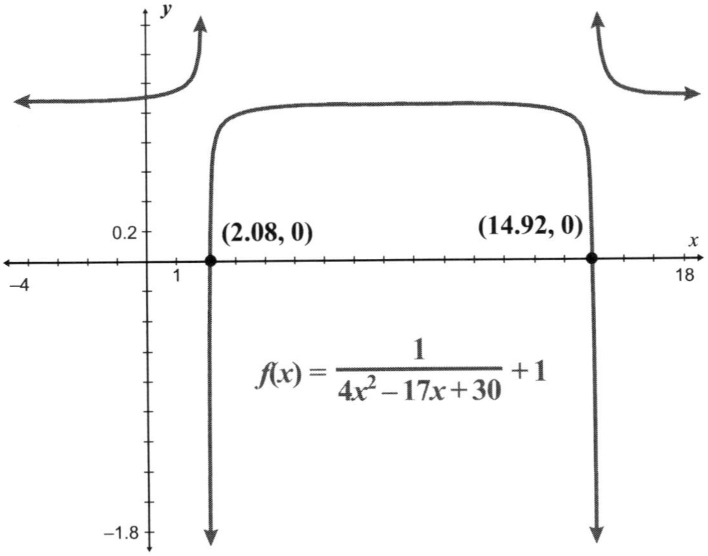

The approximate solutions are

$x \approx 2.08$ and $x \approx 14.92$

Resolution

Use a graphing window that includes all the functions' intersections.

Validation

$$\frac{1}{(2.08)^2-17(2.08)+30}+1 \overset{?}{\approx} 0 \quad \text{or} \quad \frac{1}{(14.92)^2-17(14.92)+30}+1 \overset{?}{\approx} 0$$

$0.032 \approx 0$ ✓ or $0.032 \approx 0$ ✓

ERROR: Failing to substitute back in after the equation is solved

Incorrect Process

Solve for x

$x^4 - 13x^2 = -36$

$x^4 - 13x^2 = -36$

$x^4 - 13x^2 + 36 = 0$

$z = x^2$

$z^2 - 13z + 36 = 0$

$(z-4)(z-9) = 0$

$z - 4 = 0$ or $z - 9 = 0$

$z = 4$ or $z = 9$

Correct Process

$x^4 - 13x^2 = -36$

$x^4 - 13x^2 + 36 = 0$

$z = x^2$

$z^2 - 13z + 36 = 0$

$(z-4)(z-9) = 0$

$z - 4 = 0$ or $z - 9 = 0$

$z = 4$ or $z = 9$

$x^2 = 4$ or $x^2 = 9$

$x = \pm 2$ or $x = \pm 3$

Resolution

Be sure to substitute back in, the term you removed.

Validation

Let $x = \pm 2$ or $x = \pm 3$

$x^4 - 13x^2 = -36$

$(\pm 2)^4 - 13(\pm 2)^2 \stackrel{?}{=} -36$

$16 - 52 \stackrel{?}{=} -36$

$-36 = -36$ ✓

or

$(\pm 3)^4 - 13(\pm 3)^2 \stackrel{?}{=} -36$

$81 - 117 \stackrel{?}{=} -36$

$-36 = -36$ ✓

Preparation Inventory 6.4

I can...	OR *Here's my question...*
☐ solve an equation in quadratic form.	☐
☐ solve an equation using a graphing utility.	☐
☐ validate approximate solutions to an equation.	☐
☐ solve the kinds of equations that appear in this section.	☐

Critical Thinking Questions 6.4

1. Why are the graphical solutions to an equation usually only approximate?

2. Give three reasons solving an equation symbolically might not be possible.

3. Why would you solve an equation graphically?

4. Why is the graphing window important in solving an equation graphically?

5. Why is solving an equation that has a zero on one side often easier than solving an equation with non-zero expressions on both sides?

6. Why would you use a graphical solution to an equation when you already have a symbolic solution?

Demonstrate Your Understanding 6.4

Solve each of the following equations for the indicated variable.

1. $3\ln(x) = x$ (for x)

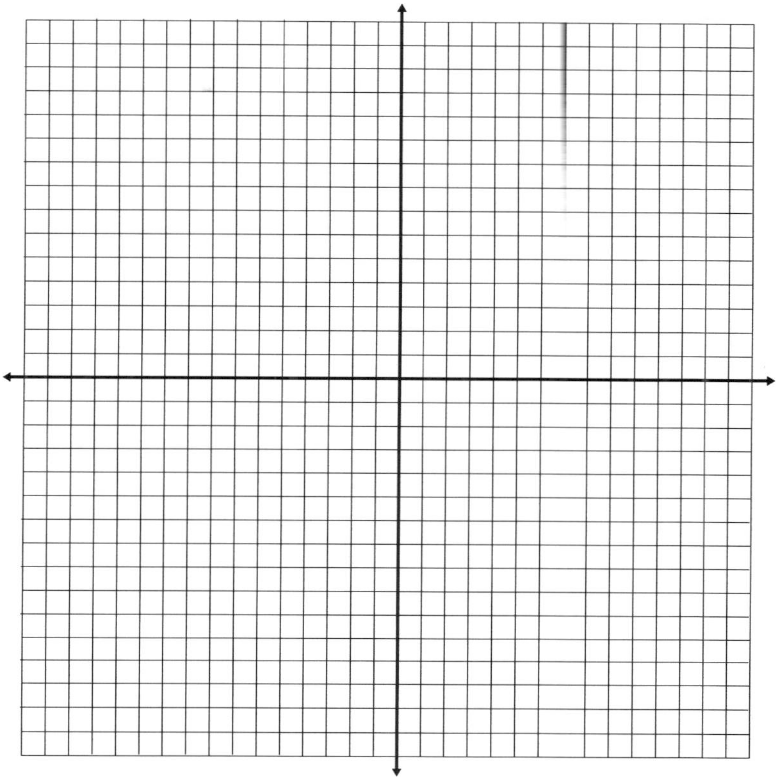

2. $x^2 - \sin(x) = 1$ (for x)

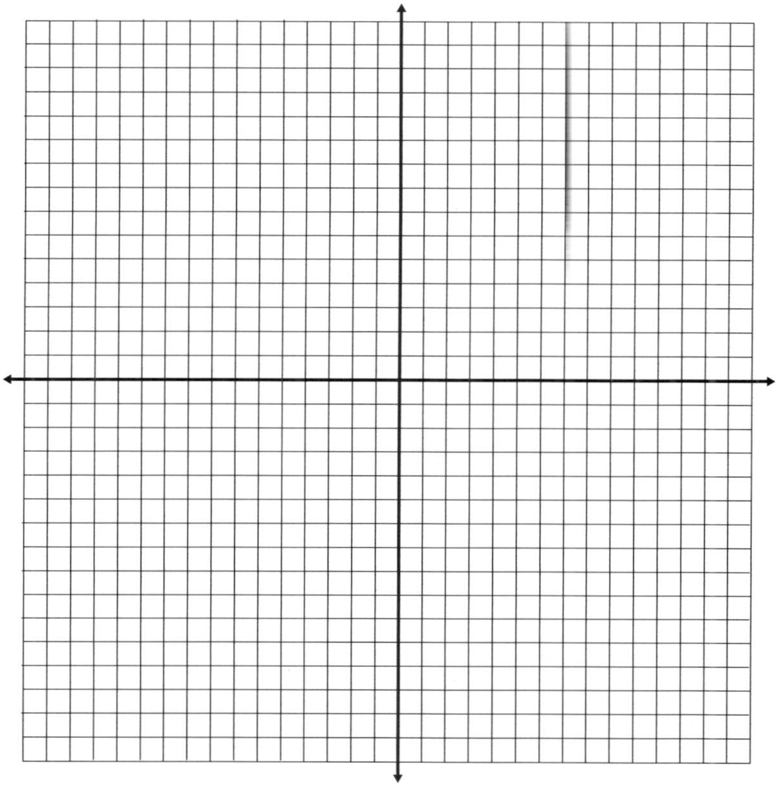

3. $x^4 + 5x^2 = 6$ (for x)

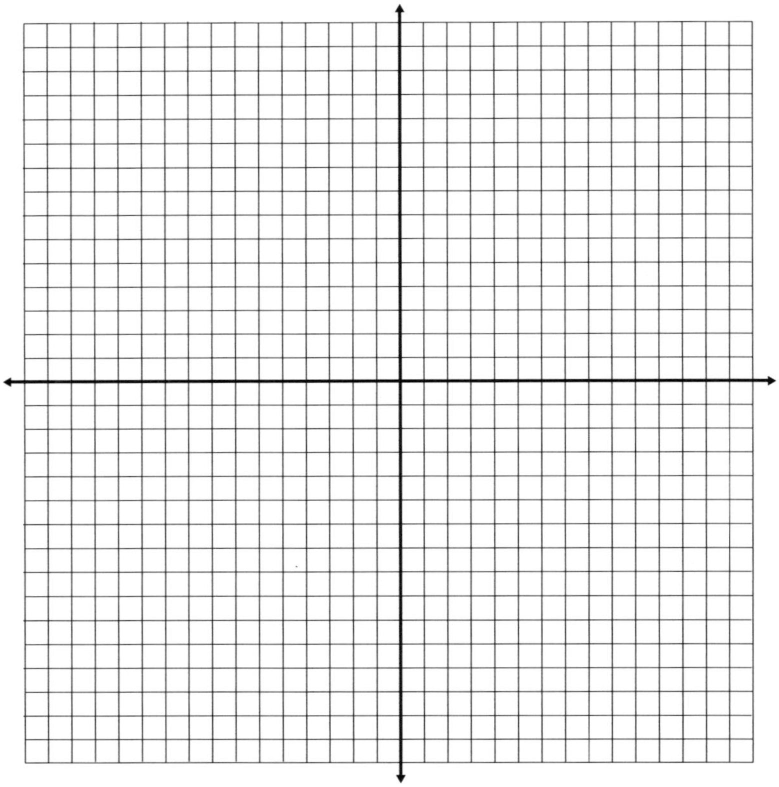

4. $3y^{1/3} - y^{2/3} = -4$ (for y)

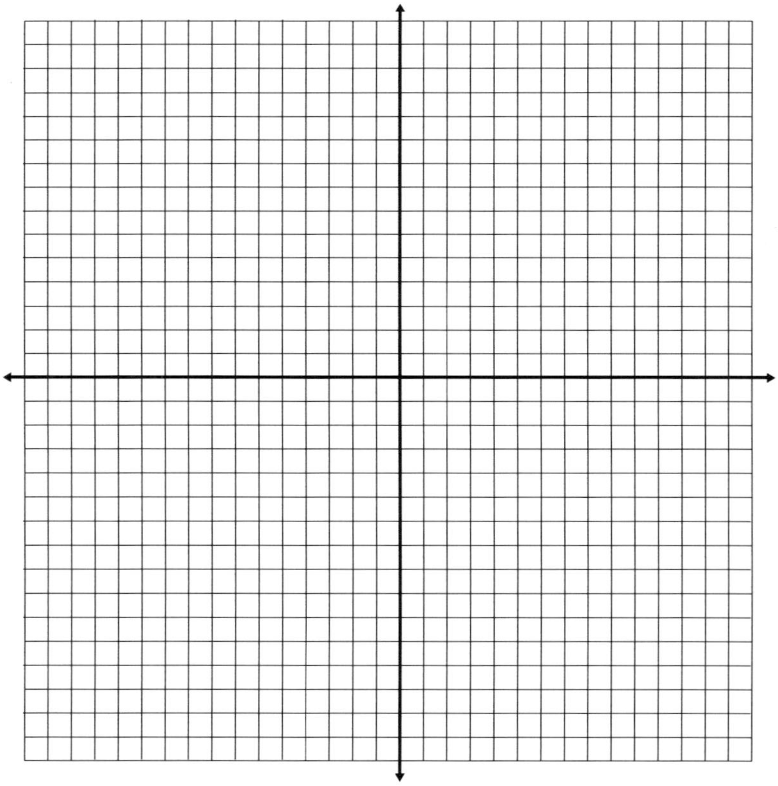

5. $p + 2\sqrt{p} = 3$ (for p)

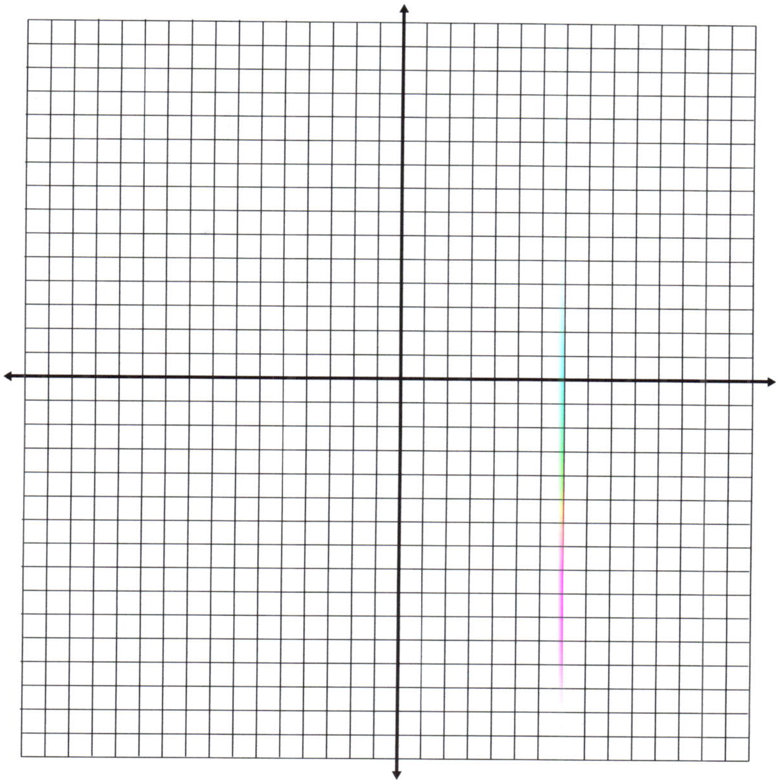

6. $3^x + 4 = 5 - 2 \cdot 3^x$ (for x)

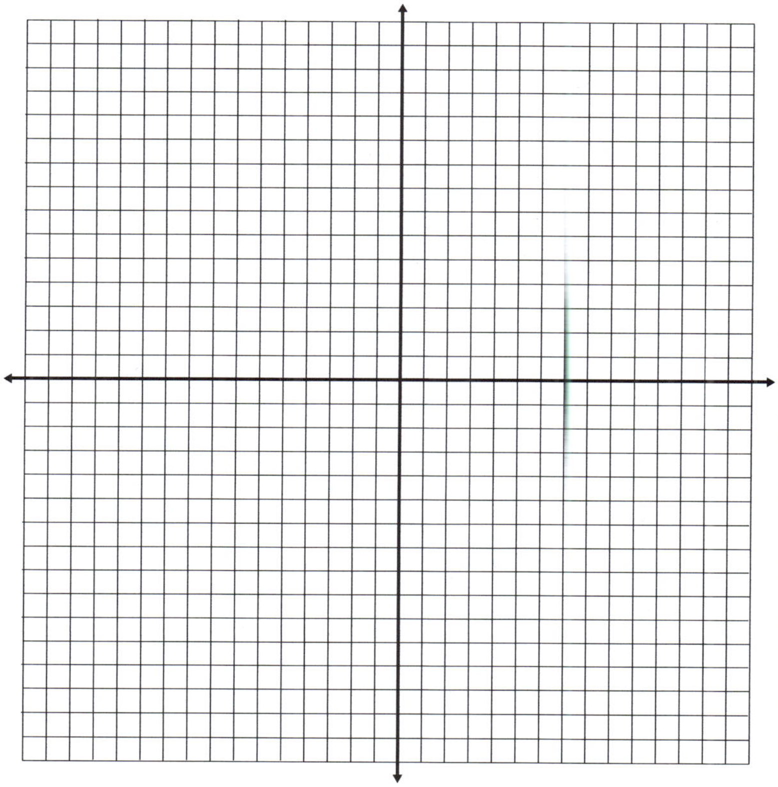

Create and Solve the Hardest Problem 6.4

Your Hardest Problem

Problem Solution

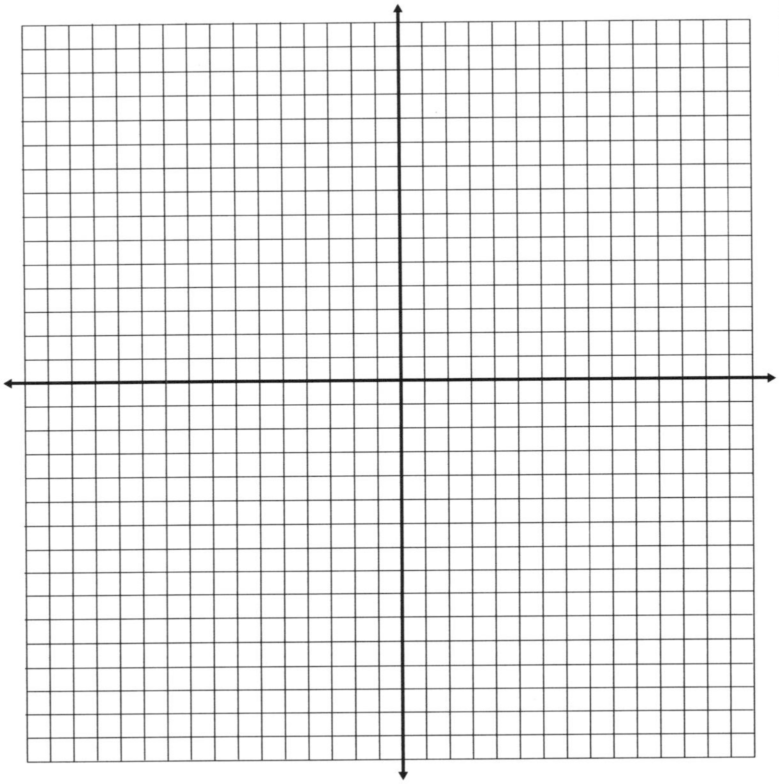

What makes a problem where you must solve an advanced equation graphically a hard problem to solve?

Solve the problem correctly in the appropriate location. Be sure to identify and describe the error you found in the worked solution!

Problem 1

Solve for x: $x^4 - 8x^2 + 12 = 0$

Describe the Error

Worked Solution *(Find the error!)*

$x^4 - 8x^2 + 12 = 0$

$\left(x^2\right)^2 - 8x^2 + 12 = 0$

Let $z = x^2$

$z = x^2$

$z^2 - 8z + 12 = 0$

$(z - 6)(z - 2) = 0$

$z - 6 = 0$ or $z - 2 = 0$

$z = 6$ or $z = 2$

Validate: $z^2 - 8z + 12 = 0$

$z = 6$

$(6)^2 - 8(6) + 12 \stackrel{?}{=} 0$

$36 - 48 + 12 \stackrel{?}{=} 0$

$0 = 0$ ✓

$z = 2$

$(2)^2 - 8(2) + 12 \stackrel{?}{=} 0$

$4 - 16 + 12 \stackrel{?}{=} 0$

$0 = 0$ ✓

Correct Process

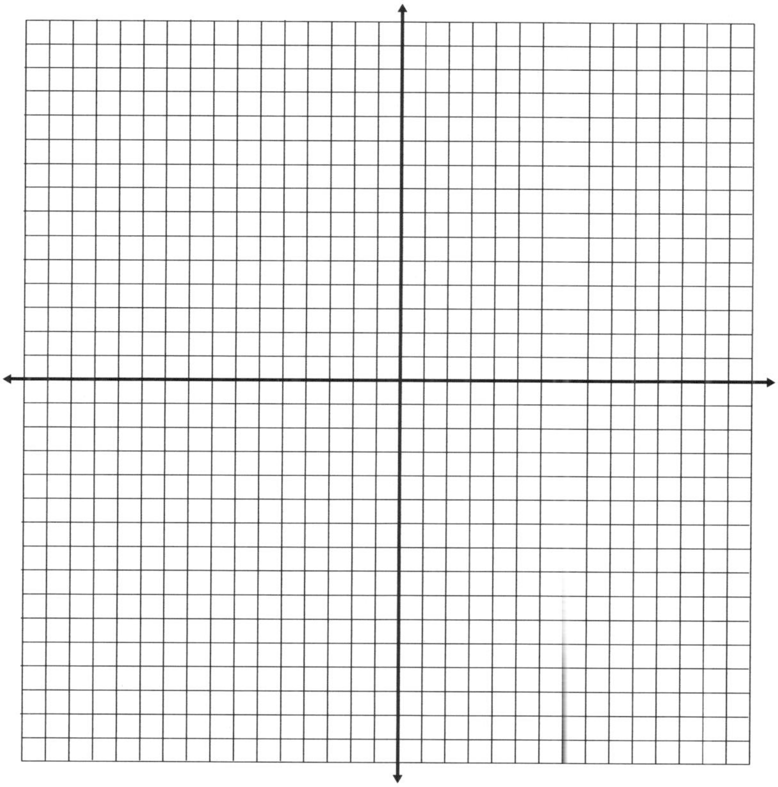

Problem 2

Solve for x: $10x = 3^x$

Describe the Error

Worked Solution *(Find the error!)*

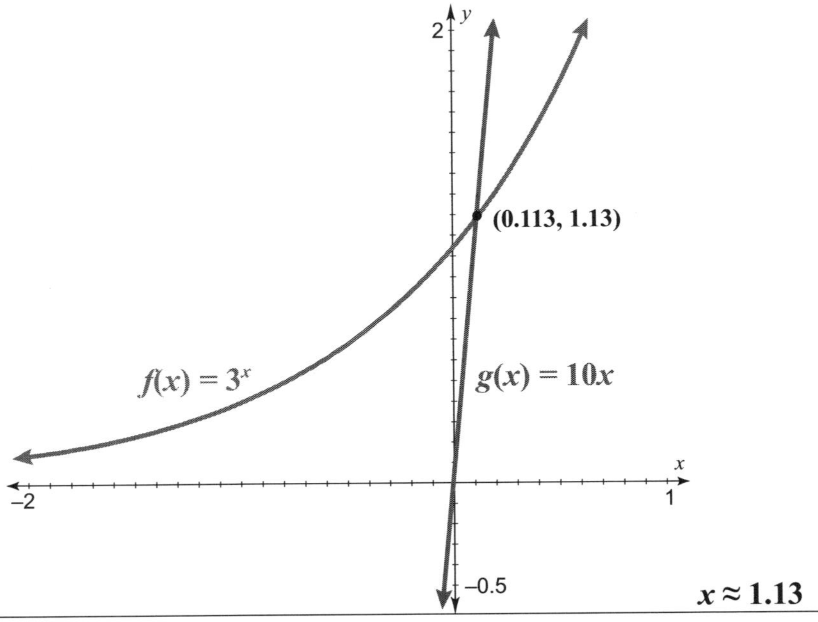

Correct Process

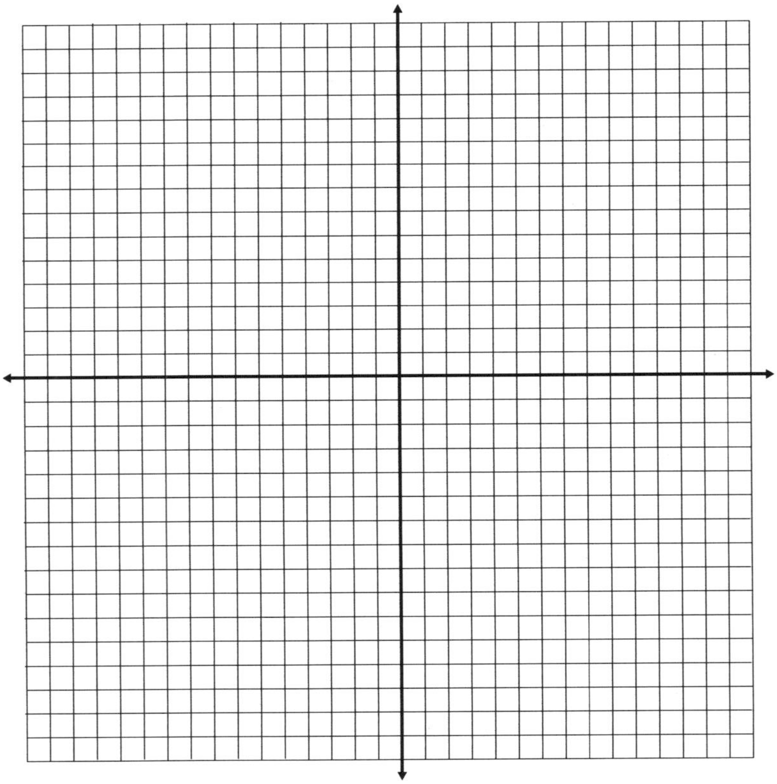

Problem 3

Solve for y: $y^{3/4} - 8y^{1/4} = 0$

Describe the Error

Worked Solution *(Find the error!)*

$$y^{3/4} - 8y^{1/4} = 0$$

$$\left(y^{1/4}\right)^3 - 8y^{1/4} = 0$$

$$z = y^{1/4}$$

$$z^3 - 8z = 0$$

$$z(z^2 - 8) = 0$$

$$z = 0 \text{ or } z^2 - 8 = 0$$

$$z = 0 \text{ or } z^2 = 8$$

$$z = 0 \text{ or } z = \pm\sqrt{8} = \pm 2\sqrt{2}$$

Correct Process

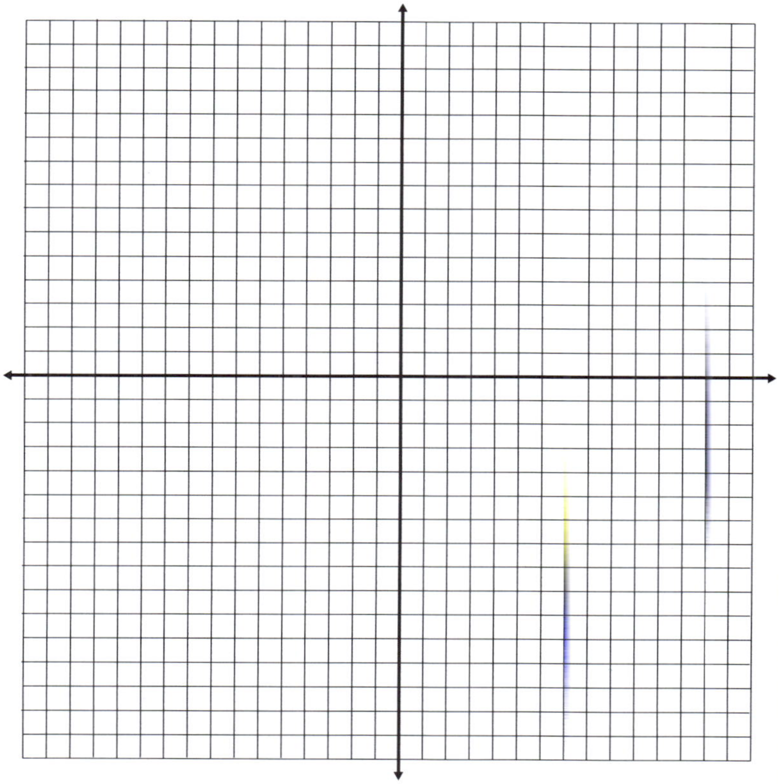

What Did You Learn About Learning Math? 6.4

1. How do you determine how to solve an equation?

2. How did you learn the various equation solving contexts?

3. What did you learn about validating solutions to equations in this section?

Concept Map — Chapter 6

Terminology from the chapter
check the terms used in your map

__absolute value equation

__eliminating the denominator

__embedded functions

__extraneous solutions

__graphing utility

__rational equation

__reduced solution set

__transformed equations

Your instructor will specify how many terms from the chapter you should use in your concept map.